ENCYCLOPEDIA OF DRUG AND DRUG DELIVERY

ENCYCLOPEDIA OF DRUG AND DRUG DELIVERY

Vol. 3

DRUG DISCOVERY

By

Dr. Kamlesh Dashora
Associate Professor
Institute of Pharmacy
Vikram University
Ujjain (M.P.)
(India)

&

Dr. Darshan Dubey
Assistant Professor
Institute of Pharmacy
Vikram University
Ujjain (M.P.)
(India)

D P H

DISCOVERY PUBLISHING HOUSE PVT. LTD.
NEW DELHI-110 002

Published by:
Tilak Wasan
DISCOVERY PUBLISHING HOUSE PVT. LTD.
4383/4A, Ansari Road, Darya Ganj
New Delhi-110 002 (India)
Phone: +91-11-23279245, 43596064-65
Fax : +91-11-23253475
E-mail: parul.wasan@gmail.com
discoverypublishinghouse@gmail.com
web : www.discoverypublishinggroup.com

***First Edition:* 2012**

ISBN: 978-93-5056-038-9 (Set)
978-93-5056-039-6 (Vol. 1)
978-93-5056-040-2 (Vol. 2)
978-93-5056-041-9 (Vol. 3)
978-93-5056-042-6 (Vol. 4)
978-93-5056-043-3 (Vol. 5)

Encyclopedia of Drug and Drug Delivery

Printed at:
Shree Balaji Art Press
Delhi

Preface

The Present title "**Encyclopedia of Drug and Drug Delivery**" is the amazing advancement of pharmaceutical studies. It provides the various fundamental aspects of drug designing to be understood adequately. It has been compiled for undergraduate, postgraduate students, research scholars, teachers, technicians, industrial microbiologists and scientists involved in pharmaceutical research and development. The basic concepts have been clearly explained and their functions are adequately highlighted. The presentation of text is simple and systematic. The selection of chapters and topics is according to the specified syllabi of several Indian Universities and are so structured as to enable the student to move easily from the fundamental to the complex. It describes the ideas and techniques that are creating or new domain of science and technology. It holds the promise of revolutionizing field ranging from medical science to pharmaceutical industries. It is our earnest hope that this title will be of great value to all our students.

Organizational and economic changes in the pharmaceutical industry have led to a marked reduction in the supply of new chemical entities. This combined with an increase in the expiration of patents and tighter regulatory control has lead many companies to rethink existing product design. The growing trend to improve the efficacy of existing drug products, to both extend parent life and allow competition with opportunistic generic companies, has allowed for a greater understanding of formulation design. This trend is set to continue for some time as economic and political pressure increases.

There have been many improvements in understanding the theoretical and practical aspects of drug discoveries. Numerous therapeutic compounds have been investigated to treat a variety of different illnesses. There are many pharmaceutical methods requiring a variety of formulation options. Advanced formulation design may be thought of as a way to optimize the mode of drug designing to a patient that will ultimately lead to a more effective therapy and approach an optimal therapeutic outcome for a particular drug. This book is designed to take the reader through various cutting-edge areas of advanced formulation design by describing therapeutic categories that may be beneficial. The reader will be introduced to some novel technologies and updated on the state of the art in areas designed to overcome poor efficacy issues.

To make the work more comprehensive and informative, the authors have consulted many authoritative books, research journals, abstracts, monographs etc., so there can be no claim to originality except in the manner of treatment.

The authors express their thanks to their friends and colleagues whose continue inspirations have initiated them to bring out this book.

The authors express their gratitude to Mr. Wasan and staff of M/s Discovery Publishing House Pvt. Ltd., for their whole hearted co-operation in the publications of this book.

The authors will remain sincerely responsible for any shortcomings of the book and be grateful to the readers for their suggestions and constructive criticism for the continuous betterment of the book. They have this opportunity to appeal to the readers to send their suggestions straightaway to this publisher.

–Authors

Contents

1

Phylogenetic Drug Discovery

The field of genomics has fundamentally changed the way research has been done since the early 1990s and is actively being incorporated into strategies for identifying new drugs and other therapeutics. The emergence of the field of genomics was catalyzed by two crucial factors: the development of automated DNA sequencing and the availability of high-performance computing infrastructure. Upon this foundation, many associated techniques have been developed or improved, such as microarraying, proteomics, and other laboratory and bioinformatic techniques. All of these "*genomics technologies*" have one aspect in common: the ability to analyze gene function in a high-throughput manner. Antimicrobial drug discovery was one of the first areas to benefit from genomics, due to relatively small genome sizes and, in many cases, systems amenable to genetic manipulation. This chapter describes how bioinformatic and genomic technologies are being leveraged for the discovery of novel antimicrobials.

STEPWISE DRUG DISCOVERY

The process of discovering and developing new antimicrobial drugs is outlined here. The first step is target identification, which in the genomic era is usually done by a bioinformatic filtering process to identify genes specifically conserved in bacteria. The next step is to validate the drug target, usually by confirming that the gene encoding the drug target is essential for viability or, in some cases, essential for virulence. A validated drug target is then used to design an assay that will be used to screen a large deck of small molecule compounds which have been designed by companies from natural products, medicinal chemistry, and combinatorial chemistry. From the screen, a number of "hits" are identified, validated, and prioritized. The hits then progress to "leads" following secondary assays, SAR (structure-activity relationship) studies, and preliminary pharmacology and toxicology studies.

At the end of the preclinical process, a drug candidate is chosen and an IND (Investigational New Drug) application is submitted to the Federal Drug Administration. Compound production is scaled up and the drug candidate then goes through clinical trials to determine safety, efficacy, and dosing. Following the successful conclusion of the clinical trials process, an NDA (New Drug Application) is filed with the FDA. Once the drug has been approved by the FDA (about 20% of INDs reach this point), the company markets and sells the drug; new indications are also sought as part of the life-cycle management process.

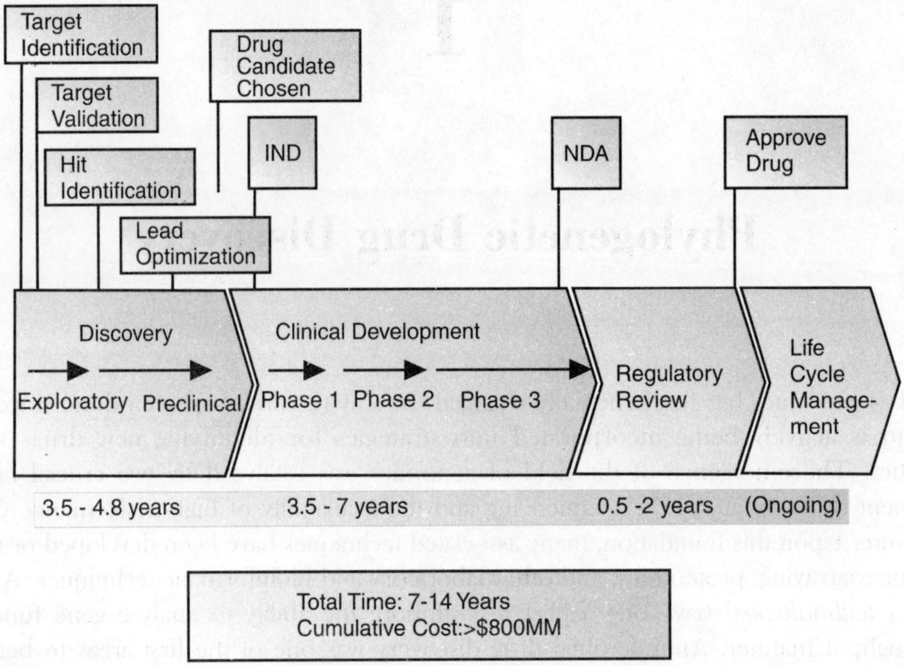

Fig. 1.1. Overview of drug discovery and development.

It should be noted that the attrition rate from target to approved drug is high: For an average large pharmaceutical company, it has been estimated that 60 targets are required to result in 20 drug candidates, which, after going through clinical trials, result in 3 new drugs. The process also takes a significant amount of time: In the past, the process from target identification to approved drug has taken an average of 14 years, although to compete more effectively drug companies are striving for an ~8-year timeline.

SEQUENCING GENOME

Automated sequencers were developed in the mid-1980s and functionally put to use in the early 1990s for *expressed sequence tag* (EST) projects. With the infrastructure for high-throughput sequencing, assembly, and data management established, the next major milestone was the publication of the first bacterial genome. This was a critical development for genome sequencing for several reasons. First, the accepted paradigm for sequencing DNA over 50 kb was a top-down, cosmid-by-cosmid approach involving a time- and resource-intensive upfront mapping stage. *Haemophilus influenzae* was sequenced by a whole-genome shotgun method, rapidly accelerating project timelines while still providing very high quality sequence data. Second, the project set a high standard for the publication of future genome sequences: an entire microbial genome without gaps, extensively annotated, and making use of web pages to include the unprecedented volume of data.

Finally, the completion of an ~2 Mb segment of DNA suggested the feasibility of a incorporating shotgun sequencing to accelerate the sequencing of the human genome; in fact, the Venter group went on to utilize the whole-genome shotgun technique to completely sequence the ~120 Mb *Drosophila* and ~2910 Mb human genomes. Whole-genome shotgun sequencing is the most rapid, cost-effective method for sequencing microbial genomes. Since 1995, hundreds of microbial genome projects have been initiated and to date, over 55 microbial genomes have been completed. This number continues to rapidly increase, and lists on the worldwide web should be referred to for the most up-to-date information.

The process of microbial genome sequencing has been reviewed in more detail elsewhere, but essentially consists of the following steps: (i) constructing a library of randomly sheared DNA, (ii) shotgun sequencing of clones for high genome coverage, (iii) assembly of the genome into contiguous sequences (contigs), (iv) closure of the gaps between contigs, and (v) gene finding and annotation. Nearly every step of a genome sequencing project involves computers or bioinformatic tools: for instance, basecalling (e.g., ABI Basecaller, sequence assembly), gene finding, and suites of tools for multiple analyses (the GCG package and the genome project management tool MAGPIE).

Moreover, at the core of every successful genome project is a relational database to archive and track everything from raw sequence chromatograms to a series of updated assemblies to properties associated with each gene (e.g., sequence, coordinates on the genome, similarity to other genes, and experimental data) Microbial genome projects can range in effort from low-pass sequencing to rapidly identify the majority of genes in a genome (often seen in industrial settings) to published whole-genome sequencing projects where every gap is closed, representing a firm framework for future experimental work since the genome has become a finite entity.

COMPUTATIONAL ANALYSIS OF GENES

Clearly, bioinformatics is essentially integrated into all aspects of microbial genomics and drug discovery. Some of the major uses of bioinformatic tools are presented below.

FASTA and BLAST

Many programs exist, and some of the most widely used include FASTA, the workhorse program BLAST, and the more sensitive (and more computationally intensive) Smith-Waterman algorithm. Most genome projects use BLAST to search the genes found within a genome; however, determining what constitutes true signal in the case of weakly similar genes is always a contentious issue. This is especially true for small genes found by genome sequencing projects that are conserved at midrange similarity among different bacteria. In general, about 30-50% of the genes identified in microbial genome sequencing projects are considered unknown, depending on the organism and the annotation criteria used. One effort to enrich for potentially meaningful similarities found among the noise is the use of iterative BLAST searching, known as Psi-BLAST. Low-coverage genome projects may, like EST projects, suffer from a higher error rate and programs that take frameshifting into account, such as Framesearch, are invaluable.

Data Base

Similarity may exist between two proteins that is only apparent when focusing on small regions, such as conserved motifs. For many years researchers have made use of the Prosite database of motifs and the Blocks database. A more recent collection of motifs, the Pfam database, was constructed using the Hidden Markov model-based HMMER suite of tools. Finally, the Collection of Orthologous Groups (COG) database contains clusters of proteins found throughout sequenced genomes that are potentially related to each other.

Analysis of Entire Genome

Once whole-genome sequence information began to appear, it became clear that tools designed to analyze entire genomes needed to be designed. The SEEBUGS suite of tools allows users to identify a concordance of genes conserved among selected organisms at any selected similarity level and includes the capability of quickly identifying "neighbors" of selected genes in other genomes and specifically subtracting out selected organisms. This program was designed to address an issue at the heart of antimicrobial drug discovery: identifying potential drug targets (gene products) that are conserved in certain groups of bacteria (e.g., gram-positive and gram-negative) but not found in humans. This allows researchers to build in up front both spectrum and selectivity for novel classes of antibiotics. Another tool that has more recently been developed for whole-genome comparisons is the Comprehensive Microbial Resource.

Clustal Package

Alignments of multiple sequences from a diverse set of organisms helps to highlight conserved regions; one of the standard tools for this is the CLUSTAL package. Conserved regions identified by the multiple sequence alignment can then be more carefully analyzed by motif searching techniques to identify additional related genes or by phylogenetic analyses. Also, regions conserved in alignments of linear amino acid sequences can be the focus of three-dimensional modeling and structural similarity searching techniques.

Preparing Phylogenetic Tree

Phylogenetic analysis takes the relationships found among a set of aligned sequences and graphically presents them as a "tree." The most widely used programs are PAUP (Phylogenetic analysis using parsimony;) by D. Swofford and the PHYLIP phylogenetic suite of programs by J. Felsenstein and collaborators. There are three commonly used methods: distance, parsimony, and maximum likelihood (ML). Each has its benefits and drawbacks. Most molecular biologists use distance methods, while parsimony and ML are generally preferred but require more expertise and computing time. The trees produced can show the evolutionary relationships of the genes under examination (a gene tree) or the species under examination (a species tree). However, it must be noted that these two types of tree are not the same thing and do not provide the same information.

Phylogenetic trees can help to classify an unknown reading frame. An excellent review of the use of ML was recently published. Before genome sequences became available, much of the phylogenetic analysis of microbes focused on taxonomic classification of microbes and the establishment of homology among genes. The flood genes from genome sequencing projects made the finding of gene family relationships even more important to understanding the function of gene products and the evolution of micro-organisms. Genome sequence information requires new computational methods that build on the comparison of predicted gene sequence, incorporating other important information such as structures, domain shuffling, expression patterns, and gene adjacency in genomes.

Three Dimensional Structures of Proteins

A number of techniques are used to examine relationships among three-dimensional structures of proteins, such as structural homology comparisons of different solved structures or the "threading" of query sequences versus known structures. Some more recent approaches use a combination of techniques, such as Psi-BLAST coupled with structural prediction programs. The field of rational drug design is of particular interest to pharmaceutical and biotechnology companies. Some techniques include virtual screening of decks of compounds or even chemical fragments for hits that fit into an important region of a protein and the use of molecular modeling techniques to optimize a hit into a lead compound [e.g., structure–activity relationship by nuclear magnetic resonance].

Proteomics

Genomics technologies such as microarraying, proteomics, and high-throughput biology techniques give vast amounts of data to be analyzed. Powerful methods for providing context to this information involve overlaying the information onto a metabolic pathway map or the assignment of role category information (e.g., cellular role, enzyme class, and gene family). Metabolic pathway modeling has resulted in some excellent knowledge bases, particularly the EcoCyc and KEGG databases. One example underscoring the power of having complete genome sequence information in hand for metabolic reconstruction and comparative genomics work is a study examining the presence/absence and conservation of citric acid cycle enzymes across 19 genomes. Role categorization methods have also been developed for organizing microbial genome data, such as the categories developed for many microbial genome projects and the Gene Ontology effort that has been used primarily in model organism sequencing projects.

Bacterial Genome

One feature unique to bacterial genomes is the grouping of genes into operons—coregulated sets of genes that are often related in function. Rigorously defining an operon is a tedious process, especially if done experimentally; however, a number of computer-based prediction methods have been developed. One particularly intriguing approach utilizes the gene clustering approach used by others but does so in combination with other data, such as mRNA expression patterns and protein-protein interactions, resulting in a powerful predictive tool.

GENOMIC TECHNOLOGY

A number of genomics technologies exist for the analysis of all genes and/or gene products in a genome. Although these tools are being used for all of the steps in the drug discovery and development process, much of the effort is focused on the use of these technologies for target identification and validation.

Transcriptional Profiling Technique

Microarraying is a high-throughput transcriptional profiling technique that is being applied increasingly to examine differential gene expression in microbial genomes. The technique is based on the hybridization of labeled RNA to miniaturized deposits of DNA on small solid supports. Two widely used formats are synthetic Affymetrix GeneChip arrays, composed of short oligonucletides photolithographically deposited onto silicon wafers, and "spotted" microarrays, usually consisting of a gene fragment (several hundred basepairs) deposited onto a glass microscope slide. Basically, RNA derived from various conditions of interest are labeled (one dye in the case of Affymetrix and two dyes for spotted arrays), hybridized to the arrays, and the signal quantitated. After the data have undergone some initial filtering and have been stored into a database, a number of software packages can be utilized for expression analysis.

In addition to programs offered by microarray manufacturers, developed in-house by companies, or published by academic groups, a number of third-party packages are used such as Genespring or Rosetta Resolver. There are many recent publications involving expression profiling of microbes; a few that have a drug discovery focus are emphasized here. In one study, the pathogen *Mycobacterium tuberculosis* was treated with the commonly used antibiotic isoniazid, and gene expression changes were monitored using microarrays. Genes observed to change in response to the drug were known to be part of the isoniazid mechanism of action, although some additional genes of unknown function were also affected.

In fact, there are many drug mechanism of action studies that have been performed in industry, but few have been published to date. One area of great interest to industrial groups working on antifungal drug discovery is the effect of azoles and other drugs on ergosterol biosynthesis. A recent publication examined the transcriptional profile of azole-treated *Saccharomyces cerevisiae* and compared these patterns to profiles of *S. cerevisiae* strains with knockouts of ergosterol biosynthetic genes. Responsive genes included those in the ergosterol pathway as well as mitochondrial genes, oxidative stress genes, and a few genes of unknown function. Also in this study, an antifungal drug of a novel chemotype was profiled and found to be similar to the azoles, thus underscoring the power of microarray-based drug profiling studies.

Another microarray experiment by pharmaceutical researchers focused on two key regulators of virulence in the gram-positive pathogen *Staphylococcus aureus* in an effort to develop treatment against pathogenesis-based targets. Using a custom-designed Affymetrix array, the researchers found genes known or expected to be regulated by *agr* and SarA, but also found known genes not thought to be directly involved in staphylococcal pathogenesis, suggesting that these global transcriptional regulators play a broader role than previously thought. DNA microarrays are also being used for other genomic experiments, such as whole-genome DNA-DNA hybridizations.

Examples of these microarray-based studies that focus on important pathogens include comparisons of BCG vaccine strains to other mycobacterial isolates, comparison of isolates of *Streptococcus pneumoniae* to all genes in the sequenced type 4 strain, and comparison of divergent lineages of *S. aureus* to the COL genome to identify pathogenicity islands and dispensable genetic material, estimated to be approximately 22% of the genome. Other techniques independent of microarrays can be used to look at differential gene expression in microbes. Techniques that have found some usage include differential display, subtractive hybridization, and quantitative polymerase chain reaction (PCR). Quantitative reverse-transcription PCR has increasingly been used as the method of choice to validate microarray results. Additionally, the technique has become robust and accurate enough to be used as a transcriptional profiling tool in its own right. For any transcriptional profiling technique using bacteria, a notable experimental distinction is the absence of a usable poly-A tail, requiring the modification of many of the molecular techniques that have been devised using eukaryotic RNA.

Protein Complement of the Genome

Proteomics is the study of the protein complement of the genome. Differential expression can be observed at the protein level for different samples by protein separation followed by staining and comparison of the signal for the separated species. The protein separation technology that has been in use for the past several decades is two-dimensional Sodium dodecyl sulfide-polyacrylamide gel electrophoresis, but other separation technologies are being developed, such as multidimensional liquid chromatography.

Similarly, Coomassie blue and silver stain are often used today, but, again, newer technologies, such as SYPRO dyes, are becoming available. Once proteins are separated and species representing differentially expressed proteins found, the next step is protein identification, which can be done by techniques such as mass spectroscopy (MS/MS), matrix-desisted laser-desorption ionization time-of-flight, and peptide sequencing. Extensive cataloging of microbial proteome components has been performed, such as a study of the *H. influenzae* proteome published 2 years after the genome sequence was determined.

Other Experimental Techniques

A number of other experimental techniques are being used to determine the function of genes discovered by microbial genome sequencing projects. One is global gene knockout campaigns using transposon mutagenesis, random and directed gene knockouts, genome footprinting, and antisense technology. Additionally, it is also possible to include regulatable promoters, gene fusions, or oligonucleotide tags when performing gene knockouts, allowing for large-scale systematic analyses of gene function. Knockouts of every gene in model micro-organisms, such as *Bacillus subtilis* or *Saccharomyces cerevisae*, have been performed by the consortia put in place to sequence these genomes. Additionally, techniques that analyze bacterial virulence, such as *in vivo* expression technology and signature-tagged mutagenesis, have relied on fairly random approaches; these techniques stand to benefit tremendously from completed genomes and the development of systematic knockouts and associated constructs. A second widely employed technique is the analysis of protein-protein interactions using genetic systems based on the yeast two-hybrid system.

Recent genomewide protein-protein interaction studies have been performed for phage, bacteria, and yeast, and dovetailing this technology with oligonucleotide microarrays should bring increased throughout. Finally, advances in screening techniques have lead to a number of "*gene-to-screen*" technologies, developed by biotechnology companies in response to the explosion in genomic information and the subsequent bottleneck in screening gene products of unknown function. Many of these techniques are based on thermal denaturation, NMR, and phage display technologies and are broadly applicable to all areas of drug discovery. Much of the work using this technology for antimicrobial drug discovery is proprietary at present, although there are some examples of recently published gene-to-screen work to identify novel antibiotics.

CONCLUDING REMARKS

The sequencing and publication of over 50 genomes since 1995 should be regarded as a monumental achievement; however, an equally impressive post-genomic effort is now building on that infrastructure, using both computational and laboratory-based high-throughput approaches to understand gene function in the context of microbial physiology. Genomics has helped change the face of microbial research within a decade from a gene-by-gene approach to a more comprehensive, "*systems biology*" approach.

The availability of microbial genome sequences and the means to rapidly make use of it will allow an unprecedented ability to develop therapies to treat infectious diseases, including antibacterial and antifungal compounds and new classes of vaccines. As the genomes of model eukaryotes and humans are elucidated, a powerful knowledge-base-leveraging microbial pathogenesis and cell biology work should enable the development of a new generation of therapies. These include therapies that target other functions in the microbial cell, such as virulence, or that address infection from the host perspective and will hopefully address the problem of pathogens resistant to anti-microbial therapy.

2

History of Drug Discovery

Today's medicine is based on traditional medicine. Traditional medicines exist in every continent of the globe and in every cultural area of the world. The most famous ones are traditional Chinese medicine in East Asia, Ayurvedic medicine in India, and formerly Galenic medicine in Europe, all of which have some resemblance to one other. Each of these traditional medicines has its own origins and an individual basic philosophy. The art of practicing *Chinese medicine* stretches back over several thousand years. The legendary culture hero, Shen-nong, is said to have tested many herbs for their medical properties. *Pen-ts'ao,* the first compilation of herbal medicines, is connected with his name.

Since Ancient times, the Chinese have divided the world into five symbolic elements: Wood, Fire, Earth, Metal and Water. Everything in the world is dominated by one of these elements, and their constant interplay, combined with those of yin and yang, explain all change and activity in nature. The positive, generative cycle proceeds as follows: Wood burns to generate Fire; Fire produces ashes which generate Earth; Earth generates Metal, which can be mined from the ground; when heated Metal becomes molten like Water; Water promotes the growth of plants, thereby generating Wood. The negative subjugable cycle is complementary to the positive, generative one. Chinese medical views regarding the vital internal organs are based on the theories of *yin* and *yang,* the five elements (which are each related to body organs and colors), and the meridians. With medicines consisting mainly of herbal drugs and minerals, Chinese doctors manipulate these natural relationships to adjust energy imbalances caused by the excess or deficiency of these forces in the body.

Chinese physicians and philosophers developed a special system of physiology describing vital organs as storage houses and vital connections as meridians that became the basis of acupuncture. In India *Ayurveda, Siddha* and *Unani* systems of medicine provide healthcare for a large part of the population. The word *Ayurveda* is composed of two parts: Ayu (= life) and Veda (= knowledge). Scholars of Ayurveda had placed the origins of this science of life at some time around 6000 BC. They were orally transmitted by successive generations. The principles were recorded in great detail in compendia, which are called *Samhitas*. Compared with modern anatomy and physiology Ayurveda is based on certain fundamental doctrines, known as the *Darshanas*, such as the seven *Dhatus*. They can be described not exactly as organs but as body constituents. The three *Doshas*: *vata*, *pitta* and *kapha* are regulators of cell function in various ways. A balance of the three doshas is essential for maintaining health. Imbalance of the doshas creates disease. Drug therapy in Ayurveda is highly individualized.

The choice of drugs as well as the dose is not only influenced by the disease process, but also by the constitution of individuals and environmental conditions which affect the balance of the doshas and therefore the response to drugs. Both traditional Chinese and Ayurvedic medicine developed further in terms of formulations. There is also the tendency to adopt the modern forms of clinical trials, but there has never been a change in paradigm as far as the basic philosophy is concerned. *Traditional European medicine* goes back to the time of Egyptian and Babylonian-Assyrian culture. About 3000 BC the Sumerians developed a system of cuneiform writing that enabled them to write on clay tablets. At the time of King Assurbanipal 2000 BC a first comprehensive Materia Medica was written containing approximately 250 vegetable drugs and 120 mineral drugs.

Around 1500 BC, the famous Egyptian papyrus *Ebers* was written, describing more than 700 drugs drawn from plants, animals, and minerals. Some are still used in our time, such as garlic and poppy seeds. Greek philosophers such as Empedocles, Aristoteles and Pythagoras, all from around 500 BC, influenced European medicine a great deal. They created the theory of the Four Elements, which were proposed to be the components of all matter, including animals and Man. These Four Elements were Water, Air, Fire and Earth. The most famous physicians were Hippocrates (around 400 BC) and Galenus, living in the second century after Christ. In medieval times the system of pathology and therapy, originating from Galenus, dominated Western medicine. The fundamental principle in so-called Galenism was the transformation of humoral pathology into a rigid dogma. The school of Hippocrates had already formulated the theory of the Four Humours (paralleling the Four Elements), the correct balance of which meant health, while every disturbance of this balance caused disease.

There were Four Humours: Blood (coming from the heart), Phlegm (supposed to come from the brain), Yellow Bile (supposed to be secreted from the liver), and Black Bile (supposed to come from the spleen and the stomach). Each of these humours had definite qualities. Blood was moist and warm; Phlegm, moist and cold; Yellow Bile, warm and dry; Black Bile, cold and dry. Furthermore, there was a definite connection between the predominance of one humour in the metabolic system and an individual's temperament, such as sanguine, phlegmatic, choleric, and melancholic. European medicine has been influenced by Arabian medicine, mainly by the physician Ali Ibn Sina, who was known in Europe under the name Avicenna. His books were translated into Latin and his *Canon Medicinae* has influenced European medicine for centuries.

Saint Hildegard of Bingen, the abbess of a monastery, was one of the most famous physicians and pharmacists in the twelfth century. She wrote many books on human nature and the use of herbal drugs. Theophrastus Bombastus von Hohenheim, also called Paracelsus and who lived in the fifteenth century, was a reformer of European medicine. One of his ideas was that in nature a remedy could always be found against every disease. The Latin sentence "*Ubi malum, ibi remedium*" was the core idea of the signature theory that the shape or the color of a plant indicates against which disease the herbal drug can be used. European medicine has undergone a great change in paradigm. In contrast to traditional Chinese medicine and Ayurveda, traditional European medicine no longer exists and is only a matter of history of medicine. Even the self-understanding of modern phytotherapy in Europe is completely based on the allopathy that means modern Western medicine.

CLINICAL APPROACHES

The most important achievements of *modern Western medicine* were made in several areas such as diagnosis, infectious diseases, endocrinology and medicinal chemistry. Virchow founded cellular pathology in the nineteenth century. The intensive use of the microscope in medicine with histological comparison of diseased and normal organs allowed the change from humoral to *cellular pathology. Medicinal chemistry* as an important science started less than 100 years ago. The active principles of plants, mostly alkaloids, were isolated and were the starting point for syntheses, such as: morphine and papaverine from *Papaver somniferum* for synthetic analgesics and spasmolytics; atropine from *Atropa belladonna* for synthetic spasmolytics; cocaine from *Erythroxylon coca* for synthetic local anesthetics; quinine and quinidine from *Cinchona succirubra* for synthetic anti-malaria drugs and antiarrhythmics; ephedrine from *Ephedra sinica* for synthetic sympathomimetics and sympatholytics including β-blockers, xanthines such as caffeine; theobromine and theophylline from *Coffea arabica, Theobroma cacao, Camellia sinensis* for vasotherapeutics such as pentoxyphylline; ergot alkaloids from *Claviceps purpurea* for semi-synthetic ergot derivatives; reserpine and ajmaline from *Rauwolfia serpentina* for synthetic antihypertensives and antiarrhythmics; physostigmine from *Physostigma venenosum* for potential antidementic drugs; glycosides from *Digitalis lanata* and *Digitalis purpurea* for semi-synthetic cardiac glycosides; anthraquinones from *Senna angustifolia* or *Rhamnusfrangula* or *Rheum officinale* for synthetic laxatives.

Pharmacological research started in Europe in the second half of the nineteenth century when its founders, Rudolf Buchheim and Oswald Schmiede berg, investigated the action of existing drugs in animal experiments. With the emergence of synthetic chemistry the pharmacological evaluation of these products for therapeutic indications became necessary. Many new drugs were discovered by this classical approach during the twentieth century. The *classical way of pharmacological screening* involves sequential testing of new chemical entities or extracts from biological material in *isolated organs* followed by tests in whole animals, mostly rats and mice but also higher animals if indicated. Most drugs in use nowadays in therapy have been found and evaluated with these methods. The chemistry and pharmacology of drugs found and used in therapy until 1970 are described in the five volumes of the book by Ehrhart and Ruschig H (eds), *Arzneimittel, Entwicklung, Wirkung Darstellung* (1972).

In the mid 1970s *receptor binding assays* were introduced as an approach for compound evaluation by the development of radioligand binding assays, based on evaluation procedures and mathematical calculations provided by Schild (1947), Scatchard (1949), Stephenson (1956), Ariens and van Rossum (1957), Arunlakshana and Schild (1959), Furchgott (1966), Cheng and Prusoff (1973), Rodbard and Frazier (1975), Lefkowitz and Williams (1977), Bennett (1978), Creese (1978),

Munson and Rodbard (1980), McPherson (1985a,b), Tallarida and Murray (1987), and Greenstein (1991). Receptor binding assays were described for various transmitters as well as assays for ion channels and neurotransmitter reporters. Characterization and classification of receptors is a continuous procedure. The use of radioligand-binding assays has facilitated the design of new chemical entities, especially as the information obtained has been used in deriving molecular models for the structure-activity relationship.

The receptor technology provides a rapid means to evaluate small amounts of compound (5-10 mg) directly for their ability to interact with a receptor or enzyme, independent of its efficacy. But as new assays are developed, it also provides a means to profile the activity of compounds against a battery of binding sites, thereby yielding an *in vitro* radioligand binding profile. Receptors have been divided into an increasing number of subclasses so that full characterization of a new chemical entity by receptor pharmacology also needs time and material. DNA sequencing technology has identified many receptors, mostly belonging to the G-protein-coupled receptor superfamily. Reverse molecular pharmacology and functional genomic strategies are recommended to identify the activating ligands for these receptors.

The reverse molecular pharmacological methodology includes cloning and expression of G-protein-coupled receptors in mammalian cells and screening these cells for a functional response to cognate or surrogate agonists present in biological extract preparations, peptide libraries, and complex compound preparations. The ligand-binding assay is a powerful tool in the search for agonists and antagonists for novel receptors, and for identification of novel classes of agonists and antagonists of known receptors. However, it does not differentiate between agonist and antagonist. Ligand-binding mass screening can be adapted for very high throughput. However, well-defined criteria have to be fulfilled to avoid blind alleys.

PHARMACOLOGICAL APPROACHES

It is universally bemoaned that development costs for new pharmaceutical drugs are exploding, whilst the output of new drugs is decreasing. *Sams-Dodd* (2006) analyzed the strength and limitations of various drug discovery approaches. The *target-based drug discovery approach*, once welcomed as a change in paradigm, has for the past 10-15 years been the dominating approach in drug discovery. However, in the past few years the commercial value of novel targets in licensing deals has fallen dramatically, reflecting the fact that the probability of reaching a clinical drug candidate for a novel target is very low. This has naturally led to questions regarding the success of target-based drug discovery and, more importantly, a search for alternatives. The *mechanism-based approach*, which corresponds to the target-based approach, screens for compounds with a specific mode of action.

The focus on a single mechanism that permits the use of high-throughput screening (HTS) and ultra-HTS technology for screening makes it highly effective for the identification of target-selective compounds. However, despite the fact that this approach is highly advantageous from a scientific and practical viewpoint, it does not translate into a high success rate for novel targets. A success rate of 3% for reaching preclinical development is not attractive. Each drug discovery project usually lasts 2-4 years, and if 33 targets on average have to be evaluated to identify one that can proceed into preclinical development, it means that at least 66 years of research is needed to produce one successful drug discovery project for a novel target. Considering that the attrition rate caused by toxicological effects and lack of clinical effects are also substantial, it means that novel unvalidated targets are not attractive for the pharmaceutical industry. The *physiology-based approach* was the first drug discovery paradigm, and this has resulted in many effective treatments. It seeks to induce a therapeutic effect by reducing disease-specific symptoms or physiological changes and screens compounds for these properties in animal models that mimic specific aspects of disease symptomatology.

The screening is conducted in isolated organ systems or in whole animals. The physiology-based approach suffers from low screening capacity and difficulty in identifying the mode of action of drugs. The *function-based approach* seeks to induce a therapeutic effect by normalizing a disease-specific abnormality. Compared with the mechanism-based approach, functional parameters represent a higher level of organism complexity because function requires the integrated action of many mechanisms. However, unlike the physiology-based approach, the parameters cannot be compared to the symptoms observed in patients. The screening capacity cannot be used for library screening.

In summary, no clear advantage of one drug discovery paradigm over the others could be demonstrated, because all of them have their specific strengths and weaknesses. The classical approaches cannot be abandoned.

Cuatrecasas (2006) analyzed the reasons for insufficiency of drug discovery and development. The low productivity is certainly not related to available budgets, which have increased 30-fold since 1970. The Food and Drug Administration (FDA) or regulatory issues are not fundamental barriers and contribute only marginally to the decline in drug productivity. Likewise, one cannot blame the current state of scientific advances, which in the last 20 years have been revolutionary.

Much more important are changes in the approach to management. Foremost among the issues that cripple drug research and development (R&D) is that while utmost creativity and innovation are required, the R&D is conducted in traditional for-profit corporations that are virtually indistinguishable operationally from those that conduct little or no R&D. Most corporations' top management does not understand the complexities of science, its mode of conduct or objectives, and runs the companies in ways that stifle creativity and innovation.

Management

In the 1970s things began to change. Modern managers entered as chief executive officers (CEOs) and other high-level executives, mostly with little or no technical experience. Freedom, spontaneity, flexibility, nimbleness, tolerance, compassion, humor, and diversity were replaced by bulky and inflexible organizational structures characterized by regimentation, control, conformity, and excessive bureaucracy.

There are still companies that try to focus on excellent science and that attract first-rate scientists. These are, however, exceptional situations. Drug R&D thrives in a creative, flexible, and nonautocratic environment. Success depends on individual freedom and inspiration rather than dogmatic leadership.

Public Companies

The ownership of public companies consists mainly of shareholders who expect rapid (and substantive) returns on their investments. This contrasts (and often conflicts) with the nature of the business objectives, which must be based on long-term investments in science and technology.

Merger and Acquisitions

The decreasing earnings have stimulated mergers and acquisitions, driven by the desire to acquire existing sales (products) while decreasing costs via layoffs. This has created conditions that catalyze further inefficiencies and suffocation of innovation. The merged megacompanies' research organizations must be integrated rapidly and redundancies eliminated, often-times in haste. Companies have become much less interested in developing drugs that will sell less than $1 billion a year. Without these "blockbusters" they cannot maintain the traditionally high profits. Berkowitz and Sachs (2002) described discovery and development of the blockbuster drug omeprazole.

Drugs of Rare Diseases

In the 1960s through the 1990s, orphan diseases were simply rare diseases for which drug companies did not wish to initiate discovery programs, for commercial reasons. It was understood that prices could not be adjusted out of line with the prevailing standards. In 1983, the US Congress passed The Orphan Drug Act to provide tax, funding, and exclusivity incentives to companies developing drugs for low-prevalence diseases (defined as fewer than 200,000 cases) that were not expected to provide investment returns. However, many drugs that have become blockbusters had early histories of major disinterest and skepticism. There are many publications on orphan drugs and drugs for rare diseases, but they concern mainly cost-effectiveness and funding of clinical trials, but not pharmacological methods. It seems necessary that experimental pharmacologists know rare diseases and think about methods to evaluate drugs against rare diseases.

Researchers are now looking back to the golden age of drug discovery in the 1950s and 1960s of the last century. What was different at that time? A few personal remarks: in the 1960s, we were looking for analgesics in a special chemical series. However, we were afraid of addictive and hallucinogenic potential. In animal experiments we found a compound almost as effective as methadone. After 2 weeks of toxicity tests, we gave it to one of our behavior cats. The results were unequivocal. So, without informing anybody, one of us took a rather high dose orally. He did not hallucinate. So we gave the compound to patients.

There were no side-effects after oral dosing, but hallucinogenic side-effects were reported after intravenous dosing. We had to drop development. Another example: we were looking for an effective and highly tolerable local anesthetic. One pharmacologist performed animal experiments, but he injected interesting compounds into his forearm to evaluate pain and the local anesthetic effect. An excellent compound was found this way. One further example: one of our colleagues was testing sedative and hypnotic compounds. After some toxicological studies, he gave the compound to a colleague in charge of first clinical trials. He took the compound himself, but then collapsed behind his desk.

Conclusion: we were not afraid to take our compounds ourselves even flaunting regulations. Thus compounds were tested by the pharmacologists themselves. If we look back to the great breakthroughs in the 1950s and 1960s, we see that many important ones came from serendipitous findings. Horace Walpole (1717–1797) coined the term serendipity in 1754 in allusion to an ancient oriental legend of the "Three Princes of Serendip". "Serendipity" in drug discovery implies the finding of one thing while looking for something else.

The diuretic action of sulfanilamide and acetazolamide as inhibitors of carbonic anhydrase was found in clinical trials. Sulfonylureas were tested as antibiotic drugs to be compared with sulfonamides. The blood sugar-lowering-effect of sulfonylureas designed as antibacterial agents was found just by chance. Then it turned out that the amino group was not necessary, resulting in tolbutamide. On the other hand, one has to keep in mind that millions of diabetics were treated successfully with sulfonylureas over decades, whereby only the release of insulin from the pancreatic β-cell, but not the molecular mechanism, was known.

The antipsychotic activity of phenothiazines (chlorpromazine) originally tested as antihistaminic drugs was found in clinical studies, as was the antidepressant effect of imipramine. The antidepressant effect of the monoamine oxidase inhibitor iproniazid was found in clinical trials. *Valproic acid* was used as an organic solvent in research laboratories for eight decades, until the observation of action against pentylenetetrazol-induced convulsions in rodents. Clinical experience emphasized therapy for absence seizures in primary generalized epilepsies. Anecdotal observations in patients with both epilepsy and migraine headaches led to the discovery of its antimigraine efficacy. More than a decade later, clinical trials established a significant efficacy of valproate in mania.

Butazolidin was used as an additive to increase the solubility of aminopyrine resulting in a combination product. The therapeutic value of butazolidin as single drug was recognized later. The blood-pressure-lowering effect of *clonidine* was found when tested as a vasoconstrictive drug for nasal application. The mechanism of action was clarified in animal experiments. *Furosemide* was found to be a potent diuretic and saluretic drug comparable to other substances. But animal experiments and studies in volunteers showed an effect at doses at which other diuretic drugs were no longer effective. Thus the name "*high-ceiling diuretic*" was coined.

The discovery of *sildenafil*, a highly selective inhibitor of phosphodiesterase-5 (PDE-5), was the result of research on chemical agents that might be useful in the treatment of coronary heart disease. Initial clinical studies on sildenafil were not promising with respect to its anti-anginal potential. However, the incidental discovery of its anti-impotence effect led to its approval for the treatment of erectile dysfunction. Most new drugs, however, were found by the *classical approach*, which has the advantage of relatively high relevance. If a compound has blood-pressure-lowering activity in hypertensive rats after oral dosage, the chances of activity in humans are high.

Measurement of dose-response curves, effects over a given period of time, and comparison of the effects after intravenous and oral administration have already given hints for pharmacokinetic data. This approach had the disadvantage that it was time-consuming and required relatively large amounts of the new compound (usually about 5 g). Furthermore, this approach provided little information about the molecular mechanisms involved in the observed effects. What were the deficiencies of the classical approach? We did not know the exact mode of action of our drugs in spite of the fact that we could treat patients effectively.

This situation has changed. Molecular biology has a significant impact on drug discovery, offering us over 30,000 genes and their protein products as potential drug targets, and revolutionizing the processes involved in seeking drug targets, testing hypotheses, and screening for activity. In addition, the techniques of combinatorial chemistry are available, which generate thousands of compounds to test against the thousands of targets. The search for new lead compounds involves intricate high-throughput screening technology with tremendous capacity, and the use of cell culture and recombinant human cells and enzymes. So we have many hypotheses and we know many biochemical steps that could be the mode of action of a new drug, but we do not know the new drug. The biochemical approach reduced the use of animals.

When animal experiments are performed at a later stage, quite often the results of *in vitro* experiments cannot be validated *in vivo*. In all countries we have regulations for the care and use of laboratory animals. Animal rights extremists have created an atmosphere where students are reluctant to perform whole-animal studies. Our experience tells us that only a research worker who loves animals and is not afraid of them should perform animal experiments. What can be recommended from experimental pharmacology to increase the efficacy of drug development?

1. Looking for *short-cuts*, even neglecting the present rules when performing self experiments.
2. To give pharmacologists the freedom, time and money to realize their own ideas.
3. Giving room for serendipity.
4. Performing animal experiments in parallel to *in vitro* studies.
5. In respect to *orphan drugs*, it seems necessary that experimental pharmacologists know rare diseases and think about methods to evaluate drugs against these diseases.
6. Pharmacologists should have training in medicine, not only in biochemistry.
7. Outstanding pharmacologists should be selected as leaders in research and development.

LATEST FINDINGS IN DRUG DISCOVERY

It was always due to the ingenuity of the chemist that new chemical entities in pharmaceutical research were created. For some period of time, computer-based molecular modeling was thought to be the most effective approach. With the increasing characterization of the three-dimensional structures of receptors and enzymes, the design of molecules that interact with these biological targets was thought to be an intellectual approach using modern computer technology. This attitude to the development of combinatorial chemistry has changed. It has been defined as the systematic and repetitive covalent connection of a set of different "building blocks" of varying structures to each other to yield a large array of diverse molecular entities. An interim analysis showed only limited success of combinatorial chemistry approaches in terms of delivering leads. The interest in natural products has been regained.

It has become possible to use solid- or solution-phase syntheses with different chemistries and scaffolds to produce libraries tailor-made for finding or optimizing a lead directed at almost any class of target. Several suggestions were made to enhance the success of the hit to lead process. Golebiowski et al. (2003) reviewed lead compounds discovered from libraries at various research institutes. Design of screening libraries targeted at macromolecules, such as synthetic and G-protein-coupled receptors, was described by Juliano et al. (2001), Crossley (2004), Jimonet and Jager (2004), and Srinivasan and Kilburn (2004). Purification strategies for combinatorial and parallel chemistry were discussed by Edwards (2003). New tools, such as molecular docking algorithms, mapping of protein-binding sites by nuclear magnetic resonance, preparation of highly enantioselective selectors for chiral HPLC, microfluidic systems, light-directed synthesis, and homology modeling of proteins, allow an unprecedented level of rational design to guide the synthesis of prospective drugs.

Spectroscopic methods, including especially mass spectrometry and to a lesser extent infrared and nuclear magnetic resonance, have been applied at different levels of combinatorial library synthesis: in the rehearsal phase to optimize the chemistry prior to library generation, to confirm library composition, and to characterize after screening each structure that exhibits positive response. Schmid et al. (2000), Heeren et al. (2004), and Zhang et al. (2005) discussed the application of mass spectrometry using high-performance Fourier transform ion cyclotron resonance mass spectrometry (FTICR-MS) methods. Blundell and Patel (2004) underlined the value of high-throughput X-ray crystallography for drug discovery.

High-Throughput Screening (HTS)

The logical development of receptor technology was high-throughput screening (HTS). This evolution was closely connected with the changes in strategy of chemical synthesis. The vast number of compounds produced by combinatorial chemistry and the possibility of testing many compounds, including natural products, in a short period of time by HTS stimulated each other. High-throughput screening is the process of testing a large number of diverse chemical structures against disease targets to identify "hits". Compared to traditional drug screening methods, HTS is characterized by its simplicity, rapidness, taking the ligand-target interactions as the principle, as well as leading to a relatively high information harvest.

As a multidisciplinary field, HTS involves an automated operation platform, a highly sensitive testing system, specific screening models (*in vitro*), an abundant components library, and a data acquisition and processing system. Various technologies, especially technologies such as fluorescence, nuclear magnetic resonance, affinity chromatography, surface plasmon resonance, and DNA microarray, are available, and the screening of more than 100,000 samples per day is possible. Fluorescence-based assays include the scintillation proximity assay, time-resolved energy transfer, fluorescence anisotropy, fluorescence correlation spectroscopy, and fluorescence fluctuation spectroscopy. With the introduction of robotics, automation and miniaturization techniques, it became feasible to screen 50,000 compounds a day with complex workstations. Full automation of all assay steps, from compound addition to data collection, ultimately allows screens to be run continuously – 24 h a day, 7 days a week.

Another progress is the development of 1536- well microtiter plate formats. Multiple types of formats are emerging beside the 96-, 384-, and 1536-, even 3456- and 9600- well plates. The trend toward assay miniaturization for high-throughput and ultra-high-throughput screening intensified the development of homogenous, fluorescence-based assays in higher density, smaller volume microplate formats. Microfluidic devices have been designed to perform continuous-flow biochemical and cell-based assays providing orders of magnitude reduction in reagent consumption. Harding et al. (1997) described the development of an automated HTS system involving two systems, the first handling isotopic assay ("hot" system), the second non-isotopic assays ("cold" system). An ultra-high-throughput screen based on fluorescence isotropy and performed entirely using 1536- well assay plates was published by Turconi et al. (2001).

Further studies on miniaturized HTS technologies were published by Wolcke and Ullmann (2001) and Brandish et al. (2006). Target validation should not be underestimated in the drug research and development process. Enzyme assays for HTS were evaluated by Goddard and Reymond (2004). Assaying enzyme-catalyzed transformation in high throughput is crucial to enzyme discovery, enzyme engineering and the drug development process. In enzyme assays, catalytic activity is detected using labeled substrates or indirect sensor systems that produce a detectable spectroscopic signal upon reaction. *High-content screening* was proposed by Giuliano et al. (1997) as a new approach to easing key bottlenecks in the drug discovery process.

High-content screening defines the role of targets in cell function by combining fluorescence-based reagents with the ArrayScan to automatically extract temporal and spatial information about target activities within cells. High-content screening integrates cell-based assays, high-resolution fluorescence microscopy, and proprietary image processing algorithms for the automated analysis of cellular and subcellular events. It delivers the potential to screen using targets that previously were minimally used or avoided due to a lack of a robust way to measure them, such as morphology changes, cellular differentiation and cytoskeletal changes, cell-to-cell interactions, chemotaxis and motility, as well as spatial distribution changes such as receptor trafficking or complex formation. Clark and Pickett (2000) underlined the importance of computational methods for the prediction of "drug-likeness", in particular for the prediction of intestinal absorption and blood–brain barrier penetration.

Several authors expressed their thoughts on the future of HTS. Valler and Green (2000) expressed some doubts on diversity screening in drug discovery and proposed focused screening using three-dimensional information about the target which may improve the hit rate 10- to 100-fold over random screening. Nevertheless, this technical progress has led to a new concept in drug discovery: "a change in paradigm". Large numbers of hypothetical targets are incorporated into *in vitro* or cell-based assays and are exposed to large numbers of compounds of natural or synthetic origin. This change in paradigm created new areas, such as bioinformatics, cheminformatics, and functional genomics.

Many compounds can elicit a positive response in a particular assay, which is called a hit. In this new terminology "hit" is a molecule with confirmed activity from the primary HTS assay, a good profile in secondary assays and a confirmed structure. The hits have to be structurally defined and may give rise to leads. "Lead" is a hit series for which the structure-activity relationship is shown and activity demonstrated both *in vitro* and *in vivo*. The implementation of these methods is being complemented by an increase in the use of automation and robotics. Automation may be defined as the use of stand-alone instrumentation (work stations) that performs a given task, whereas robotics uses a robot arm or track systems to move microplates between these instruments.

These technologies are continuously improved. High-throughput screening methods are also used to characterize metabolic and pharmacokinetic data about new drugs. Kansy et al. (1998)

recommended parallel artificial membrane permeation assays for the description of passive absorption processes early in HTS. High-throughput screening assays have been used for the assessment of cytochrome P450 isoenzyme metabolism using fluorogenic Vivid substrates. A large pharmaceutical company has the potential to screen up to 100,000,000 compounds per year. In any case, the leads or derivatives of them have to be tested in more complex models, isolated organs or animals. Only compounds active in classical pharmacological tests can be taken into preclinical and clinical development.

Since only a few drug candidates, originally screened in HTS, have reached the market, the searching for higher productivity of HTS is continuing, especially because development costs for new drugs have increased tremendously. With the completion of the human genome sequencing, bioinformatics will be an essential tool in target discovery, and the *in silico* analysis of gene expression and gene function will be an integral part of it, facilitating selection of the most important targets for a disease under study. Optimistic papers on the value of virtual screening of virtual libraries were written by Green (2003) as well as by Davies et al. (2006). Some doubts were expressed on the philosophy of research management in the pharmaceutical industry neglecting the influence of creative scientists, failing to understand biological complexity, and overestimating the outcome of the "change in paradigm". Nevertheless, the race to test more and more compounds in a shorter time in ultra-HTS with more automation is continuing.

False positive in HTS

As for every screening procedure, HTS suffers from type 1 and type 2 errors. Type 1 errors are false positives. In HTS, a poor candidate or an artifact gives an anomalously high signal, exceeding an established threshold. Type 2 errors are the false negatives. In HTS, a perfectly good candidate compound is not flagged as a hit, because it gives an anomalously low signal. Moreover, a low degree of relevance of the test may induce a high failure rate of type 2. Much more attention is given in the literature to false-positive results in HTS than to type 2 errors. Some of the false positives are promiscuous compounds that act noncompetitively and show little relationship between structure and function. McGovern et al. (2002, 2003) and Seidler et al. (2003) found that one mechanism of inhibitor promiscuity is aggregate formation: individual molecules group together to form particles 30-1000 nm in diameter and these aggregates are the active inhibitory species.

Mathematical calculations of HTS

The ability to identify active compounds ("hits") from large libraries accurately and rapidly is the ultimate goal of HTS assays, which depends on the suitability and quality of the assay. The Z-factor and the Z′ factor were introduced to characterize the reliability of high-throughput assays. The Z-factor is a dimensionless, simple statistical characteristic, which reflects the assay signal dynamic range and the data variation associated with the signal measurements. The Z-factor is defined as a screening window coefficient, which uses the formula

$$Z = 1 - \frac{3\sigma_s + 3\sigma_c}{|\mu_s - \mu_c|}$$

in which μ_s and μ_c denote the means of the population and control (usually background) signals, respectively, and the signal standard deviations are denoted as μ_s and μ_c, respectively. This coefficient takes into account the assay signal's dynamic range, the data variation associated with the sample measurement in the presence of the test library, and the data variation associated with the reference control measurement. The Z′-factor has been defined as the ratio of the separation band to the dynamic range of the assay, based on the positive control and negative control data of the assay. It takes the formula

$$Z' = 1 - \frac{3\sigma_{c+} + 3\sigma_{c-}}{|\mu_{c+} - \mu_{c-}|}$$

in which μ_{c-} and μ_{c+} denote the means of the negative control signal and positive control signal, respectively. The standard deviations of the signals are denoted as σ_{c-} and σ_{c+}, respectively. The Z′-factor is a simple, dimensionless, and characteristic parameter for the quality of each assay. Values of 0.7 or higher are considered to indicate a valuable assay. The Z-factor and the Z′-factor have been used by various authors to characterize the quality of assays.

Fundamental Changes

Fundamental changes in assay technologies have facilitated the development of high-throughput screening (HTS). Zehender et al. (2004) recommended Speed-Screen as the "missing ling" between genomics and lead discovery. SpeedScreen is a label-free, in- solution, affinity-based selection methodology for HTS. The protocol compromises in-solution affinity selection, followed by size exclusion chromatography in combination with microbore-liquid-chromatography/electrospray-ionization mass spectroscopy. In *multiplexing assays,* more than one component is studied in a single sample. Advantages of this strategy include economy, since theoretically the same sample may be used for many measurements. Also, because the exact same sample is used for different assays, sample errors caused by heterogeneity are minimized. Nieuwenhuijsen et al. (2003) published a dual luciferase multiplexed HTS platform for protein-protein interactions. Maley et al. (2004) used multiplexed real-time reverse transcriptase-polymerase chain reaction (RT-PCR) for HTS applications.

Fluid dispensing systems

Several systems for fluid dispensing are in use: the piezo- and inkjet, the "air displacement", and the "pin-tool" systems. Piezoelectric and inkjet systems were the first introduced to the market. There are three types of inkjet dispensers: thermal, solenoid, and piezoelectric. All of these use some means of compressing a liquid against a small orifice to create sufficient linear velocity to eject the fluid in the form of a drop. The "pintool" systems use pins (needles) to transfer liquid. Up to 1536 needles are immersed into the fluid and withdrawn, taking up a few nanoliters and transferring the fluid into the wells of a microtiter plate.

Utilization of syringes and syringe needles as a contact tool enables multiple dispensing from a single fill. Generally, syringe dispensing has a higher degree of accuracy and repeatability compared with pin dispensing tools. Microfluidic arrays for high-throughput submicroliter assays using capillary electrophoresis were described by Gibbons (2000). Capillary electrophoresis is an analytical technique that can achieve rapid high-resolution separation of water-soluble components present in small sample volumes. Separations are based on the electrically driven movement of ions. In HTS, analytes are separated according to differences in electrophoretic mobility, depending on the charge-to-size ratio. The most common "no contact" delivery format is piezoelectric acoustic delivery. This technique delivers a powerful sonic or ultrasonic pulse, which accelerates a tiny droplet from the surface of liquid in the stock vessel, through the air and onto the dry capture of a microplate. Hsieh et al. (2004) used disposable piezoelectric ejectors for liquid dispensing in ultrahigh-throughput microarray screening.

In vitro biochemical assays

Cell-based assays are increasingly attractive alternatives to *in vitro* biochemical assays for HTS. They fall into three broad categories: second messenger assays that monitor signal transduction following activation of cell-surface receptors; reporter gene assays that monitor cellular responses at the transcription/translation level; and cell proliferation assays to monitor the overall growth/no growth response of cells to external stimuli. Hynes et al. (2003) described a luminescence-based assay for screening the viability of mammalian cells, based on the monitoring of cell respiration by means of a phosphorescent water-soluble oxygen probe that responds to changes in the concentration of dissolved oxygen by changing its emission intensity and lifetime. Yan et al. (2002) described a cell-based HTS assay system for monitoring G-protein-coupled receptor activation using β-galactosidase enzyme complementation technology. This technology uses a pair of inactive β-galactosidase deletion mutants as fusion partners to the protein targets of interest. To monitor G-protein-coupled receptor activation, stable cell lines expressing both G-protein-coupled receptor and β-arrestin- β-galactosidase fusion proteins are generated.

Following ligand stimulation, β-arrestin binds to the activated G-protein-coupled receptor, and this interaction drives functional complementation of the β-galactosidase mutant fragments. G-protein-coupled receptor activation is measured directly by quantitating restored β-galactosidase activity with the chemoluminescent Gal-Screen reagent. Cell-based reporter assays are used where human receptors are transfected into null cell lines either alone, or as part of receptor systems constructed to show alterations in light production (luciferin-luciferase) or light transmission (melanophore), which can be measured independently of radioactivity within minutes. Pathirna et al. (1995) developed a transient transfection HTS assay to identify nonsteroidal human progesterone receptor modulators using luciferase expression. Zhu et al. (2004) found a good correlation between high-throughput pregnane X receptor (PXR) transactivation and binding assays.

Another reporter gene, green fluorescent protein, can be measured by fluorescence techniques in real time so that kinetic parameters can be determined from a single well of a microtiter plate. Ting et al. (2001) described genetically encoded fluorescent reporters of protein tyrosine kinase activities in living cells. Gonzalez and Negulescu (1998) described intracellular detection assays for HTS which employ either fluorescent or luminescent readouts. The method is particularly suitable for measuring calcium mobilization. Using receptor selection and amplification technology (R-SAT), it is possible to rapidly quantify specific pharmacological responses as a function of changes in cell proliferation rates.

NIH3T3 cells (or other cell lines) are transfected with the plasmid coding for the receptor of interest and pSV-β-Gal. Cells are then transferred to 96-well microtiter plates and test compounds are added. Several days later, β-Gal activity is measured via a simple, inexpensive colorimetric assay. As a result of compound treatment, the changes in enzyme activity are a function of the proliferative activity of the cells. This functional assay can quantitatively differentiate between full agonists, partial agonists, neutral antagonists, and negative antagonists (inverse agonists). An interesting aspect of this technology is that many genes in a related family can be combined in the same cell population for convenient assay. Bailey et al. (2002) discussed the applications of transfected cell microarrays in high-throughput drug discovery.

Transfected cell microarrays are a complementary technique in which array features comprise clusters of cells overexpressing defined cDNAs. Complementary DNAs cloned in expression vectors are printed on microscope slides, which become living arrays after the addition of a lipid transfection reagent and adherent mammalian cells. Dupriez et al. (2002) and Le-Poul et al. (2002) described aequorin-based functional assays for G-protein-coupled receptors, ion channels, and tyrosine kinase receptors. AequoScreen is a cellular-based functional assay in which the cells are loaded with the apoaequorin cofactor coelenterazine, diluted in assay buffer, and injected into plates containing the samples to be tested. Kunz-Schughart et al. (2004) recommended the multicellular spheroid model and the use of 3-D cultures for HTS.

Gaspari et al. (2004) described the quantification of the proliferation index of human dermal fibroblast cultures with the ArrayScanTM high-content screening reader. Isailovic et al. (2005) presented a high-throughput method for measuring single-cell fluorescence spectra. Upon excitation with a 488 nm argon-ion laser many bacterial cells were imaged by a 20 × microscope objective while they moved through a capillary tube. Fluorescence was dispersed by a transmission diffraction grating, and an intensified charge-coupled device (ICCD) camera simultaneously recorded the zero and the first orders of the fluorescence from each cell. *Flow cytometry* was recommended for high-throughput, high-content screening by Edwards et al. (2004). Flow cytometry is a platform for quantitative multi-parameter measurements of cell fluorescence. Homogeneous discrimination between free and cell-bound fluorescent probe eliminates wash steps to streamline sample processing. Waller et al. (2004) used flow cytometry as a technique for studying GPCR assembly, pharmacology, and screening.

Ion channels

Ion channels represent a class of membrane-spanning protein pores that mediate the flux of ions in a variety of cell types. To date, >400 ion channels have been cloned and characterized, and some of these channels have emerged as attractive drug targets. Zheng et al. (2004b) reviewed the technologies that are currently used for the screening of ion channels. The technologies discussed are binding assays, ion flux assays, fluorescence-based assays, and automated patch- clamp instrumentation. Gill et al. (2003) described flux assays in HTS of ion channels in drug discovery. Technologies based on flux assays are available in a fully automated high-throughput format for efficient screening. This application offers sensitive, precise, and reproducible measurements giving accurate drug rank orders matching those of patch-clamp data. *Automated patch-clamp systems* useful for HTS techniques, such as Qpatch, have been recommended by several authors.

Biochemical detection

Fluorescence-based assay technologies play an increasing role in HTS. When compared to other biochemical and cell-based techniques, fluorescence has a significant advantage over such methods as isotopic labeling, colorimetry, and chemiluminescence.

Sensitivity

It is relatively simple for modern instrumentation to reliably detect a signal being emitted from a single fluorescent molecule. In biological applications, this level of sensitivity means molecules that may be present only in very small numbers can be easily detected. In the case of imaging application, such as high-content screening, their intracellular location can be determined. This high level of sensitivity also means that transient biological events can be detected very quickly, hence enabling the measurement and understanding of events that occur very rapidly inside living cells. The inherent sensitivity of fluorescence technology also permits the use of very low concentrations of fluorescent label.

Specificity

Fluorescent probes are readily available for labeling virtually any biomolecule, structure, or cell type. Immunofluorescent probes can be directed to bind not only to specific proteins but also to specific conformations, cleavage products, or site modifications, such as phosphorylation. Individual peptides and proteins can be engineered to autofluorescence by expressing them as AFP chimeras inside cells. Such high levels of specificity enable the use of several different fluorescent labels - each emitting at a unique color - and the subsequent understanding of the complex interactions that occur among and between subcellular constituents.

Accuracy

Reagents used for fluorescence labeling are well understood and their performance has been characterized under a wide variety of biological environments.

Flexibility

Fluorescence technologies have matured to the point where an abundance of useful dyes is now commercially available. Many dyes have been developed and optimized for labeling specific cell compartments and components, as well as for labeling proteins (antibodies, ligands, toxins, etc.), which can be used as probes. Other fluorescent sensors have been designed to report on biological activities or environmental changes (pH, calcium concentration, electrical potential, proximity to other probes, etc.). Multiple fluorescent labels can be used on the same sample and individually detected quantitatively, permitting measurement of multiple cellular responses simultaneously. Many quantitative techniques have been developed to harness unique properties of fluorescence, including:

1. Fluorescence resonance energy transfer (FRET)
2. Fluorescence polarization or anisotropy (FP)
3. Time-resolved fluorescence (TRF)
4. Fluorescence lifetime measurements (FLM)
5. Fluorescence correlation spectroscopy (FCS)
6. Fluorescence photobleaching recovery (FPR)
7. Molecular beacons (MB)

Mere et al. (1999) reported miniaturized *fluorescence resonance energy transfer* (FRET) assays and microfluidics as key components for ultra-HTS. An ultra-HTS system (UHTSS) platform integrates several microfluidic dispensing devices, which have been developed to deliver nanoliter to microliter volumes at high speed into the appropriate wells of 384-well NanoWellTM assay plates. The activity of the compounds dispensed into the assay wells is quantified via a highly sensitive dual-wavelength emission fluorescence detector. Pope et al. (1999) discussed the advantages and disadvantages/limitations of fluorescence techniques for the establishment of miniaturized homogeneous screening assays, including prompt intensity assay (FLINT), anisotropy assay (FA), prompt energy transfer assay (FRET), fluorescence correlation spectroscopy (mass-dependent FCS), and fluorescent correlation spectroscopy (mass-independent FCS).

Nagai et al. (2004) used FRET technology to develop genetically encoded fluorescent indicators for various cell functions. Chin et al. (2003) described miniaturization of cell- based beta-lactamase-dependent FRET assays to ultrahigh-throughput formats to identify agonists of human liver X receptors. Wallrabe and Periasami (2005) reviewed imaging protein molecules using Förster (or fluorescence) resonance energy transfer (FRET) and fluorescence lifetime imaging measurements (VLIM). Dynamics of fluorescence fluctuations in green fluorescent protein observed by *fluorescence correlation spectroscopy* (VCS) were described by Haupts et al. (1998).

Auer et al. (1999) recommended fluorescence correlation spectroscopy for lead discovery by miniaturized HTS. By monitoring interactions "of single molecules in femtoliter volumes, fluorescence correlation spectroscopy offers the highest potential as the detection technique at the nanoscale. Schwille et al. (2000) reported that fluorescence correlation spectroscopy reveals fast optical excitation-driven intramolecular dynamics of yellow fluorescent proteins. Haupts et al. (2000) compared macroscopic versus microscopic fluorescence techniques in (ultra)-HTS. Rudiger et al. (2001) reviewed single-molecule detection technologies in miniaturized HTS and studied binding assays for G-protein-coupled receptors using fluorescence intensity distribution analysis (FIDA) and fluorescence anisotropy. As a confocal detection technology, FIDA inherently allows reduction of the assay volume to the microliter range and below without any loss of signal.

Palo et al. (2002) introduced a method that combines the benefits of both fluorescence intensity distribution analysis (FIDA) and fluorescence lifetime analysis. It is based on fitting the two-dimensional histogram of the number of photons detected in counting time intervals of given width and the sum of excitation to detection delay times of these photons. Referred to as fluorescence intensity and lifetime distribution analysis (FILDA), the technique distinguishes fluorescence species on the basis of both their specific molecular brightness and the lifetime of the excited state and is also able to determine absolute fluorophore concentrations. Jager et al. (2003a) reviewed fluorescence techniques for high-throughput drug discovery. In particular, the family of fluorescence techniques, Fluorescence Intensity Distribution Analysis (FIDA), which is based on confocal single-molecule detection, has opened up a new field of HTS applications.

Eggeling et al. (2005) recommended the combination of Förster resonance energy transfer (FRET) and two-color global fluorescence correlation spectroscopy (2CG-FCS) for high-throughput applications in drug screening. Sportsman and Leytes (2000) discussed miniaturization of homogenous assays using *fluorescence polarization* (FP), which can detect changes in molecule size as well as quantify the binding of small molecules to larger molecules. Polarization of fluorescence occurs when a fluorescent molecule is illuminated with plane-polarized light, provided that the molecule does not move during the course of fluorescent lifetime. For typical fluorescent molecules such as fluorescein, the polarization is not observed if the molecule is rotating rapidly in solution under conditions typically used in biological assays.

Thus, if the polarization of fluorescence of a fluorescein-labeled ligand is measured, a low polarization will be observed when the ligand is free in solution, and a high polarization will be observed when the ligand is bound to a macromolecule such as a specific receptor or antibody. Banks et al. (2000) described fluorescence polarization assays for HTS of G-protein-coupled receptors in 384-well microtiter plates. Parker et al. (2000) used fluorescence polarization to develop HTS assays for nuclear receptor-displacement and kinase inhibition. This method is a solution-based, homogeneous technique, requiring no immobilization or separation of reaction components. As an example, the authors described the fluorescence polarization-based estrogen receptor assay based on the competition of fluorescein-labeled estradiol and estrogen-like compounds for binding to the estrogen receptor.

Li et al. (2000) developed an assay for measuring the activity of an enzyme that transfers multiple adenine-containing groups to an acceptor protein, based on fluorescence polarization technology in a 1536- well plate format. Texas red (rhodamine) was covalently conjugated to the adenine of the donor substrate through a C_6 spacer arm. As a result of the transfer of the adenine-containing moieties to the acceptor protein substrate, the rotational correlation time of the Texas red conjugate increased, hence increasing the degree of fluorescence polarization. Owicki (2000) compared the perspectives of fluorescence polarization and anisotropy in HTS. Qian et al. (2004) used fluorescence polarization to discover novel inhibitors of Bcl-xL, a factor preventing apoptosis.

Lu et al. (2004) developed a fluorescence polarization bead-based coupled assay to target activity/conformation states of a protein kinase. This assay is based on the principle of the Immobilized Metal Assay for Phosphochemicals (IMAP), a fluorescence polarization assay based on the affinity capture of phosphorylated peptides. Kolb et al. (1997) and Grepin and Pernelle (2000) described the evolution of *homogeneous time resolved fluorescence* (HTRF) technology for HTS. HTRF uses the europium (Eu^{3+}) ion caged into polycyclic cryptate (Eu-cryptate) as a donor. Laser excitation at 337 nm transfers the energy from the Eu-cryptate complex to an allophycocyanin acceptor molecule (XL665), the APC. This results in the emission of light at 665 nm over a prolonged time (milliseconds). Uncoupled or free XL665 emits at 665 nm, but with a short decay.

The light emission is recorded in a time-resolved fashion over a 400-μs period, starting 50 μs after the excitation pulse, so that the autofluorescence from the media and the short-lived fluorescence of the free APC are not recorded. Eu-cryptate emits at 620 nm and is discriminated from XL665 by wavelength. The labeling of macromolecules for HTRF assay can be done in a number of ways, depending on the type of screen, using direct, indirect or semi-direct types of labeling. Direct labeling is characterized by having the (Eu)K and XL665 covalently bound to the molecule or molecules of interest.

Indirect labeling is when the (Eu)K and XL665 are bound to macromolecules through secondary interactions, such as biotin-streptavidin interaction or antibody binding. Other indirect methods, such as (Eu)K or XL665-labeled lectin, can be used for binding to membranes. The semi-direct method is a combination of direct and indirect methods and is frequently used in HTRF. HTRF immunoassays generally use a direct labeling strategy in which two antibodies against different antigenic sites are used. One antibody is labeled with (Eu)K and another is labeled with XL665. When both bind to the antigen, energy transfer occurs and the long-lived XL665 signal is generated.

Europium-labeled recombinant protein G as a fast and sensitive immunoreagent for time-resolved immunofluorometry had been described by Markela et al. (1993). Merk et al. (2004) described time-resolved fluorescence measurements using a microlens array and area imaging devices. For applications in the TRF mode the Plate::Vision 96-microlens array reader was compared with the LEADseeker Generation IV multimodality imaging system. Selvin (2002) discussed the principles and biophysical applications of lanthanide-based probes. Using luminescent lanthanides, instead of conventional fluorophores, as donor molecules in resonance energy transfer measurements offers many technical advantages and opens up a wide range of applications including farther measurable distances with greater accuracy, insensitivity to incomplete labeling, and the ability to use generic relatively large labels.

Karvinen et al. (2002) developed a homogenous time-resolved fluorescence quenching assay (LANCE) for caspases. The assay uses a peptide labeled with both a luminescent europium chelate and a quencher. Cleavage of the peptide by caspase-3 separates the quencher from the chelate and thus recovers europium fluorescence. To study the interaction of two proteins, the JUN FOS dimerization was used as a model assay. The JUN protein was biotinylated. Streptavidin:(Eu)K was used to bind that label to the JUN-biotin protein via SA- biotin binding.

The FOS peptide was directly labeled with XL665. Signal is only generated when JUN-biotin:SA-(Eu)K and FOS-XL665 dimerize to form the activated AP1 transcription complex. The binding of ligands to membrane receptors and the competitive displacement by unknown molecules is one of the most common screening assays in drug discovery. To test the applicability of HTRF for receptor binding, the interaction of epidermal growth factor (EGF) with its receptor (EGFR) was studied. To generate a HTRF signal, a semi-direct labeling method was developed. EGF was labeled directly with (Eu)K, and a monoclonal antibody against a nonbinding portion of the EGF receptor was used.

The anti-EGRF antibody was labeled with XL665. The IC_{50} of displacement could be determined. The phosphorylation of enzymes, proteins and receptors is a major mechanism of cell regulatory pathways and is therefore a frequent target for drug discovery. The EGFR-intrinsic tyrosine kinase activity was used to develop an HTRF assay. Biotin-labeled (glutamine-alanine-tyrosine)$_n$ (poly GAT) was used as a substrate. The EGF receptor was isolated from A431 cells after partial purification of a cell homogenate. After the phosphorylation reaction, the HTRF signal was generated by adding streptavidin-XL665 and (Eu)-K-antiphosphotyrosine antibody. To establish the biological function of HTRF in this assay, inhibition of tryphostin-47 was assayed.

Homogeneous *time-resolved fluorescence resonance energy transfer* (TR-FRET) *assays* represent a highly sensitive and robust HTS method for the quantification of kinase activity. Moshinski et al. (2003) reported a widely applicable high-throughput TR-FRET assay for the measurement of kinase autophosphorylation. Inhibition of the vascular endothelial growth factor receptor 2 (VEGRF-

2) was used as prototype. Glickman et al. (2002) published a comparison of ALPHAScreen, TR-FRET and TRF as assay methods for FXR nuclear receptors. Newman and Josiah (2004) described utilization of fluorescence polarization and time-resolved fluorescence resonance energy transfer assay formats for structure-activity relationship studies. Src kinase was used as a model system.

Beasley et al. (2003) described the evaluation of compound interference in immobilized metal ion affinity-based fluorescence polarization (IMAP) detection with a four million member compound collection. IMAP is a non-separation-based, antibody-independent FP assay that can be applied to many types of protein kinases and phosphatases. In this technology, a fluorescently labeled peptide substrate is phosphorylated and then captured on immobilized metal nanoparticles. The binding of the phosphorylated peptide to the nanoparticles is detected using FP. Loomans et al. (2003) reported HTS with immobilized metal ion affinity-based fluorescence polarization detection as a homogeneous assay for protein kinases. Klumpp et al. (2006) discussed the development of homogeneous, miniaturized assays for the identification of novel kinase inhibitors from very large compound collections.

In particular, the suitability of TR-FRET based on phospho-specific antibodies, an antibody-independent fluorescence polarization (FP) approach using metal-coated beads (IMAP technology), and the determination of adenosine triphosphate consumption through chemiluminescence was evaluated. The *Acumen Explorer* is a laser-scanning system that measures fluorescence and luminescence, combining the ability to detect sub-micron events and features with ultra-fast acquisition and data management. Different parts of the cell can be localized by labeling with Hoechst 33342, Alexa 488, Alexa 546, Alexa 594, or Cys 5. Many different colors (readouts) can be used simultaneously. The *fluorometric imaging plate reader* (FLIPR) permits rapid, kinetic measurements of intracellular fluorescence. Simultaneous measurements in 96 wells and in real time can be made every second: before, during and after the addition of test compounds. The detection optics of FLIPR are based on cooled charge-coupled device (CCD) technology. With each kinetic update, the system takes a picture of the bottom of a microplate, recording a signal for all the individual cells simultaneously.

Enhanced sensitivity for cell-based assays is accomplished via an optical detection scheme, which allows for signal isolation on a cell monolayer. The system is ideal for monitoring intracellular calcium fluxes that occur within seconds of activation of G-coupled receptors. Using calcium-sensitive dyes such as Fluo-3 permits the derivation of full dose–response (or dose–inhibition) curves in a matter of minutes. Hodder et al. (2004) reported miniaturization of intracellular calcium functional assays to the 1536-well plate format using a fluorometric imaging plate reader. An intracellular calcium functional assay against the rat muscarinic acetylcholine receptor subtype 1 Gprotein-coupled receptor was miniaturized and executed in modified instruments. Robinson et al. (2004) used a fluorescence microplate reader to compare ratiometric and nonratiometric Ca^{2+} indicators for the assessment of intracellular free Ca^2 in a breast cancer line.

Bednar et al. (2004) used fluorescence detection of calcium flux for kinetic characterization of novel NR2B antagonists. Swartzman et al. (1999) developed a simple, homogeneous bead-based immunoassay for use with *fluorometric microvolume assay technology* (FMAT). Koltermann et al. (1998) proposed rapid assay processing by integration of *dual-color fluorescence cross-correlation spectroscopy* (RAPID FCS) as an ideal tool for ultra-HTS when combined with nanotechnology, which can probe up to 10^5 samples per day. Auer et al. (1999) used fluorescence correlation spectroscopy for lead discovery by miniaturized HTS. Winkler et al. (1999) presented *confocal fluorescence coincidence analysis* as an alternative to HTS. Confocal fluorescence coincidence analysis extracts fluorescence fluctuations that occur coincidentally in two different spectral ranges from a tiny observation volume of less than 1 fl.

This procedure makes it possible to monitor whether an association between molecular fragments labeled with different fluorophores is established or broken, providing access to the characterization of a variety of cleavage and ligation reactions in biochemistry. Jager et al. (2003b) published a modular, fully integrated ultra-HTS system based on *confocal fluorescence analysis techniques*. The goal was to achieve high-data quality in small assay volumes (1-4 µl) combined with reliable and unattended operation. Two new confocal fluorescence readers have been designed. One of the instruments is a four-channel confocal fluorescence reader, measuring with four objectives in parallel. The fluorescence readout is based on single-molecule detection methods, allowing high sensitivity at low tracer concentrations and delivering an information-rich output.

The other instrument is a confocal fluorescence imaging reader, where the images are analyzed in terms of generic patterns and quantified in units of intensity per pixel. Both readers span the application range from assays with isolated targets in homogenous solution or membrane vesicle-based assays (four-channel reader) to cell-based assays (imaging reader). Results from a comprehensive test on these assay types demonstrate the high quality and robustness of this screening system. Heilker et al. (2005) recommended confocal fluorescence microscopy for HTS of G-protein-coupled receptors. For HTS, Schmid et al. (1998) proposed a method to reversibly attach receptor proteins via an affinity tag to a quartz surface and subsequently detect with high sensitivity the real-time binding of ligands by *total internal reflection fluorescence*.

The CytoFluor Fluorescence Multi-Well Plate Reader is a versatile fluorescence intensity measurement instrument that detects and quantifies fluorescent molecules over a five-log dynamic range. The system has a sensitivity of better than 8 fmol/well for fluorescein. Ramm (1999) described the technique of *image-based screening in the LEADseeker system*, which contains a cooled CCD camera and a telecentric lens with integral epifluorescence illumination. *Laser scanning imaging systems* have been developed to measure cellular and subcellular quantitation of fluorescence in whole cells. *Chemiluminescence* is a photometric technique that is applicable to HTS. Detection of chemiluminescence is a convenient adjunct to fluorescence, since most plate readers capable of measuring fluorescence will measure luminescence as well.

This technique has been used predominantly with luciferase reporter genes in cell-based assays and in high-sensitivity enzyme-linked immunosorbent assays (ELISA), employing chemolumogenic substrates for alkaline phosphatase and horseradish peroxidase. The Luc-Screen assay system with extended-glow light emission was designed for sensitive detection of the firefly luciferase reporter enzyme. Liochev and Fridovich (1997) recommended lucigenin luminescence as a measure of intracellular superoxide dismutase activity in *Escherichia coli*. The NorthStar HTS workstation has been developed as a chemiluminescence analyzer for the HTS laboratory. Featuring online injection, a broad dynamic range, and an extremely low crosstalk, the NorthStar HTS workstation allows the routine analysis of more than 500,000 samples a day. Grepin et al. (2001) presented an assay for precise and direct quantification of specific endogenous mRNAs in cell lysates.

The technology is based on the bioluminescence detection of multiple biotinylated ssDNA probe/endogenous mRNA hybrids, which are captured onto a streptavidin-coated multi-well plate. Using the Xpress-Screen technology, an assay was developed aimed to monitor the induction of endogenous cFos mRNA expression in NGF-treated PC12 cells and the miniaturization of the assay in a 384-well format, which is adapted to HTS. Pommereau et al. (2004) compared the performance of two HTS assays for a serine/tyrosin kinase: a microplate-based, bioluminescent assay that uses the luciferin/luciferase system to monitor ATP consumption, and a microfluidic assay that measures the change in mobility in an electric field of a fluorescently labeled peptide upon phosphorylation. O'Brien et al. (2005) described homogenous, bioluminescent protease assays using caspases-3 as a model that are significantly faster and more sensitive than the fluorescent caspases-3 assays.

David et al. (2002), Gopalakrishnan et al. (2002), and Anderson et al. (2004) used *microarrayed compound screening* (μARCS), a well-less, high-density, ultra-HTS format, for various assays. In μARCS, the reagents are incorporated into agarose gels and layered over compounds arrayed on polystyrene sheets, mimicking the pipetting steps in a microplate assay. μARCS is a very flexible format and eliminates the requirement for complex liquid handling instruments and avoids problems associated with evaporation and plate edge effects. *Bioluminescence resonance energy transfer* (BRET) was developed as a technique to study protein-protein interactions. BRET is a natural phenomenon, observed in marine organisms, in which energy transfer occurs between luminescent donor and fluorescent acceptor proteins. Oxidation of cœlenterazine by *Renilla* luciferase produces light with a wavelength of 480 nm. In the sea pansy *Renilla*, the close proximity of a green fluorescent protein allows a non-radiative energy transfer that results in light emission at 509 nm by the green fluorescent protein.

In the BRET technology, the first protein partner is fused to *Renilla* luciferase, whereas the second protein is fused to a fluorescent protein, e. g., yellow fluorescent protein. If the two partners do not interact, only one signal, emitted by the luciferase, can be detected after addition of its substrate cœlenterazine. If the distance between the two partners is less than 10 nm (100 Å) and the partners interact, resonance energy transfer occurs between the luciferase and the yellow fluorescent protein and an additional signal can be detected. The method was used to study the dimerization of G-protein-coupled receptors. Based on BRET technology, Boute et al. (2002) developed a test to monitor the activation state of the insulin receptor. BRET technology was recommended for HTS assays. A BRET2 assay was recommended by Bertrand et al. (2002). *Renilla* luciferase was used as the donor protein, while the green fluorescent protein 2 was used as the acceptor protein.

In the presence of the cell-permeable substrate DeepBlueC, *Renilla* luciferase emits blue light at 395 nm. If the green fluorescent protein 2 is brought into close proximity to *Renilla* luciferase via a specific biomolecular interaction, it will absorb the blue light energy and re-emit green light at 510 nm. Vrecl et al. (2004) described the development of a BRET2 screening assay using β-arrestin 2 mutants. Pfleger and Eidne (2004) used BRET for the detection of real-time interactions involving G-protein-coupled receptors. Heding (2004) described the BRET7 receptor/beta-arrestin assay in drug discovery and screening that can be used to screen all seven-transmembrane receptors independent of their signaling pathway.

Electrochemiluminescence has been recommended as a diagnostic and research tool. Electrochemiluminescence is a means of converting electrical energy into radiative energy. It involves the production of reactive intermediates from stable precursors at the surface of an electrode. These intermediates then react under various conditions to form excited states that emit light. Golla and Seethal (2004) reported a sensitive, robust, high-throughput electrochemi-luminescence assay for rat insulin. The introduction of the *scintillation proximity assay* (SPA) obviated the need to separate bound from free radioactivity in the conventional radioligand-binding assay. One version of the scintillation proximity assay utilized polyvinyltoluene microspheres or beads into which a scintillant has been incorporated.

When a radiolabeled ligand is captured on the surface of the bead, the radioactive decay occurs in close proximity to the bead, and effectively transfers energy to the scintillant, which results in light emission. Zheng et al. (2004a) reported high-throughput cell- based screening using the scintillation proximity assay for the discovery of inositol phosphatase inhibitors. *FlashPlate technology* has been described by Brown et al. (1997). FlashPlates are white 96-well polystyrene microplates coated on the inside with solid scintillant in a polystyrene polymer. The polystyrene surface provides a hydrophobic surface for absorption of proteins, such as antibodies and receptors. When the binder molecule is coated on the FlashPlate, the reaction of the mixture of radiolabeled tracer and unlabeled ligand of interest with the binder molecule, and detection of the bound radioactivity can be done in one step. Due to the microplate surface scintillation effect, only the bound radioactivity will be detected by a scintillation counter. Dillon et al. (2003) used a FlashPlate assay for the identification of PARP- 1 inhibitors.

The use of *high-performance microphysiometry* in drug discovery has been discussed by Alajoki et al. (1997). Microphysiometry measures the extracellular acidification rate. It is governed by catabolism, via the excretion of acidic end-products, such as carbon dioxide and lactic acid. In addition, it is substantially influenced in a complementary fashion by the regulation of intracellular pH, a process sensitive to physiological changes such as receptor activation. The light-addressable potentiometric sensor (LAPS) uses a light-generated alternating current to probe the electrical potential at the interface between an electrolyte and the insulated surface of an appropriately biased silicon chip; the surface potential depends on pH in a Nernstian fashion. This system has been optimized for HTS. Microarray compound screening (μARCS) was used by David et al. (2002), Gopalakrishnan et al. (2002), and Anderson et al. (2004). Abassi et al. (2004), Atienza et al. (2005), and Yu et al. (2006) described dynamic monitoring of cell adhesion and spreading and real-time monitoring of morphological changes in living cells on *microelectronic sensor arrays*. Special microplates are integrated with electronic sensor arrays.

A promising probe for quantitative genomic studies is the *molecular beacon* (MB), a single-stranded DNA molecule composed of a hairpin-shaped oligonucleotide that contains both a fluorophore and a quencher group. Molecular beacons act like switches that are normally closed – when the fluorophore and the quencher are together, fluorescence is "off". Conformational changes open the hairpin, separate the fluorophore and the quencher, and the fluorescence is turned "on". These sensitive fluorescent DNA probes can be used for real-time biomolecular recognition of unseparated target species.

Molecular beacons are useful for studies with proteins, molecular beacon aptamers. Santangelo et al. (2004) described dual FRET molecular beacons for mRNA detection in living cells. *Surface plasmon resonance* is a method for characterizing macromolecular interactions. It is an optical technique that uses the evanescent wave phenomenon to measure changes in refractive index very close to a sensor surface. The binding between an analyte in solution and its ligand immobilized on the sensor surface results in a change in the refractive index. The interaction is monitored in real time and the amount of bound ligand and rates of association and dissociation can be measured with high precision. Since there is no need for fluorescent or radioisotopic labels, surface plasmon resonance biosensors are amenable to characterizing unmodified biopharmaceuticals, studying the interaction of drug candidates with macromolecular targets and identifying binding partners.

Genetic Background

The concept of an altered response based on genetic background is not new. In 510 BC, Pythagoras recognized that some individuals developed hemolytic anemia with fava bean consumption. In 1914, Garrod expanded this observation, stating that enzymes detoxified foreign agents so that they were excreted harmlessly. However, some people lack these enzymes and experience adverse effects. Hemolytic anemia due to fava bean consumption was later determined to occur in glucose-6-phosphate-deficient individuals. The term pharmacogenetics was first used by Friedrich Vogel in 1959. The dawn of pharmacogenetics continued by combining Mendelian genetics with observed phenotypes.

In 1932, Snyder performed the first global study of ethnic variation and deduced that taste deficiency was inherited. He proposed that the phenylthiourea nontaster phenotype was an inherited recessive trait and that the frequency of occurrence differed between races. Other genetic differences, such as aldehyde dehydrogenase and alcohol dehydrogenase, were discovered. These deficiencies are common in Asians and cause them to have less alcohol tolerance. Similarly, polymorphisms in the *N*-acetyl transferase enzyme are also segregated by ethnicity and correlate to the latitude of the country. These genetic differences were originally thought to be caused by genetic variance, but it was not until the advent of molecular biology that disease states could be carefully analyzed. The following dogma:

$$\text{gene} \rightarrow \text{protein} \rightarrow \text{biochemical process} \rightarrow \text{disease state}$$

became the model for examining human diseases. Following this scheme, sickle cell anemia was the first trait to reveal that a single point mutation can change protein structure and lead to a disease phenotype. Single nucleotide polymorphisms (SNPs) are responsible for many diseases. SNPs can change receptors, transport proteins, and drug-metabolizing enzymes. The advances in genetic technology to detect polymorphisms have caused an explosion in pharmacogenetic research and many of these discoveries have been employed in clinical practice. Many thousands of polymorphisms can be determined simultaneously in a patient. These SNPs are selected as markers evenly distributed in the genome, in the hope that functionally relevant polymorphisms can be associated with specific markers by virtue of their proximity to the chromosome.

Such genome-wide association studies are used in the discovery of susceptibility genes for diseases, such as asthma and prostate cancer, but they are equally suitable for determining the genes in drug response. Disorders such as Huntington's disease or cystic fibrosis are associated with defects in a single gene product identified by positional cloning. This involves the isolation of a gene with the information on its sub-chromosomal localization and the phenotypic expression of a mutation of this gene that results in a particular disease state. Positional cloning is based on the unique, inheritable DNA fingerprint associated with each individual, from which the inheritance of identified polymorphisms can be tracked, along with the inheritance of the disease.

While this approach has been more or less successful in diseases involving a single gene defect, more complex diseases, such as Alzheimer's disease, cancer, asthma or cardiovascular diseases, appear to result from the influence of multiple gene defects. The next step is to identify the function of the protein expressed by the DNA using functional genomics. This technique involves the use of sequence comparisons with known proteins, using computer databases, transgenic expression or expression in surrogate systems; gene knockouts in mice, or relatively simple organisms such as *Drosophila melanogaster, Caenorhabditis elegans,* Zebrafish or *Saccharomyces cerevisiae.* A two-hybrid system to identify molecular binding partners can be used to identify proteins that are associated with the gene product of interest. Progress in genomics technology, especially the elucidation of the human genome, has created an unique opportunity to significantly impact on the pharmaceutical drug development processes.

Genetic polymorphism in drug-metabolizing enzymes, transporters, receptors, and other drug targets are linked to individual differences in the efficacy and toxicity of many medications. Pharmacogenomic studies can elucidate the inherited nature of these differences, thereby enhancing the possibilities of drug discovery and providing a stronger scientific basis for optimizing drug therapy on the basis of each patient's genetic constitution. Receptor polymorphism is also the reason for differences in drug effects among patients. Two genetic variants of the angiotensin-converting enzyme (ACE) are described, depending on an insertion (I-form) or deletion (D-form) of a base pair at position 287 in the gene, which are equally distributed in the Caucasian population. Individuals with the D/D-form express ACE levels much higher than I/I individuals. They have an increased risk for myocardial infarction and may respond better to therapy with ACE inhibitors.

Another example is the association between β_2-adrenergic receptor polymorphisms and asthma. There are three polymorphisms that alter receptor function, which may influence the therapeutic success. Genomics will introduce a new dimension in drug research. The number of molecular targets modified by the complete armamentarium of modern drugs is not greater than 500. The number of genes that contribute to multifactorial diseases might not be very high; the numbers reported for different forms of diabetes and hypertension are 5-10 per disease. There are 100-150 disease entities that present an epidemiological and economical problem to industrialized societies. If one assumes 10 contributing genes for 100 multifactorial diseases (including different forms of cancer, asthma, diabetes, hypertension, atherosclerosis and osteoporosis), one arrives at 1000 "*disease genes*" that dispose patients to the most important multigenetic conditions.

It appears reasonable to assume that each of these disease genes, or rather proteins that are specified by the disease genes, connects with at least 5-10 proteins that are feasible levels for drug intervention. On the basis of these calculations, one can assume that 5000-10,000 can be used as targets for drug interventions. Lennon (2000) discussed the methods for high-throughput gene expression analysis applicable to drug discovery, including differential display of eukaryotic mRNA, the modification thereof called restriction enzyme analysis of differentially expressed sequences (READS), expression sequence tags (EST) methodology, serial analysis of gene expression, filter arrays, and DNA microarrays. Celis et al. (2000) gave an excellent survey on microarray technology.

Gene expression analyses and genetic polymorphisms are important not only for determining predisposition to disease and for drug discovery, but also for predicting the incidence of adverse drug reactions. In the past, several drugs had to be withdrawn from the market due to rare, but severe, adverse events. Some drug metabolic enzyme variants have been found to cause severe adverse events. They form qualitatively or quantitatively different metabolites with toxicological implications. Most of the enzymes involved in both drug metabolism and the elimination of many therapeutic agents are members of the cytochrome P450 (CYP) family that includes more than 30 isoforms. One drug-metabolizing enzyme, resulting in numerous drugs being withdrawn from the market, is CYP3A, which is involved in the oxidative biotransformation of up to 50% of clinically important therapeutic agents.

The expression of CYP3A is regulated by genetic and non-genetic factors that can result in a 5- to 20-fold interindividual variability in metabolic clearance. Another enzyme, CYP2D6 (debrisoquine hydroxylase), metabolizes one-quarter of all prescribed drugs and is inactive in 6% of the Caucasian population. A principal molecular defect in poor metabolizers is a single base pair mutation in exon 5 of CYP2C19. These enzymes are being screened at the earliest stages of the drug development process. The single nucleotide point mutation profile of an individual can indicate a predisposition for adverse side-effects. Furthermore, drug transport protein polymorphisms and their drug interactions may provide another valuable tool in screening for potential toxic effects of drugs. The pharmacogenomic approach should reduce the number of compounds failing during late clinical development. A detailed knowledge of the genetic basis of individual drug response is of major clinical and economic importance and can provide the basis for a rational approach to drug prescription.

In the sense of a "*personalized medicine*", prospective genotyping will lead to patients being prescribed drugs that are both safer and more effective. Winkelmann et al. (2001) initiated the Ludwigshafen Risk and Cardiovascular Health (LURIC) study, which aims to provide a resource for the study of environmental and genetic risk factors and their interactions, and of the functional relationship between gene variation and biochemical phenotype (functional genomics), or response to medication (pharmacogenomics) and long-term prognosis of cardiovascular disease. A vision for the future of genomics research was given by Collins et al. (2003). *Rolling-cycle amplification* (RCA) is a system for amplification, which may be used for detection of proteins or nucleic acid targets. Rolling-circle amplification driven by DNA polymerase can replicate circularized oligonucleotide probes with either linear or geometric kinetics under isothermal conditions.

In the presence of two primers, one hybridizing to the positive strand and the other to the negative strand of DNA, a complex pattern of DNA strand displacement ensues that generates 10^9 or more copies of each circle in 90 min, enabling detection of point mutations in human genomic DNA. Using a single primer, RCA generates hundreds of tandemly linked copies of a covalently closed circle in a few minutes. I-matrix-associated, the DNA product remains bound to the site of synthesis, where it may be tagged, condensed, and imaged as a point light source. The target DNA can be attached to an antibody, biotin, or other protein probe. Vanhecke and Janitz (2005) discussed functional genomics using high-throughput RNA interference. *RNA interference* (RNAi) describes the post-transcriptional silencing of gene expression that occurs in response to the introduction of double-stranded RNA into cells. To facilitate large-scale functional genomics studies using RNAi, several high-throughput approaches have been developed based on microarray or microwell assays.

The RNAi silencing mechanism is based on the fact that long double-stranded RNA is processed to short interfering RNA (siRNA) duplexes by the enzyme Dicer, which exhibit RNaseIII-like activity. A single strand of the siRNA is incorporated into the group of cytoplasmic proteins to form an RNA-induced silencing complex (RISC). Activated RISC, guided by the antisense siRNA strand, performs endonucleolytic cleavage of target mRNA. Thus produced mRNA fragments are rapidly degraded by cytoplasmic nucleases. Rondinone (2006) dicussed the therapeutic potential of RNAi in metabolic diseases. Aza-Blanc et al. (2003) described the identification of modulators of TRAIL-induced apoptosis via RNAi-based phenotypic screening. TRAIL is a TNF super- family member that induces selective cytotoxicity of tumor cells when bound to its cognate receptors.

Using RNAi, Kamath et al. (2003) performed a systematic functional analysis of the *Caenorhabditis elegans* genome. Mousses et al. (2003) describe a microarray analysis in cultured mammalian cells. Paddison et al. (2004) described a resource for large-scale RNA-interference-based screens in mammals; Berns et al. (2004), a large-scale RNAi screen in human cells to identify new components of the p53 pathway. *Promega Protocols* and *Applications Guide* is a published survey on RNA interference. The use of siRNA in therapy and drug target discovery is discussed by Cocks and Theriault (2004), Hannon and Rossi (2004), Jones et al. (2004), Kurreck (2004), Meister and Tuschi (2004), Siuod (2004), Bartz and Jackson (2005), and Catterjee-Kishore (2006). Xin et al. (2004) discussed high-throughput siRNA-based functional target validation. They initiated a project with a functional cell-based screen for a biological process of interest using libraries of small interfering RNA (siRNA) molecules. In this protocol, siRNAs function as potent gene-specific inhibitors. SiRNA-mediated knockdown of the target gene is confirmed by TaqMan analysis, and genes with impacts on biological functions of interest are selected for further analysis. Once the genes are confirmed and further validated, they may be used for HTS to yield lead compounds.

Tang (2005) discussed two classes of short RNA molecules, siRNA and microRNA (miRNA), which are incorporated into related RNA-induced silencing complexes (RISCs), termed siRISC and miRISC, respectively. They are distinct complexes that regulate mRNA stability and translation. *MicroRNAs* (miRNAs) are generated via a two-step processing pathway to yield ~22 nucleotide small RNAs that regulate gene expression at the post-transcriptional level. Initial cleavage is catalyzed by Drosha, a nuclease of the RNase III family, which acts on primary miRNA transcripts (primiRNAs) in the nucleus. MicroRNAs play important regulatory roles in animals and plants by targeting mRNAs for cleavage or translational repression. The microRNAs of *Caenorhabditis elegans* are described by Lim et al. (2003).

Denli et al. (2004) showed that Drosha exists in a multi-protein complex, the microprocessor, and begins the process of destructing that complex into its constituent components. Along with Drosha, the microprocessors also contains Pasha (partner of Drosha), a double-stranded RNA binding protein. Han et al. (2004) dissected the mechanisms of action of human Drosha by generating mutants and beycharacterizing its interacting partner, DGCR8. Berezikov et al. (2005) described phylogenetic shadowing and computational identification of human microRNA genes. Barad et al. (2004) reported microRNA expression detected by oligonucleotide microarray system establishment and expression profiling in human tissue.

Liang et al. (2005) described a miRNA profiling microarray, in which miRNAs are directly labeled at the 3′ terminus with biotin and hybridized with complementary oligo-DNA probes immobilized on glass slides, and subsequently detected by measuring the fluorescence of quantum dots labeled with streptavidin bound to miRNAs through streptavidin–biotin interaction. *Locked nucleic acids* (LNA) were reported in 1998. LNA is a nucleic acid analog containing one more LNA nucleotide monomers with a bicyclic furanose unit locked in an RNA-mimicking sugar confirmation. LNA nucleotides display unprecedented hybridization affinity toward complementary single-stranded RNA and complementary single- or double-stranded DNA. They are considered as a versatile tool for therapeutics and genomics. Tolstrup et al. (2003) reported the development of a software package, OligoDesign, which provides optimal design of LNA oligonucleotide capture probes for gene expression profiling.

Valoczi et al. (2004) described sensitive and specific detection of microRNAs by Northern blot analysis using LNA-modified oligonucleotide probes. Elmen et al. (2005) reported LNA-mediated improvements in siRNA stability and functionality. Thomassin et al. (2004) presented MethylQuant, a sensitive method for quantifying methylation of specific cytosines within the genome. The presence of a locked nucleic acid at the 3′ end of the discriminative primer provides the specificity necessary for accurate and sensitive quantification. Nucleic acid "*lariats*" have been of great interest since their discovery. The synthesis of lariat RNA and DNA via template-mediated chemical ligation of Y-shaped oligonucleotides was reported by Carriero and Damha (2003). Deoxyribozymes that synthesize branched and lariat RNA were studied by Wang and Silverman (2003). RNAse III-mediated degradation of unspliced premRNAs and lariat introns was reported by Danin-Kreiselman et al. (2003). Coombes and Boeke (2005) evaluated the detection methods for large lariat RNAs.

Computational Approach

The term "proteome" was coined by Wilkins et al. (1995) to describe the protein complement of the genome. The term proteomics was first used to describe the complexity of proteins expressed in an organism using two-dimensional gel electrophoresis (2-DE) followed by quantitative analysis. 2-DE remains the highest resolution protein separation method available, but the ability to concentrate and identify low abundance proteins has always been an extremely difficult problem. Mass spectrometry (MS) has been an integral part of approaching this problem. Although improvements in 2-D gel technology have been made since its introduction, three enabling technological advances have provided the basis for the foundation of the field of proteomics. The first advance was the introduction of large-scale nucleotide sequencing of both expressed sequence tags (ESTs) and genomic DNA. The second was the development of mass spectrometers able to ionize and mass-analyze biological molecules and the widespread introduction of mass spectrometers, capable of data-dependent ion selection for fragmentation (MS/MS) (i. e., without the need for user intervention). The third was the development of computer algorithms able to match uninterpreted (or partially interpreted) MS/MS spectra with translations of the nucleotide sequence databases, thereby tying together the first technological advances. The mass spectrometer instruments are named for their type of ionization source and mass analyzer. To measure the mass of molecules, the test material must be charged (ionized) and desolvated. The two most successful mechanisms for ionization of peptides and proteins are matrix-assisted laser desorption ionization (MALDI) and electrospray ionization (ESI).

In MALDI the analyte of interest is embedded in a matrix that is dried and then volatilized in a vacuum under ultraviolet laser irradiation. This is a relatively efficient process that ablates only a small portion of the analyte with each laser shot. Typically, the mass analyzer coupled with MALDI is a time-of-flight (TOF) mass analyzer, that measures the elapsed time from acceleration of the charged (ionized) molecules through a field-free drift region. The other common ionization source is ESI, in which the analyte is sprayed from a fine needle at high voltage toward the inlet of the mass spectrometer (which is under vacuum) at a lower voltage. The spray is typically either from a reversed-phase HPLC column or a nanospray device, similar to a microinjection needle. The ions formed during this process are directed into the mass analyzer, which could be a triple-quadrupole, an ion trap, a Fourier-transform ion cyclotron resonance (FT-ICR), or a hybrid quadrupole TOF (Qp-TOF) type.

Although unambiguous identification of a protein cannot always be derived from the masses of a few of its peptides in the tandem mass spectrometer, peptide ions from the first mass spectrometer run are fragmented and identified in a second run to yield a more valuable commodity of a peptide sequence. Annis et al. (2004) and Whitehurst et al. (2006) proposed affinity selection-mass spectroscopy (AS-MS) as a useful high-throughput assay system for the discovery and characterization of all classes of integral membrane protein ligands, including allosteric modulators. The goal of proteomics is a comprehensive, quantitative description of protein expression and its change under the influence of biological perturbations, such as disease and drug treatment. A combination of mRNA and protein expression patterns has to be simultaneously considered to develop a conceptual understanding of the functional architecture of genomes and gene networks.

New methods are created such as automated proteomics platforms, combining two-dimensional electrophoresis, automated spot picking and mass spectrometry. Improvements in quality, ability and utility of large-scale tertiary and quarternary protein structural information are enabling a revolution in rational design, having a particular impact on drug discovery and optimization. Disease proteomics will give a better understanding of disease processes, develop new biomarkers for diagnosis and early detection of disease, and accelerate drug development. One may expect that with the new approach of drug research, including combinatorial chemistry, genomics, pharmacogenomics, proteomics, and bioinformatics, unprecedented results will be generated. The difficulties arising in proteomic experiments from data analysis are discussed by Patterson (2003). Duncan and Hunsucker (2005) discussed the value of proteomics as a tool for clinically relevant biomarker discovery and validation. A critical review on the impact of "OMIC" technology in drug development is given by Bilello (2005).

Hybridization Reaction

Array technology is based on the RNA and DNA hybridization reaction previously used in Northern and Southern expression analysis. However, these traditional approaches allowed the detection of only one single gene. In contrast, gene arrays and microarrays allow the analysis of hundreds or thousands of genes simultaneously. Gene chips can be global, containing the entire genome of one species on the slide. Besides the miniaturization, the greatest leap of technology was the development of fluorescence-labeled nucleotides, which can be detected by laser scanning. They have now replaced the use of radioactively labeled DNA. Two different methods are used in array production: chips are loaded with either synthetic oligonucleotides or with a DNA fragment isolated directly from the respective gene.

A specific process allows the synthesis of oligonucleotides directly onto the glass surface. Normally, every gene is represented by about 20 oligonucleotides from different regions of the gene. Each oligonucleotide is 25-30 base pairs in length and is synthesized next to a control oligonucleotide with a mismatch.Whole genome arrays were pioneered by Affymetrix using a proprietary chip technology but are now available from other suppliers as well. The development of DNA chips was pioneered by researchers at Stanford University. Along with the Stanford Microarray Database a number of additional centers voted for an open online access to their array databases. Killion et al. (2003) created the Longhorn Array Database (LAD), which is an open-source, MIAME-compliant implementation of the Stanford Microarray Database.

The standard MIAME (minimum information about a microarray experiment) was proposed by Brazma et al. (2001). Kamb and Ramaswami (2001) presented a simple method for statistical analysis of intensity differences in microarray-derived gene expression data which does not require multiple replicates. The difference-averaging method enables determination of variances as a function of signal intensities by averaging over the entire dataset. Novoradovskaya et al. (2004) recommended Universal Reference RNA (URR) as a standard for microarray experiments, a useful tool for monitoring intra- and inter-experimental variation. A variety of microassays were used to determine the percentage of spots hybridizing with URR and producing signal above a user defined threshold (microarray coverage).

Microarray coverage was consistently greater than 80% for all arrays tested. Regardless of using oligonucleotides or DNA arrays, one always needs two RNA samples for chip analysis. The RNA can be isolated from treated versus untreated cell lines, from the serum of an animal before or after drug administration, or from tumor versus normal tissue. Standard protocols require about 100 µg total RNA, from which poly A^+ RNA is extracted. Next, the RNA is converted by reverse transcription into cDNA in the presence of fluorescence-labeled dyes. Commonly, the sample is labeled with Cy-5 and the control with Cy-3. Both cDNAs are combined and hybridized against the array.

Since DNA-DNA binding reactions are thermodynamically slow, the chip is incubated overnight at temperatures above 60°C. Chip analysis is performed by scanning with two lasers at 562 nm

(Cy-3) and 644 nm (Cy-5). The overlay of both scans shows genes that are deregulated. By convention, overexpressed genes are shown in green, repressed genes in red and unchanged expression in yellow. Data are evaluated by scatter blot analysis, where the relative expression of each gene is indicated by logarithmic axes. Sophisticated bioinformatic protocols are used for further annotation and data comparison. The best approach to microarray analysis is to begin with a small number of the elements in the microarray known to be a pattern and ask questions of the other elements in the microarray; i. e., perform instantaneous scientific experiments regarding whether each of the other elements in the microarray is related to the known pattern.

Emerging technologies and perspectives for microarray applications are reviewed by Gebauer (2004). Epstein and Walt (2003) described fluorescence- based fibre optic arrays as a universal platform for sensing. Optical fibres provide a universal sensing platform as they are easily integrated with a multitude of different sensing schemes. Such schemes enable the preparation of a multitude of sensors including high-density oligonucleotide arrays and high-throughput cell-based arrays. Imaging fibre bundles comprised of thousands of fused optical fibres are the basis for an optically connected, individually addressable parallel sensing platform. Fibre optic imaging bundles possess miniature feature sizes (3- to 10-μm-diameter fibres), allowing high-density sensor packing ($\sim 2 \times 10^7$ sensors per cm^2). Ge (2000) described the development of a universal protein array (UPA) system that provides a sensitive, quantitative, multipurpose, effective and easy technology to determine not only specific protein-protein interactions, but also specific interactions of proteins with DNA, RNA, ligands and other small chemicals.

The application of biochip and microarray systems in pharmacogenomics is discussed by Jain (2000). MacBeath et al. (1999) discussed the possibilities of printing small molecules as microassays and detecting protein-ligand interactions en masse. Firestein and Pisetsky (2002) published guidelines for studies using DNA microarray technology. The criteria for evaluating microarray studies are: (1) reproducibility must be demonstrated, including rigorous evaluation of the run-to-run variability of each gene; (2) detailed statistical analysis is required; including appropriate corrections for repeated multiple measurements; (3) homogeneous cell populations should be optimally studied, to reduce the complexity of analysis; (4) a non-array method must confirm changes in the expression of key genes. Ziauddin and Sabatini (2001) developed a microarray-driven gene expression system for the functional analysis of many gene products in parallel.

Mammalian cells are cultured on a glass slide printed in defined locations with different DNAs. Cells growing on the printed areas take up the DNA, creating spots of localized transfection within a lawn of non-transfected cells. By printing sets of complementary DNAs cloned in expression vectors, microarrays are made, whose features are clusters of live cells that express a defined cDNA at each location. Feldman et al. (2002) described advantages of mRNA amplification for microarray analysis and recommended routine mRNA amplification for all cDNA microarray-based analysis of gene expression. Nielsen et al. (2005) developed a ready-to-spot polymer microarray slide, which is coated with a uniform layer of reactive electrophilic groups using anthraquinone-mediated photo-coupling chemistry. The slide coating reduces the hydrophobicity of the native polymer significantly, thereby enabling robust and efficient one-step coupling of spotted 5′-amino-like oligonucleotides onto the polymer slide.

The coated polymer microarray slide represents a robust and cost-effective platform for pre-spotted oligonucleotide arrays. Luo and Geschwind (2001) reviewed microarray applications in neuroscience with reference to applications in neurologic diseases in humans and the use of animal models. Bunney et al. (2003) reviewed new strategies to discover candidate vulnerability genes in psychiatric disorders by microarray technology. Templin et al. (2002) discussed the possibilities offered by protein microarray technology. Recent developments in the field of protein microarrays show applications for enzyme-substrate, DNA-protein and different types of protein-protein interactions. The authors discuss theoretical advantages and limitations of any miniaturized capture-molecule-ligand assay system. Salisbury et al. (2002) described peptide microarrays for the determination of protease substrate specificity using fluorogenic peptidyl coumarin substrates.

The technology, the materials and the methods used for antibody microarrays are described by Lal et al. (2002) and Kusnezow et al. (2003). Gopalakrishnan et al. (2002) described the application of Micro Arrayed Compound Screening (microARCS) to identify inhibitors of caspase-3. In this format, 8,640 discrete compounds are spotted and dried onto a polystyrene sheet, which has the same footprint as a 96-well plate. A homogeneous time- resolved fluorescence assay format (LANCE) was applied to identify the inhibitors of caspase-3 using a peptide substrate labeled with a fluorescent europium chelate and a dabcyl quencher. The caspase-3 enzyme was cast into a thin agarose gel, which was placed on a sheet containing test compounds. A second gel containing caspase substrate was then laid above the enzyme gel to initiate the reaction. Caspase-3 cleaves the substrate and separates the europium from the quencher, giving rise to a time-resolved fluorescence signal, which was detected using a ViewLux charge-coupled device imaging system.

Protein chip technology has been addressed by Zhu and Snyder (2003). Protein microarray technology has shown its great potential in basic research, diagnostics, and drug discovery. It has been applied to analyze antibody-antigen, protein-protein, protein-nucleic acid, protein-lipid and protein-small molecule interactions, as well as enzyme-substrate interactions. Progress in the field of protein chips includes surface chemistry, capture molecule attachment, protein labeling and detection methods, high-throughput protein–antibody production, and applications to analyze entire proteomes. Lenigk et al. (2002) recommended plastic biochannel hybridization devices as a new concept for microfluidic DNA arrays. Lee et al. (2004) described high-throughput screening of novel peptide inhibitors of an integrin receptor from the hexapeptide library by using a protein microarray chip. Novel peptide ligands with high affinity to the integrin were identified from the peptide libraries with this chip-based screening system by a competitive inhibition assay in a simultaneous and high-throughput fashion.

Hsieh et al. (2004) described an ultra-high-throughput microarray generation and liquid dispensing using multiple disposable piezoelectric ejectors. Lim et al. (2005) used microarray analysis to show that some microRNAs downregulate a large number of target mRNAs. Howbrook et al. (2003) reviewed developments in microarray technologies including gene expression analysis, DNA microarrays, protein and peptide microarrays, glycomics, tissue and cell microarrays as well as progress in microarray technology, such as microfluids, molecular imprintation, and nanotechnology. Gosalia and Diamond (2003) developed a method that performs enzymatic assays in nanoliter volumes in the liquid phase with minimal evaporation, no cross- contamination, and high reproducibility, and that has the capability to rapidly assemble multicomponent reactions with minimal usage in a microarray format. Chemical compounds within individual nano- liter droplets of glycerol are microarrayed onto glass slides at 400 spots/cm^2.

Using aerosol deposition, subsequent reagents and water are metered into each reaction center to rapidly assemble diverse multi-component reactions without cross-contamination or the need for surface linkage. This proteomics technique allows the kinetic profiling of protease mixtures, protease–substrate interactions, and high-throughput screening reactions. Kusnezow and Hoheisel (2002) discussed promises and problems of antibody microarrays and compared this approach with the achievements of DNA microarrays. Rondelez et al. (2005) developed microfabricated arrays of femtoliter chambers that allow single molecule enzymology. The microchip ensures that the chambers are uniform and precisely positioned.

Rougemont and Hingamp (2003) provided a method for automatically determining relevant gene clusters among the many genes monitored with microarrays. Forster et al. (2004) described triple-target microarray experiments as a novel experimental strategy. There are two main dye-labeling strategies for microarray studies based on custom-spotted cDNA or oligonucleotide arrays: (1) dye-labeling of a single target sample with a particular dye, followed by subsequent hybridization to a single microarray slide, (2) dye-labeling of two different target samples with two different dyes, followed by subsequent co-hybridization to a single microarray slide. The two dyes most frequently used for either method are Cy3 and Cy5. The authors proposed and evaluated a novel experimental set-up utilizing three differently labeled targets co-hybridized to one microarray slide. In addition to Cy3 and CY5, Alexa 594 was incorporated as a third dye label. Sharma et al. (2004) developed software called "ArrayD" that offers various alternative design solutions for a microassay given a set of user requirements.

Watson et al. (2004) published methods for high- throughput validation of amplified fragment pools of bacterial artificial chromosome (BAC) DNA for constructing high-resolution CHG assays. The development of array based on comparative genomic hybridization (CGH) technology provides improved resolution for detection of alterations in genomic DNA copy number. In array CHG, generating spotting solution is a multi-step process where BAC clones are converted to replenishable PCR-amplified fragment pools (AFP) for use as spotting solution in a microarray format on glass substrate.

Current methods for identifying initial BAC clones can be adapted to verify the identity of AFP spotting solutions used in printing assays. Of these methods, AFP sequencing is considered to be the most efficient for large-scale identification of spotting solution in a high-throughput manner. Daly et al. (2005) presented a statistical method based on propagation of error to evaluate concentration estimation errors in an ELISA microarray process. Do and Choi (2006) gave a review of normalization techniques of microarray data from single-labeled platforms such as the Affymetrix GeneChip rray to dual-labeled platforms, e. g., spotted array, focusing on their principles and assumptions.

Ma and Horiuchi (2006) recommended chemical microarray as a new tool for drug screening and discovery in order to find new ways in HTS to profile the activity of large numbers of chemicals against hundreds of biological targets in a fast, low-cost fashion. Uttamchandani et al. (2005) reviewed recent advances and applications of small molecule microarrays. Small molecule microarrays (SMM) are of value in high-throughput exploration, both in the identification of biologically significant natural and synthetic small molecules and in harnessing their vast potential towards medicinal and diagnostic applications. Sobek et al. (2006) recommended microarray technology as a universal tool for high-throughput analysis of biological systems and reviewed the recent developments in surface chemistry and derivatization.

Modern Biomedical Research

Magnetic resonance imaging (MRI) and spectroscopy (MRS) have become established technologies in modern biomedical research, providing relevant information on the various stages of the drug discovery and development process and are an important tool in drug discovery and metabolism. Several successful and potential applications of MRI and MRS in stroke, rheumatoid arthritis, osteoarthritis, respiratory diseases, oncology and cardiovascular diseases are described. Strengths of MRI are: (1) *high soft tissue contrast*, which is governed by a multitude of parameters; (2) *noninvasiveness*, which is relevant when studying chronic diseases and allows for the translation of study protocols from animals to humans; and (3) the *high chemical specificity* of MRS, allowing the identification of individual analytes based on compound-specific resonance frequencies, an important prerequisite for the *in vivo* study of tissue metabolism.

The main disadvantage of MRI/MRS is the *inherently low sensitivity* due to the low quantum energy involved as compared to optical spectroscopy, leading to a low degree of spin polarization. Only nuclei with high natural abundance and high intrinsic magnetic moment, such as protons, provide sufficient signal intensity for imaging applications. In addition, exogenous contrast agents (CA, paramagnetic or superparamagnetic compounds) with different levels of specificity are often used to modulate local signal intensities. MRI is a multidimensional technique providing cross-sectional images as well as full three-dimensional (3D) data of body regions. Bammer et al. (2005) gave a review on the foundations of advanced MRI. The mathematical and physical basics as well as the medical applications of MRI are described by Hornak (2007). Using time domain NMR techniques, slice selection with shaped pulses, special encoding by pulsed field gradients, gradient echo detection and Fourier transform methods, highly resolved images of anatomy can be generated in almost real time.

The most frequently used parameters are the spin-lattice relaxation time (T_1) and the spin-spin relaxation time (T_2). MRI is routinely used as a tomographic imaging technique, where anatomical pictures are created of 1-mm-thick tissue sections. The contrast differences between brain structures in most MRI techniques are determined by the different densities and diffusion of protons, as well as differences in relaxation times. T_2 images are sensitive to water and, because all pathological alterations in brains are associated with altered distribution of tissue water (edema), this technique is highly useful for visualizing the spatial distribution of lesions. Contrast in T_1 images is determined mainly by different lattice densities. Dense structures, such as compact white matter, have low T_1 values, whereas relatively loose structures, such as gray matter or lesions, have higher T_1 values.

To distinguish inflammatory active from inactive lesions, the paramagnetic dye gadolinium-DTPA is intravenously injected (0.1-0.3 mmol kg^{-1}) and, in areas of increased blood-brain barrier permeability, leaks into the brain parenchyma, causing local enhancement of the T_1-weighted signal intensity. Srijkers et al. (2007) reviewed the current status of MRI contrast agents. Sillerud and Larson (2006) reviewed screening methods for drug discovery based on nuclear magnetic resonance (NMR), which were boosted by the introduction of sophisticated computational techniques for reducing the time for the acquisition of the primary NMR data for multidimensional studies. Further non-invasive radiological techniques, such as BOLD-MRI (blood-oxygen-level-dependent MRI), positron emission tomography (PET) and ^{31}P MRS, have been evaluated.

With BOLD-MRI, brain function has been successfully probed and has initiated the widely practiced field of functional MRI (fMRI). Mirsattari et al. (2005) described physiological monitoring of small animals during fMRI. Sauter et al. (2002) applied fMRI to study functional recovery following cytoprotective treatment in a rat model of human embolic stroke. In addition, spectroscopic imaging with *in vivo* magnetic resonance ^{1}H-spectroscopy MRI, known as MRSI, enables the quantification of relative concentrations of specific metabolites, such as intramyocellular lipid content, at predetermined locations in organs. Another important MRI technique is magnetization transfer ratio (MTR) imaging. The MTR of a given tissue is defined as the ratio of free protons to protons bound to tissue macromolecules.

Among other methodologies in functional imaging is the use of electron paramagnetic resonance (EPR), a technique almost identical in principle and practice to NMR. EPR addresses unpaired spins, and in a given magnetic field the resonance frequency of an ensemble of electrons is 658 times higher than that of protons. There are two approaches of EPR imaging (ERPI), namely the continuous-wave (CW) modality (constant frequency with a field sweep) or the time-domain (termed Fourier transform, FT) modality (acquisition of an impulse response of the system at constant field). Devasahayam et al. (2004) described tailored sinc pulses for uniform excitation and artifact-free radiofrequency time-domain EPR imaging.

A truncated sinc [sin $(x)/x$] pulse, tailored to compensate for the Q-profile (RF frequency response) of the resonator, was shown to yield images from phantom objects as well as *in vivo* images, with minimal distortion. Another very closely related modality known by two alternate names, PEDRI (proton electron double resonance imaging) and OMRI (Overhauser-enhanced magnetic resonance imaging), utilizes the Overhauser effect, by which protons are polarized in the presence of stable free radicals to provide highly enhanced MR images at very low fields.

The principle behind OMRI is the saturation of free radical resonance by continuous (or pulsed) irradiation at the electron resonance frequency, which leads to an Overhauser enhancement of the proton magnetization (also known as dynamic nuclear polarization, DNP) via mainly the time-dependent dipolar interactions between the unpaired electrons and the water protons. This may lead to a high enhancement factor. This technique has been shown to produce useful functional proton images that can be used to perform non-invasive quantitative *in vivo* oximetry. The sensitivity of OMRI can be additionally improved by field cycling.

Krishna et al. (2002) used OMRI for tumor oximetry and co-recording of tumor anatomy and tissue oxygen concentration. Post (2003) used exchange-transferred nuclear Overhauser effect spectroscopy (et-NOESY) to probe the conformation of a ligand while bound to its macromolecular receptor.

CHARACTERISTIC ERRORS IN SCREENING

Any screening procedure has a characteristic error rate. This is inevitable because in high-throughput screening it is necessary to sacrifice some accuracy or precision to achieve the requisite speed. Thus when a large number of compounds is carried through a particular screen, some of the compounds will be classified incorrectly. A screen may be used in an absolute sense, so that compounds that pass a certain criterion are termed positives, whereas those that fail to meet the criterion are termed negatives. Compounds that pass, but should have failed, are false positives. In general, false positives are tolerable, if they are not too numerous, because they will be rectified later. Compounds that fail, but should have passed, are false negatives. False negatives are lost forever if the failure eliminates them from further testing. All screening procedures are based on assumptions of analogy. They have different degrees of relevance or predictability. Studies in phase II clinical trials will predict with high probability the results in large clinical trails. But even here there is the possibility of false-positive or false-negative results.

The relevance of a test is much less in early pharmacological tests, such as used in high-throughput screening. Generally, the relevance is inversely proportional to the simplicity of the test. In any case, one is confronted with the problem of false-positive results (type I errors) and false-negative results (type II errors). In each step two sources of error for false-positive results have to be taken in account:

1. a = error of the first type due to the model
2. α = error of the first type due to statistics

For definition: In the error of the first type a compound is considered to be active, but is actually ineffective. This type of error will be clarified during further development, after negative clinical trials at the latest. However, there are also two sources of error for false-negative results:

1. b = error of the second type due to the model
2. β = error of the second type due to statistics In the error of the second type a compound is considered to be ineffective, but is actually effective.

This type of error will never be clarified; an effective drug has just been missed. Perhaps another investigator will test this compound under different aspects. The statistical errors derive from the fact that a pharmacological test is performed only several times or in a limited number of animals. One can specify the probability that a decision made is incorrect, i. e., a drug candidate is erroneously identified as effective when it is actually ineffective. Usually this risk is set to 5% ($P < 0.05$) and is called the statistical error of the first type or type I error. The error of the second type or type II error is connected to the type I error by statistical rules.

Usually, screening is performed sequentially. Tests in high-throughput screening are followed by tests in isolated organs, then in small animals, and special tests in higher animals, until the compound is recommended for further development and for studies in human beings. From each step errors of type I, but also from type II, arise. As a consequence, many effective compounds will be lost. There are two ways to circumvent this obstacle: (1) to increase the number of compounds entering the screening procedure dramatically, hope for a reasonable number of true positives, and accept a high rate of false-negative results as followed in the ultra high-throughput screening, or (2) to perform tests with high relevance, meaning tests with high predictive value in whole animals at an early stage. The literature on high-throughput screening includes some publications dealing with false-negative results. Zhang et al. (1999, 2000) studied the role of false-negative results in high-throughput screening procedures. They presented a statistical model system that predicts the reliability of hits from a primary test as affected by the error in the assay and the choice of the hit threshold.

The hit confirmation rate, as well as false-positive (representing substances that initially fall above the hit limit but whose true activity is below the hit limit) and false-negative (representing substances that initially fall below the hit limit but whose true activity is in fact greater than the hit limit) rates have been analyzed by computational simulation. The Z-factor and the Z′-factor were introduced to characterize the reliability of high-throughput assays. The problem of type II errors, i. e., false-negative results, also exists in many other physiological and pharmacological studies. For example, Pollard and Howard (1986) re-investigated the staircase test, a well-accepted primary screening method for anxiolytics, and found several false-negative results for clinically active anxiolytics.

3

Discovery of Plasma Biomarkers

By coupling isoelectric focusing (IEF) in the first dimension with sodium dodecyl sulfate-polyacrylamide gel electrophoresis (SDS-PAGE) in the second dimension, two-dimensional polyacrylamide gel electrophoresis (2-DE) enables the separation of complex protein mixtures according to each protein's isoelectric point (p*I*) and molecular weight. Depending on the size of the immobilized pH gradient (IPG) and SDS-PAGE gel used, 2-DE has the potential to resolve up to 10,000 proteins simultaneously, however this is unusual, and typically single 2-DE gel images generally contain 1000 to 2000 distinguishable protein spots. This typical proteome coverage can be increased if combined with additional prefractionation steps or spread over several narrow pH range IPG gels. Furthermore, 2-DE delivers a protein pattern map, which reflects changes in protein expression levels, isoforms, cleavage products, and posttranslational modifications. 2-DE is by no means the only platform for protein separations and there are inherent technical limitations. 2-DE provides good coverage of proteins with ap*I* between 3 and 11 and a molecular weight of 10 to 200 kDa; proteins possessing properties outside these ranges are typically underrepresented. In addition, hydrophobic proteins are rarely resolved and identified on 2-DE protein maps without resorting to specialized methods.

Proteins found in low concentrations within the samples are also underrepresented, but through high-abundant protein depletion and prefractionation techniques, they can be brought above the detection threshold. 2-DE is labor intensive, difficult to automate, and, despite significant methodology improvements, a degree of technical proficiency is still required to routinely produce high-quality 2-DE protein maps. Despite these limitations, the overall advantages and relative accessibility of 2-DE technology has made it an attractive methodology for biomarker discovery, and higher-throughput automated methods can be applied for biomarker validation and screening. Biomarkers are being used throughout the drug discovery and development processes to understand fundamental biological processes and relationships.

Specific biomarkers for a disease state can be monitored for patient prognosis and response to therapy and serve as tools in the development of therapeutics. Biomarkers can also be used to stratify patients and to monitor adverse effects. Clinical biomarkers hold significant potential to make drug development process more effective and efficient. However, successful biomarker discovery through protein profiling is challenging because of the dynamic range of protein expression seen in eukaryotic tissues, ranging from 10^1 to 10^6 copies per cell.

To illustrate this point, plasma biomarker discovery involves searching for extremely low abundance proteins, which comprise less than 1% of the total plasma proteome, whereas the 20 most abundant proteins represent more than 99%. The presence of high-abundant plasma proteins such as albumin and IgGs mask the detection and identification of low-abundant proteins in subsequent proteomic analysis. Removal of these abundant proteins is essential for the detection of less abundant proteins that may prove to be informative diagnostic or therapeutic markers. One technology developed to remove high-abundant plasma proteins is the Multiple Affinity Removal System (MARS).

MARS combines the specificity of antibody-antigen recognition and the efficiency of high-performance liquid chromatography (HPLC) for the simultaneous and reproducible removal of six abundant human plasma proteins (albumin, transferrin, IgG, IgA, haptoglobin, and antitrypsin). As a result, the lower-abundant, more biologically relevant proteins can be brought into detectable levels by increasing their loading onto 2-DE gels for analysis. Within this chapter, to illustrate the power of 2-DE, we provide comprehensive step-by-step instructions for the application of 2-DE to human plasma samples to determine whether specific plasma proteins or protein expression patterns could serve as biomarkers for the development of a specific adverse effect (AE) that arises upon treatment with an experimental drug.

REQUIREMENTS

Descriptions of specific equipment used in this experiment are given, however any model of comparable capability can easily be substituted.

1. Agilent 1100 HPLC system and a 4.6 × 50 mm Multiple Affinity Removal LC Column-Human 6 for plasma fractionation. Sorvall RC-5C PLUS Superspeed Refrigerated Centrifuge equipped with a SH-3 000 rotor for the concentration and buffer exchange of fractionated plasma.
2. VMax microplate reader for determination of total protein concentration.
3. C75T Ultrasonic water bath sonicator and a 5415D tabletop centrifuge for protein solubilization.
4. Milli-Q Synthesis A10 water purification system.
5. DryStrip Reswelling Tray for in-gel rehydration and equilibration of IPG strips.
6. Multiphor II Isoelectric Focusing (IEF) System equipped with a MultiTemp III Thermostatic Circulator and EPS 3501 XL power supply for IPG strip isoelectric focusing.
7. Investigator 2-DE Electrophoresis System for SDS-PAGE separation.
8. ImageScanner II flat-bed scanner for gel image digitization (14 bits, 250 ppi).
9. Software: Z4000 Large-Scale 2-DE Gel Image Analysis System and Statistical Analysis System (SAS) for image and statistical analysis.

Reagents Required

1. HPLC mobile phases: Buffers A and B.
2. Plasma dilution buffer: 1 Complete EDTA-free Protease Inhibitor Cocktail tablet in 25 mL Buffer A.
3. Non-Interfering Protein Assay Kit.
4. Immobiline DryStrip 18 cm pH 4-7 IPG gels.
5. Rehydration buffer: 5 M urea, 2 M Thiourea (Sigma), 2% CHAPS (Sigma), 2% SB3-10 (Sigma), 30 mM DTT (Sigma), and trace amount of bromophenol blue (Sigma).
6. Equilibration buffer 1: 50 mM Tris (Sigma) pH 8.8, 6 M urea, 30% glycerol (Sigma), 4% SDS (Sigma), 2% DTT, and trace bromophenol blue.
7. Equilibration buffer 2: 50 mM Tris (Sigma) pH 8.8, 6 M urea, 30% glycerol (Sigma), 2.5% iodoacetamide (Sigma), 2% DTT, and trace bromophenol blue.
8. SDS-PAGE lower tank buffer 10X stock: 2.1 M Tris pH 8.9.
9. SDS-PAGE upper tank buffer 10X stock: 1 M Tris, 1 M Tricine (Sigma), 1% SDS.
10. Investigator pre-cast homogenous 10% Tricine slab SDS-PAGE gels.
11. Investigator Silver Stain Kits.

PROCEDURE

High-abundant protein removal from crude human plasma was performed using Agilent Technologies' Human MARS. Briefly, crude human plasma was diluted with plasma dilution buffer and filtered through 0.22-µm spin filters by centrifugation at 16,000 × *g* for 1 min. All chromatographic fractionations were performed at room temperature (22°C) on an Agilent 1100 HPLC system with automated sample injection. The flow-through and eluted fractions were manually collected.

1. Set up Buffer A and Buffer B as the only mobile phases.
2. Purge lines with Buffer A and Buffer B at a flow rate of 1.0 mL/min for 10 min without the MARS column attached.
3. Set up liquid chromatography (LC) method and run two blanks by injecting 100 µL Buffer A without the MARS column attached.
4. Attach the 4.6 × 50 mm MARS column and equilibrate with Buffer A for 4 min at a flow rate of 1 mL/min.
5. Dilute crude plasma sample 5X in plasma dilution buffer (e.g., 50 µL human plasma with 200 µL of plasma dilution buffer).
6. Remove particulates from the sample by filtering through an 0.22-µm spin filter for 1 min at 16,000 × *g* at room temperature.
7. Run LC method. In this experiment, four independent 50-µL injections were made for each AE and non-AE sample.

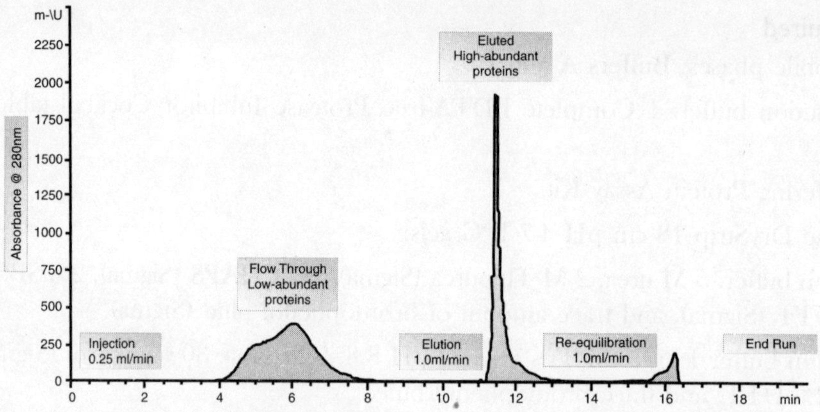

Fig. 3.1. Representative chromatogram of the affinity removal of high-abundant proteins from human plasma.

(a) Inject 50 μL diluted plasma at a flow rate of 0.25 mL/min.

(b) Collect the flow-through fraction which appears between 3 and 9 min and store at 4°C until all four injections for that particular sample are completed.

(c) Collect eluted fraction which appears between 9 and 12 minutes and store at 4°C until all four injections for that particular sample are completed.

(d) Regenerate column with Buffer A for 7.4 min at 1 mL/min.

8. Repeat step 7 three more times (total of four injections per sample).
9. Pool, concentrate, and buffer exchange flow-through fractions.

(a) Pool collected flow-through fractions from injections 1, 2, and 3 for each sample in a 4-mL, 5000 MWCO Spin concentrator.

(b) Centrifuge at ~4000 × *g* for 6 min at 4°C.

(c) Add flow-through fraction from injection 4 for each sample to the concentrator and centrifuge at ~4000 × *g* until approximately 500 μL remains in the filter compartment.

(d) Add 3.5 mL of 2 M urea to concentrator and repeat centrifugation until approximately 500 μL remains in the filter compartment.

(e) Repeat step 9(d) three times to ensure Buffer A has been sufficiently replaced with 2 M urea.

10. Collect the concentrated/buffer exchanged samples and store at –20°C for later used.

Protein Assays

The protein concentration of each pooled sample was estimated using the Non-Interfering Protein Assay Kit. This assay works in the presence of commonly used laboratory agents (reducing agents, detergents, amines, urea, etc.). The kit uses a proprietary Universal Protein Precipitating Agent for removing interfering agents, such as urea, from protein samples such that only clean protein remains in the tube to be assayed.

1. Thaw samples to be assayed on ice.

2. Prepare a standard curve (0 to 50 μg) by pipetting the appropriate volume of the BSA standard supplied with the kit into duplicate wells of a 96-deep well plate.
3. Add 10 μL of each sample to duplicate wells of the plate.
4. Add 0.5 mL UPPA-I into each well, cover the plate with the sealing mat (VWR), and vortex.
5. Incubate the plate for 3 min at room temperature.
6. Add 0.5 mL UPPA-II into each well, cover the plate with the sealing mat, and vortex.
7. Centrifuge the plate at 8000 × *g* for 7 min to sediment precipitated protein.
8. Discard the supernatant by inverting the plate into an appropriate waste receptacle then softly blotting the plate on absorbent tissue paper. Allow the liquid to completely drain out of each well.
9. Add 100 μL Copper Solution and 0.4 mL water into each well, seal the plate, and vortex to completely dissolve the protein precipitate.
10. Add 1 mL Reagent II into each well–for simultaneous mixing, Reagent II should be added by rapidly shooting into the well.
11. Incubate the plate at room temperature for 20 min.
12. Remove 200 μL of solution and pipette into the corresponding wells of a 96-shallow well assay plate (VWR).
13. Read absorbance on a plate reader at 480 nm.
14. Plot absorbance against protein concentration of the standard curve, and determine concentration of unknowns.
15. Once protein concentrations have been determined, aliquot 150 μg of each sample into 1.5-mL Eppendorf tubes and store at –20°C.

Process of Rehydration

In preparation for IEF, 150-μg aliquots of each sample were diluted to a total volume of 350 mL in rehydration buffer. The samples were loaded onto Immobiline DryStrip 18 cm pH 4-7 IPG gels during an overnight in-gel rehydration/sample loading step.

1. Thaw samples on ice.
2. Dilute each sample to a total volume of 350 mL in rehydration buffer.
3. Incubate at room temperature for 1 h with a 1-min water bath sonication every 15 min.
4. Clear any insoluble material through centrifugation at 14,000 rpm for 10 min at room temperature.
5. Level the Immobiline DryStrip Reswelling tray.
6. Pipette the entire 350 μL sample into the grooves of the Immobiline DryStrip Reswelling tray.
7. Peel off the protective cover sheets from the IPG strips, and position them such that the gel is in contact with the solution.

8. Cover each IPG strip with ~2 mL of Immobiline DryStrip Cover Fluid.
9. Allow the strips to rehydrate overnight at room temperature.

IEF (Isoelectric Focusing) System

IEF was performed utilizing a Multiphor II IEF System at 20°C under the following voltage gradient: 0 to 500 V over 1 min; increased to 3500 V over 1.5 h and held at 3500 V for approximately 73,000 V· h.

Generating the multiphor II unit

1. Set the thermostatic circulator to 20°C and begin cooling.
2. Pipette approximately 10 mL Cover Fluid on cooling plate. Position the DryStrip tray on the cooling plate so the anodic electrode connection of the tray is positioned at the top of the plate near the cooling tubes.
3. Pipette 15 mL Cover Fluid into the tray. Place the DryStrip aligner, 12-groove-side up, into the tray on top of the Cover Fluid.
4. Cut two IEF electrode strips to a length of 110 mm. Place the electrode strips on a clean surface such as a pile of Kimwipes. Soak each electrode strip with 0.5 mL water. Blot with Kimwipes to remove excess water.

Loading IPG strips for IEF

1. Remove the rehydrated IPG strips from the reswelling tray and remove excess Cover Fluid by blotting on a Kimwipe, avoiding contact with the gel surface.
2. Immediately transfer the rehydrated IPG strips to adjacent grooves of the aligner. Place the strips with the acidic end at the top of the tray near the anode. Align the IPG strips so that the anodic gel edges are lined up.
3. Place the moistened electrode strips across the cathodic and anodic ends of the aligned IPG strips. The electrode strips must at least partially contact the gel surface of each IPG strip.
4. Align each electrode over an electrode strip, ensuring that the marked side corresponds with the side of the tray giving electrical contact. When the electrodes are properly aligned, press them down to contact the electrode strips.
5. Cover the IPG strips with 100 mL Cover Fluid.
6. Place the lid on the Multiphor II unit.
7. Connect the leads on the lid to the power supply. Ensure that the current check on the EPS 3501 XL power supply is switched off and begin IEF using the following gradient: 0 to 500 V over 1 min; increased to 3500 V over 1.5 h and held at 3500 V for approximately 73,000 V· h.

Process of Equilibration

The focused IPG strips were immediately equilibrated as follows: reduction through 15 min incubation in Equilibration buffer 1; followed by alkylation for 15 min in Equilibration buffer 2. This is a modification of the method described by Yan and co-workers.

1. Remove the focused IPG strips and blot off excess Cover fluid on a Kimwipe.
2. Place in adjacent wells of the DryStrip Reswelling Tray and cover strips with 50 mL Equilibration buffer 1. Transfers tray to an orbital shaker and lightly agitate for 15 min at room temperature.
3. Aspirate off Equilibration buffer 1 and add 50 mL Equilibration buffer 2. Transfer tray to an orbital shaker and lightly agitate for 15 min at room temperature.

SDS-PAGE

Once equilibrated, strips were loaded onto Investigator Pre-cast 10% Tricine slab gels. Separation upon apparent molecular weight was conducted using the Investigator 2-DE Electrophoresis System at 15 W/gel for approximately 5 h.

Role of gels and buffers in SDS-PAGE

1. Prepare 10 L 210 mM Tris pH 8.9: dilute 1 L of 10X lower tank buffer with 9 L of water. Add to lower tank and allow chill to 4°C.
2. Prepare 3 L 100 mM Tris, 100 mM Tricine, 0.1% SDS: dilute 300 mL of 10X upper tank buffer with 2.7 L water. Place in the cold box and cool to 4°C.
3. Load the pre-cast gels into the gel rack and insert gaskets.
4. Add enough cooled upper tank buffer to cover the tops of the glass plates

Loading IPG strips

1. Remove strips from the equilibration buffer and blot off excess buffer.
2. Place the strip into the slits of the gel plates with the acidic end to the left.
3. Use the gel installer to push the strips down to make good contact with the SDS-PAGE gel.
4. Add the remaining upper tank buffer slowly to avoid disrupting the IPG strip position.
5. Place the lid on the tank and connect the lead to the power supply and start the 15 W/gel current. Run until the bromophenol dye front is approximately 1cm from the gel bottom.

Visualization of Protein Pattern

Protein patterns were visualized through silver staining using the Investigator silver stain kit, which is a modification of the method described by Rabilloud. The procedure was modified for nondestructive staining by omission of the glutaraldehyde in Fixative 2 solution.

Method of staining

1. Fixative 1: Add 1.6 L ethanol to a 4-L beaker, then slowly add 400 mL acetic acid, bring up to 4 L with de-ionized water (final concentration of 40% ethanol and 10% acetic acid).
2. Fixative 2: Add contents of Fixative 2 bottle to a 4-L beaker and add 1.2 L ethanol, bring up to 4 L with de-ionized water, and mix (final concentration 8 mM potassium tetrathionate, 829 mM sodium acetate, and 30% ethanol).
3. Silver nitrate solution: Add content of silver bottle to a 4-L beaker and add 1 mL of formaldehyde solution, bring up to 4 L with de-ionized water, and mix (final concentration of 11.4 mM silver nitrate and 0.009% formaldehyde).

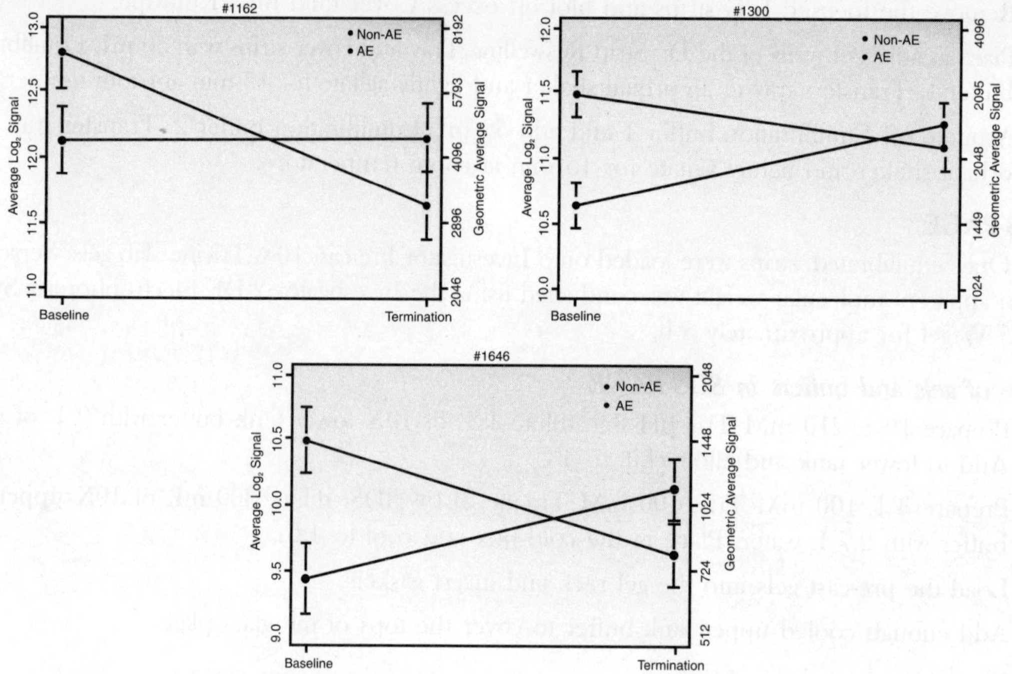

Fig. 3.2. Plots depicting "interaction" between condition and time.

4. Developer solution: add ~3.5 L de-ionized water to a 4-L beaker and stir briskly. While stirring, add contents of Developer bottle slowly until it is dissolved. Add 600 μL formaldehyde solution bring up to 4 L with de-ionized water, and mix (final concentration of 63 mM sodium thiosulfate, 0.2 M potassium carbonate, and 0.005% formaldehyde).
5. Stop solution: Add contents of Stop bottle to a 4-L beaker, add 80 mL acetic acid, bring up to 4 L with de-ionized water, and mix (final concentration of 0.4 M Tris and 2% acetic acid).

Silver staining

1. Disconnect the gel unit from the power supply and disassemble the unit. Transfer the gels to trays containing 400 mL Fixative 1 solution. Gently agitate gels on a rotary shaker for 1 h at room temperature.
2. Decant off Fixative 1 solution and add 400 mL Fixative 2 solution. Gently agitate gels overnight.
3. Decant off Fixative 2 and add 400 mL de-ionized water. Gently agitate gels for 15 min. Repeat three times with fresh water each time.
4. Place gels in 400 mL silver nitrate solution and agitate for 1 h.
5. Decant off silver solution and wash gels in 400 mL de-ionized water for 1 min.
6. Decant off water and added 400 mL developer solution. Agitate gels until spots appear, ~8 min.
7. Decant off developer solution and add 400 mL stop solution for 10 min, then add 400 mL de-ionized water.

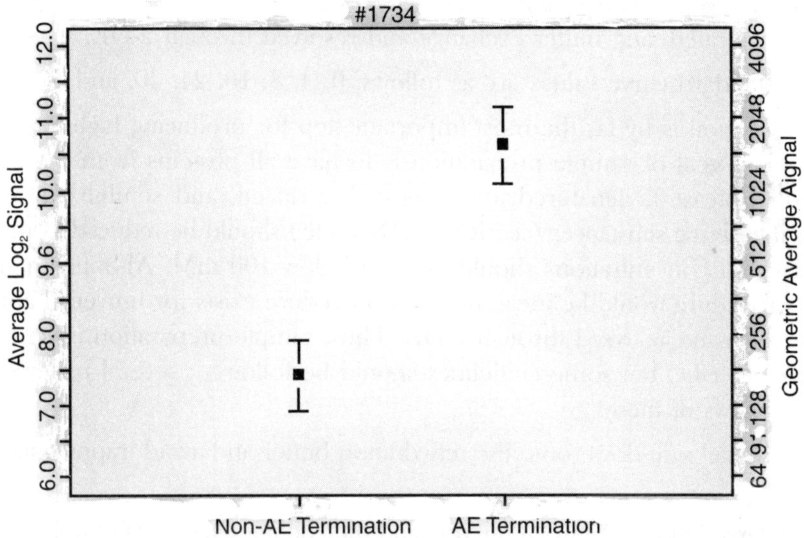

Fig. 3.3. Graph depicting differential expression of the specified protein spots between non-AE and AE samples.

Data Analysis

After electrophoresis and staining, gel image analysis is conducted to quantify protein spots, which allows gel-to-gel protein expression comparisons. This is usually accomplished with the help of commercially available 2-DE analysis software packages. Although various software packages differ in price, degree of automation, analytical speed, and ease of use, they are generally labor intensive and require considerable time for familiarization.

In general, 2-DE image analysis follows a series of steps: scanning the gel to digitize the protein pattern, filtering the image to remove background and artifacts, detecting spots, matching spots across gels, normalization, differential analysis, and statistical analysis. Spot signals were measured by densitometric scanning. Image analysis was conducted using Z4000 Large-Scale 2D Gel Image Analysis System.

The data with the raw signal values matched by spot was exported from Z4000 as a text file for further data analysis using SAS. The data was transformed to log (base 2) scale and normalized to have the same average log signal value across gels. ANOVA with mixed effects model (by spot) was used to perform statistical analysis of differential expression between conditions of interest allowing for correlation between observations from the same subject. For a pair of compared experimental groups, spots with p values less than 0.005 and at least two observations for both groups were identified as differentially expressed.

Annotation

1. All solutions should be prepared in water of 18.2 MΩ cm and total organic content of less than five parts per billion.
2. Multiple injections of each sample were made in order to recover enough depleted protein to run several 2-DE gels. Consult the column certification of analysis to verify the columns capacity.

3. Eluted fractions were retained and stored at –80°C for future analysis. However, several were pooled, concentrated, and buffer exchange and resolved through 2-DE.
4. Our typical standard curve values are as follows: 0, 4, 8, 16, 24, 40, and 50 μg.
5. Sample preparation is by far the most important step for producing high-quality reproducible 2-DE gels. The goal of sample preparation is to have all proteins from the tissue of interest deactivated (proteases), denatured, disaggregated, reduced, and solubilized. In addition, the amount of interfering substances (i.e., RNA, DNA, etc.) should be reduced as much as possible; in particular, salt concentrations should be kept below 100 mM. Although a simple one-step extraction procedure would be ideal, no such procedure exists for universal application to the many types of tissue analyzed through 2-DE. Thus, sample preparation is generally tissue and protein class specific, but some guidelines should be followed; several reviews have provided detailed overviews of these.
6. Place IPG strip gel side down onto the rehydration buffer and avoid trapping air bubbles under the strip.
7. Remove any large bubbles between the tray and the cooling plate; small bubbles can be ignored.
8. The presence of air bubbles between the strip positions under the DryStrip aligner will not affect the experiment. Avoid getting IPG Cover Fluid on top of the aligner at this point, as it interferes with visualization of the grooves.
9. SDS-PAGE buffers and gel box assembly needs to be completed before IEF to allow time for the buffers to cool and reach 4°C. Typically, the buffers and gel unit are assembled shortly after IEF has been started and allowed to cool overnight.
10. This method is quantitative when the development time is kept to 10 min or less.

4

Pharmacokinetic Models

Design of successful drug development candidates requires balancing a number of different characteristics simultaneously including intrinsic activity, biopharmaceutical properties, synthesis, stability, and many others. Independently, each of these can be considered a barrier to drug performance or development success. Among the most important of these processes determining *in vivo* performance are absorption, distribution, metabolism and excretion, collectively referred to as ADME. The structure and physicochemical characteristics of the drug are important determinants of these processes, as are the characteristics of the physiological mechanisms. Further complicating the issue is the interrelationship of many of these processes, frequently in antagonistic ways. Increasing solute lipophilicity, for example, can decrease aqueous solubility, which frequently compromises oral absorption. Also, it can increase metabolic clearance, thus making difficult sustaining the pharmacologically relevant systemic exposure.

In contrast, permeability, in many cases, increases with increasing lipophilicity, favoring absorption. The actual result will be determined by the relative contributions of these two competing phenomena. Permeability is also an important determinant of distribution, metabolism and excretion. Given these multiple dependencies upon specific drug structural characteristics, it can be seen that simultaneously optimizing all of these processes can present a formidable challenge. Occasionally, one specific, clearly dominant problem, can be identified in a particular drug discovery program. In this case, focusing on the specific property can result in successful solution of the problem and identification of a promising drug candidate.

The recent interest in development of computational structure property relationships has resulted in the discovery of promising tools to aid in these problem solving and property optimization objectives. More generally, the interrelationship of multiple solute properties must be considered simultaneously in order to achieve the desired outcome. A simple example of this is the interrelationship of solute solubility and permeability in determining oral absorption potential. Briefly, bi-directional Caco-2 cell permeability data and aqueous solubility data were measured for a series of relatively homologous antimicrobial agents at Pharmacia. All compounds were dosed both intravenously and orally in rats in order to calculate total clearance and oral bioavailability. The clearance data supported the expectation that these compounds had minimal first pass metabolism such that bioavailability was directly related to the fraction of solute absorbed.

In attempting to predict absorption of these compounds, it was found that no significant correlation existed between bioavailability (fraction absorbed) and either solubility or permeability alone. While high permeability solutes did seem to be well absorbed, so were many of the low permeability solutes. What was common among these poorly permeable, well absorbed compounds was high aqueous solubility, suggesting that this property may help to compensate for the poor *in vitro* permeability. One simple model for interrelating solubility and permeability quantitative data to predict absorption is the Maximum Absorbable Dose (MAD) model. MAD is a hypothetical value that estimates the amount of an infinite dose that could be absorbed, given a combination of solubility and absorption rate (calculated from permeability coefficient). In the present case, since a finite dose was actually given, the fraction of the dose absorbed can be estimated by dividing the MAD value by that actually dosed.

Since MAD calculates a theoretical mass of solute assuming an infinite dose, it is possible to predict masses absorbed greater than than atually dosed. In this case, the predicted fraction absorbed is assigned a value of 1. In general, the correlation between predicted fraction dose absorbed and observed bioavailability was good except for the most highly permeable, lowest solubility compounds. On the assumption that for such solutes aqueous solubility may underestimate the intestinal solubility arising from bile salts and other surfactants present in the lumen, solubility of these compounds were measured in simulated intestinal fluid (SIF) and predicted fraction absorbed recalculated. The lesson from this example is that solubility and permeability must be considered together in prioritizing these candidates with respect to absorption potential and, for highly permeable solutes, solubility in biologically relevant media may need to be considered.

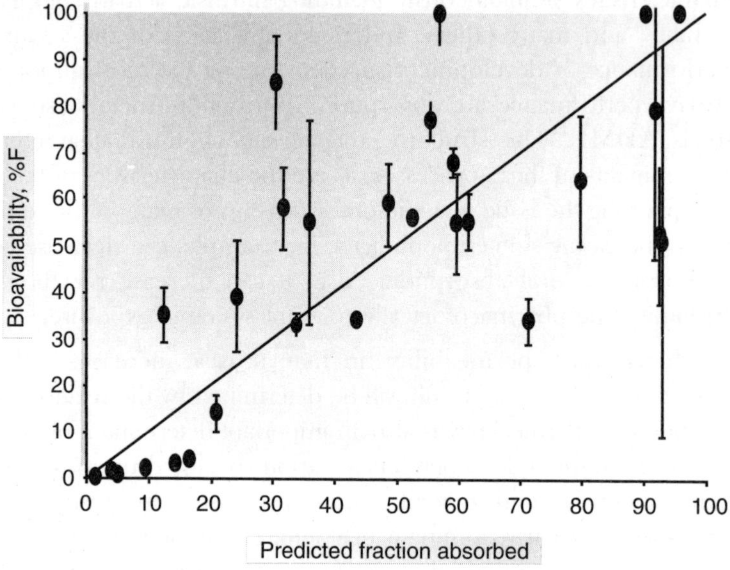

Fig. 4.1. Oral absorption of oxazolidinone antibiotics in the rat.

ESTIMATING LEAD OPTIMIZATION

While the MAD model is a useful strategy for interrelating solubility and permeability in predicting oral absorption, a more general approach to integrating solute structural characteristics into an anatomically and physiologically relevant approach for modeling absorption and disposition of xenobiotics is physiology-based pharmacokinetic (PB-PK) modeling. In PB-PK models the body is described as a system of tissue compartments perfused by known blood flows and characterized by their volumes, structures and compositions. Elimination processes are described using physiologically relevant approaches as described by Rowland, *et al.* (1973).

Disposition of the drug as a function of time is described by a series of mass balance equations, written as a system of differential equations, and the system is fitted to the available plasma and/or tissue concentrations. Historically, such PB-PK models are used on well characterized drugs or toxicants. They are rarely used on new chemical entities, probably due to the need of individual structural models for tissues (e.g. perfusion vs permeability-rate limitations) and the substantial amount of information required for their parameterization (tissue partition coefficients, tissue clearances, plasma and tissue protein binding, etc.). In an attempt to extend PB-PK modeling to use in a discovery setting, Poulin and Theil (2000, 2002) described a basic PB-PK model for rat, which was effective in predicting the pharmacokinetics of ethoxybenzamide, propranolol and diazepam, based on a very limited input (log P, pKa, intrinsic CL, protein binding, blood to plasma ratio).

Such a model clearly has potential for simulating the outcome of the first-time-in-animal study. Following Poulin and Theil's lead we developed a rodent PB-PK model to support lead optimization in the discovery setting. The basic model consisted of differential equations describing 13 tissue compartments: muscle, bone, skin, adipose, brain, heart, lungs, spleen, gut, liver, kidneys and the arterial and venous blood pools. Drug mass balance in a generic tissue compartment is described as:

$$\frac{dC_T}{dt} = C_T\left(\frac{-Q_T}{\frac{P_{t:P}}{B:P}\cdot V_T}\right) + Q_T \cdot C_{input}\left(\frac{1-E_T}{V_T}\right)$$

where C_T is the drug concentration for the tissue T, C_{input} is drug concentration in input to the tissue T, Q_T is the blood flow, V_T is the tissue volume, $P_{t:p}$ is the tissue to plasma partition coefficient, B:P is the blood to plasma ratio, and E_T is the extraction ratio. The system of differential equations was solved simultaneously for providing tissue and plasma concentration-time curves. Physiological data on blood flows (Q_T) and volumes (V_T) were obtained from the literature.

Partition into tissue compartments was assumed to be perfusion-rate limited. The partition coefficients (including B:P) were calculated using lipophilicity measurements (logP, logD), protein binding and the tissue composition model described by Poulin and Theil (2000, 2002). Finally, composition data were taken from the literature. Elimination was restricted to the liver and kidney compartments. The hepatic extraction ratio E_H was calculated from intrinsic clearance obtained from liver preparations and assuming the well-stirred model for hepatic clearance.

Renal clearance was modeled assuming glomerular filtration only, as $GFR \cdot f_{u,p}$, where GFR is the glomerular filtration rate and $f_{u,p}$ is the fraction unbound to plasma proteins. As an example of this approach, we modeled 45 Pharmacia compounds which had been dosed intravenously to rats. The input parameters were logP = 2.78, pKa = 7.3, $f_{u,p}$ = 0.065, CLint (estimated from rat hepatocytes) = 40.4 mL/min/kg. Blood: plasma ratio was estimated according to Poulin and Krishnan (1995). As can be seen from the Figure, good correspondence between predicted and observed plasma levels with time were found. This particular example represents a good prediction from the data and validates the general assumptions and interrelationships of the inputs in the model for this compound. While such a model, using only computational and/or *in vitro* inputs commonly obtained in a discovery setting, can be useful in prioritizing or ranking compounds prior to the first in animal evaluation, an additional application in helping to focus a lead optimization program can also be envisioned.

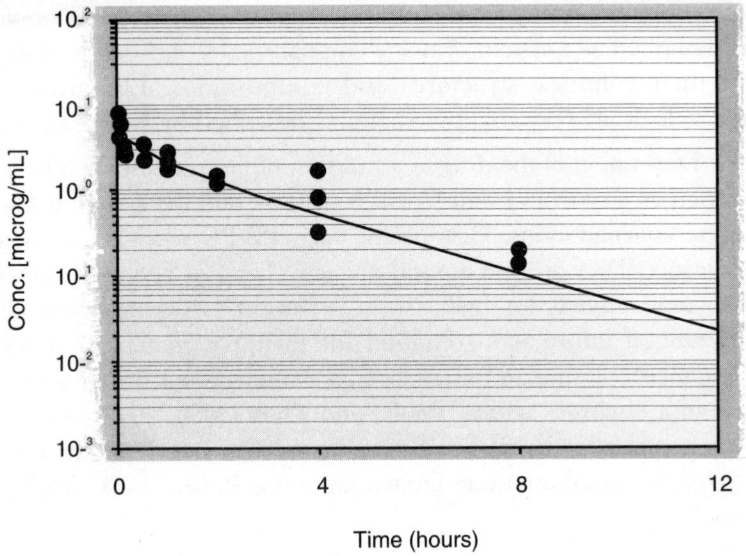

Fig. 4.2. Physiology based pharmacokinetic (PB-PK) model prediction for a model compound.

In this case, the PB-PK model is used to predict blood-level time profiles from the *in vitro* and physicochemical characteristics of the lead. The compound is then dosed to the animal in order to validate the model. If the resulting validation is reasonably successful, as in the example shown above, the PB-PK model can be used to identify those properties most important in optimizing the *in vivo* performance by means of a sensitivity analysis. These most sensitive parameters can then be focused on in the refinement program. Figure 4.3 illustrates this approach for a potential drug candidate. Consider the lead compound to have logP = 6, $f_{u,p}$ = 0.01 and CLint = 70 mL/min/kg. This combination of properties results in a fairly rapid elimination from the systemic circulation after administration. Decreasing logP to a more favorable value of 3 actually increases clearance, probably resulting from decreased tissue and/ or B:P partitioning. On the other hand, decreasing clearance, while

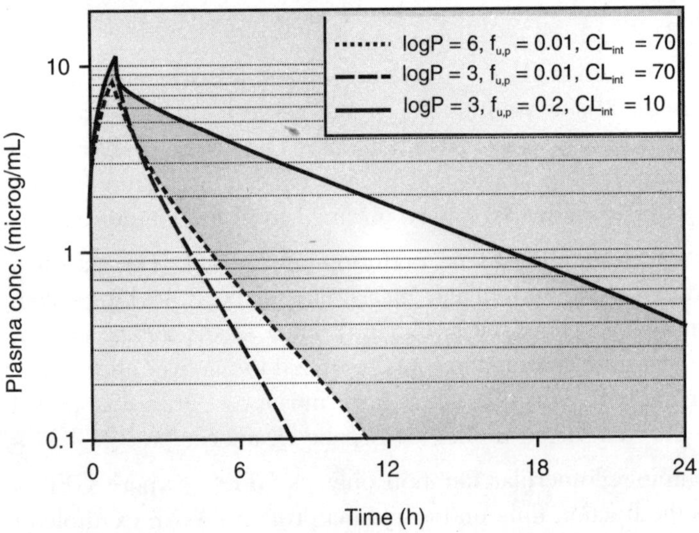

Fig. 4.3. Sensitivity analysis reflecting relative influence of logP, fraction unbound in plasma ($f_{u,p}$) and clearance on predicted blood level time profiles for a prototype lead candidate.

retaining the more favorable logP value, significantly increases predicted exposure. In this last example, the unbound fraction ($f_{u,p}$) was also allowed to increase.

Based on these considerations, the most important parameter contributing to *in vivo* performance is clearance, and a target value for an improved analog can be identified, helping to focus the chemistry effort. An enhancement to the model to add oral absorption capability has also been explored. Sophisticated compartmentalized absorption and transit (CAT) models have been developed which allow the user to model absorption into portal blood of solutes administered as solutions or solid dosage forms. In these models the intestinal tract is segmented into a series of compartments with associated physiological characteristics of the animal species of interest such as lumenal volume, surface area, pH and residence or transit time.

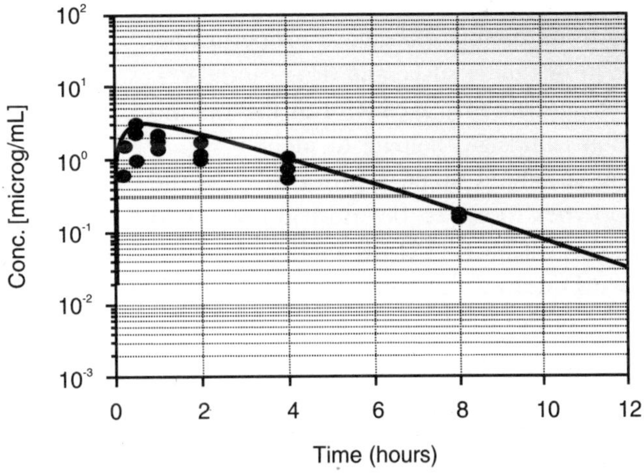

Fig. 4.4. Comparison of predicted (solid line) and measured (symbols) blood level time profiles after oral administration of a prototype lead candidate.

The solute parameter inputs are aqueous solubility as a function of pH, intestinal permeability, particle size, if administered as a suspension, and dissolution rate. The model then calculates time dependent absorption as the solute moves through the sequential compartments based on these permeability/solubility-dissolution characteristics. This profile can then be used as the input venous-blood concentration parameter to the basic PB-PK model. An example of this application is shown in Figure 4.4. In this figure, blood level time profiles are predicted and then compared to those actually observed for the model compound from Figure 4.2. The solute parameters used in the CAT model were solubility of 29.4 mg/ml at pH 1.8, 4.4 mg/ml at pH 4.6, 2.8 mg/ml (pH 5.7) and 0.06 mg/ml (pH 7.3).

Permeability was measured in Caco-2 cells ($26.0 \pm 6.1 \times 10^{-6}$ cm/sec in Ap->Bl direction and $18.0 \pm 1.0 \times 10^{-6}$ cm/sec in Bl -> Ap direction) scaled to the rat intestine, as described previously. These values were then used in Gastro-Plus assuming a solution dosage form. Again, as in the case of the intravenous model, the agreement is very good for this particular compound, capturing, in general, C_{max}, T_{max} and the overall clearance profile.

Here also, the resulting validated model can be used to perform sensitivity analysis for parameters most important to *in vivo* performance. Both permeability and clearance properties contribute to overall blood-level time profiles, with clearance having the most significant impact for this particular biopharmaceutical property profile. On the other hand, if the starting point were a more problematic combination of properties, these tools can give guidance in the direction in which to focus the optimization program and some boundary conditions on the magnitude of improvement necessary to achieve the desired end. In this regard, sensitivity analysis on fraction unbound or logP had little influence on the shape of the curves, except in the case of extreme logP values, suggesting these would not be profitable directions to explore synthetically.

OBJECTIVES OF DEVELOPING STRUCTURE-BASED MODELS

There has been significant activity in recent years toward the objective of developing structure-based predictive models of important biopharmaceutical properties for use in designing successful drug candidates. Such models may be useful in the optimization campaign, helping to identify promising structural analogs improving on the properties found to be most important in the context of the PB-PK model. The major ADME events can be considered in terms of the solute biopharmaceutical processes determining them. These biopharmaceutical processes can in turn be related to solute structural and/or physicochemical parameters of most importance in the specific property. This then forms the basis for development of predictive structure property relationships (SPR's) which can be used in lead optimization. In the next few sections we will review some characteristics of different SPR's in the context of specific lead optimization strategies. The intent is not to comprehensively review the SPR literature, which is fairly extensive, but rather consider general issues which should be addressed in the process of implementing any given property model.

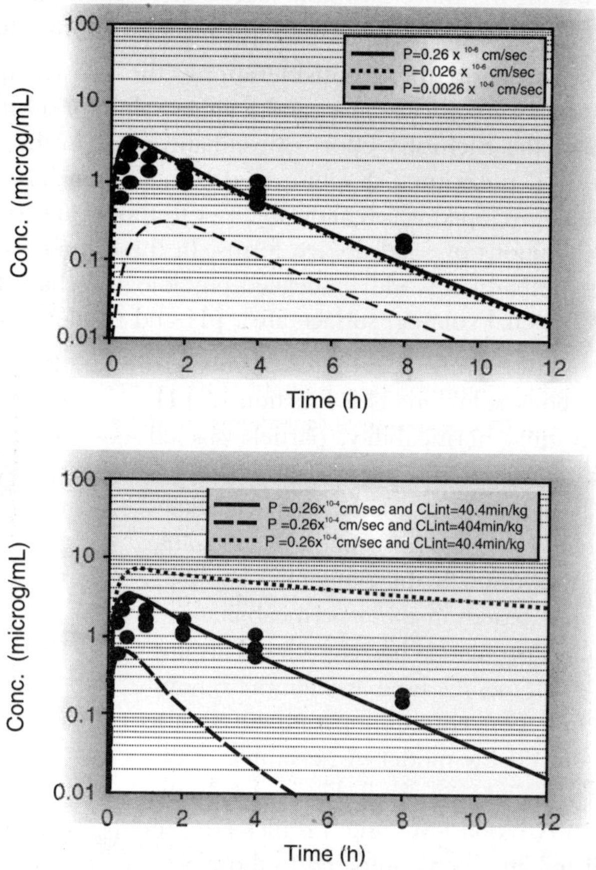

Fig. 4.5. Sensitivity analysis reflecting influence of intestinal permeability (panel A) and clearance (panel B) on predicted blood level time profiles.

Solubility

Solute solubility, as has been shown, is an important parameter in the oral absorption process. Generally, aqueous solubility is evaluated early in the lead selection process and minimally acceptable limits identified in order to consider taking a compound forward. However, in the various environments a drug will encounter *in vivo*, purely aqueous buffer is not one of them. Especially in the case of the intestinal lumen, various solubilizing agents are present which, in many cases, may improve the effective solubility of a solute *in vivo* compared to that predicted from the purely aqueous point of view. Similarly, lipids and proteins in blood may serve to increase the circulating solute concentration far beyond that expected for certain "insoluble" compounds. In spite of these caveats, aqueous solubility is an important reference point which needs to be considered and has been the subject of considerable modeling efforts. Developing structure-based solubility models from a mechanistic perspective requires specific consideration of the solvation process at a molecular level.

The forces involved in solubility are solute-solute, solvent-solvent and solute-solvent interactions. These are the same considerations that apply in solute partitioning between two immiscible liquid phases. To a first approximation then, aqueous solubility of a liquid can be modeled as a partitioning of a solute between itself, or an organic solvent surrogate, and water. Consistent with this simple model, it has been possible to correlate aqueous solubility directly with octanol-water partition coefficients for liquid solutes. In the case of solids, solute-solute forces existing in the crystal lattice complicate the prediction of solubility. At the present time, no reliable methods are available for estimating the energetics of these solid state, solute-solute interactions which must be overcome in order for the solute to release from the crystal and dissolve in water. An alternative approach to predicting solubility of solids is use of empirical, correlative models derived from collections of experimental data.

Numerous examples of such models have been published, differing generally in the composition of the training data set, descriptor representation of the solutes, and statistical method for establishing the correlation. Since these are empirical models, they are generally more accurate if the structure of the query molecule is similar to those in the training set. Thus these models strive to employ sufficient structural diversity to be useful in the entire chemistry space of interest to the medicinal chemist. Further, given the correlative nature of the model, the accuracy can be no better than the experimental accuracy of the training set data, which is generally assumed to be about 0.5 log units. An additional complication in the development and implementation of structure-based models is the influence of solute ionization state of the solute on its apparent solubility at a given pH. Since many drugs are either weakly acidic or basic, this is an important consideration.

The solubility of such substances depends upon the intrinsic solubility of the neutral species and relative concentration of neutral and charged species at the pH of interest. These concentrations depend upon the relative values of the pKa(s) of the solute and pH of medium. From a practical perspective this means that in training a model with such solutes, only the intrinsic solubility should be used. If the model is developed from apparent solubility, the relationship of pKa with the pH of data measurement must be taken into consideration, adding considerable complexity to the model. Further, if the model predicts a low intrinsic solubility, this may be somewhat misleading if the solute is expected to be highly charged at physiological pH.

In such a situation, it is desirable to incorporate a pKa estimation algorithm with an intrinsic solubility model in order to predict apparent solubility at a relevant pH. Finally, as with most statistical models, performance is measured by the model's prediction capability and this depends on the space on which the model was trained. With commercial packages, the user generally does not know on which chemical classes the model was trained. When large errors are observed between the predicted and the experimental values it is assumed that the model is not useful. This is one of the main reasons that local solubility models are developed. In this situation, local is defined as representing a relatively focused, homogeneous region of chemical space while global models are developed over much greater structural diversity. In using such models it is important to understand the limitations of the model in terms of applicability and accuracy in different chemistry and property spaces.

Toward that end, a study was undertaken at Pharmacia to compare accuracy and bias in prediction for three commercially available solubility models and a model developed at Pharmacia over different regions of chemistry spaces and three different solubility levels. The four solubility models were: ACD Labs PhysChem v. 6.0, QikProp v. 2.1, Cerius2 and the in-house model. The in-house solubility model was trained on literature and in-house data. The training sets for the three commercial models were not known. In the study, six data sets containing a total of 713 compounds were utilized. Four of the sets were local collections of data developed from internal discovery programs. They represent relatively homogeneous structural features generally arising from optimization around a biologically relevant template. One was a structurally diverse collection of Pharmacia data from a variety of discovery program teams and was used for continuous testing of the "in-house" solubility model.

Finally, the last set, data set 5, contained 85% simple, monofunctional organic compounds and the remaining 15% drugs and pesticides. The experimental solubility was compared to the predicted solubility obtained from each of the four solubility models. RMSPE (root mean squared prediction error) represents the accuracy of the prediction model in these specific data sets and is assumed to reflect the similarity of the structures in the training, model, space to the query structure. Only in the case of data sets 4 and 6, were all the compounds predicted by all four models. In the other cases, one or more of the models failed to predict solubility for some of the structures in the sets. From the results presented in the table, the in-house model has the smallest RMSPE in three out of the six data sets (1, 2, 3). This is, in part, due to the fact that the training set of this model includes compounds that are similar, but not identical, to structures in those data sets. Similarly, QikProp and Cerius2 perform comparably to the in-house model for data sets 1 and 3 respectively. None of the models perform particularly well with data set 6 suggesting this is an especially challenging region of chemistry space.

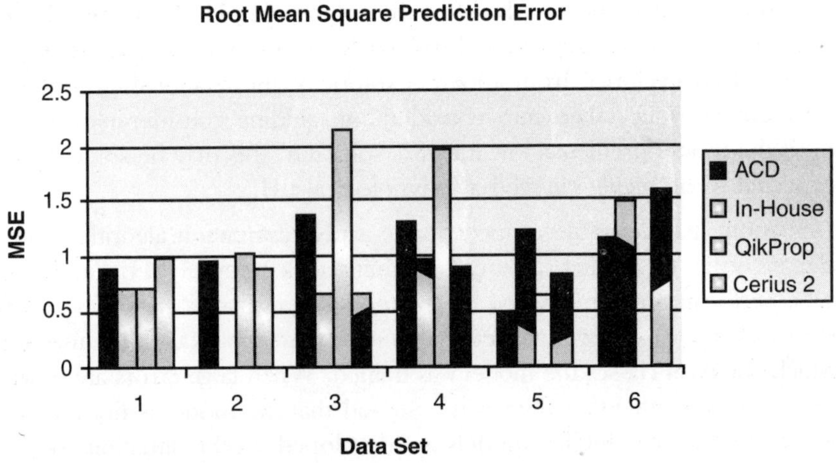

Fig. 4.6. Prediction accuracy of several global aqueous solubility models.

The implications of these results is that, while all the models are accurate in specific applications, none of them perform well over all structural diversity present in the test sets. From the perspective of prioritizing structures to synthesize based on predicted solubility properties, the existence of such biases should be taken into consideration. The residual plots differentiated by solubility range. Residual is the difference between experimental value and predicted value for each of the models and shows any bias to under or overpredict in the model. ACD and QikProp appear to be unbiased over all solubility ranges while the in-house model is over-predicting at all solubilities and Cerius2 over-predicts for low solubility compounds and under-predicts high solubility structures. Biases were observed in all the data sets, with no consistent trends with respect to the different models.

Again, from the objective of using these models prospectively in making synthetic decisions, such biases should be acknowledged and taken into consideration. With respect to lead optimization programs to overcome potential pharmacokinetic problems, accuracy in predicting the impact of subtle structural changes in a fairly homogenous structural space is desired. The results of this analysis clearly show that the performance of the different models is dependent upon the chemistry space and solubility range within that space, both with respect to accuracy and bias. While bias itself does not necessarily preclude the utility of a model, recognition that it exists and in what direction can help the user make more informed decisions with the predictions. At the least, these considerations suggest that the user of any model should characterize and validate the model in an intended application before general use.

Determinant of Absorption

Permeability is an important determinant of absorption and a number of other ADME processes dictating the *in vivo* performance of a drug. However, permeability itself is a complex process containing contributions from a number of different transport mechanisms, the relative contributions of which depend upon the structural features of the drug. After oral administration, a drug must move from the intestinal lumen through an unstirred water layer and mucus coat adjacent to the epithelial cell surface. Movement across the epithelial layer takes place by two independent routes-transcellular flux, *i.e.* movement across the cell, and paracellular flux, or movement between adjacent epithelial cells, restricted by the presence of tight junctions between the cells. The solute then encounters a number of microenvironments including a basement membrane, interstitial space and capillary wall in accessing the mesenteric circulation.

Any and all of these microenvironments can be considered a resistance to solute movement with an associated permeability coefficient. Further, the influence of drug structure with permeability in these different domains will be different. For example, unstirred water layer permeability is inversely related to solute size while paracellular permeability is dependent upon both size and charge. In the latter case, the characteristics of the paracellular "pore" result in size restricted diffusion as the size of the solute approaches that of the paracellular space. Further, cations are more permeable than neutral species, which in turn are more permeable than anions, consistent with the negative charge characteristics of the paracellular space. With respect to transcellular permeability, the relationship of solute structure with permeability again depends upon the mechanism involved. Historically, a passive diffusion pathway has been assumed for most solutes. However, an increasing number of active absorptive and secretory processes in intestinal epithelial cells are being identified for which many common drugs are substrates. These same considerations apply to tissue distribution and excretion, both of which are also dependent upon passive and active transport mechanism.

With respect to structure/transport or permeability relationships, while active transport involves specific interactions between solute and transporter, passive diffusion is dependent upon solute partitioning into the cellular plasma membrane and diffusion coefficient within the membrane. Considerable structure-property relationship (SPR) work has been published over the years to describe the passive transport mechanism, still believed in many cases to be the most significant pathway responsible for intestinal permeability and oral drug absorption. A very successful model for predicting passive permeability is the so called passive/diffusion model in which the cellular barrier is reduced conceptually to a homogenous, single biomembrane. Both partitioning and diffusion are influenced by the physicochemical and structural characteristics of the drug.

Factors influencing plasma membrane partitioning are solute size, lipophilicity, hydrogen bonding potential and charge characteristics, while diffusion is dependent upon size or total molecular surface area properties. In general, non-polar surface area favors partitioning while polar, hydrogen bonding functionality opposes partitioning. With respect to diffusion, an inverse relationship with size is found, similar to the situation with paracellular permeability. These multiple influences on permeability are manifested in a number of different ways. If intestinal permeability of a number of homologous, non-actively transported solutes is measured as a function of membrane partitioning, or more commonly, an organic solvent partition coefficient such as octanol as a surrogate, a sigmoidal relationship is frequently observed. For solutes with little or no membrane affinity, permeability is low, resulting primarily from paracellular diffusion of the solute between cells.

As the propensity of the solute to partition into the cell membrane increases, permeability also increases as a result of the significant increase in surface area of the transcellular pathway relative to the paracellular route. This increase in permeability will approach a plateau value beyond which further increases in partition coefficient do not result in increased permeability. This is the so-called aqueous boundary layer limited situation where diffusion across the cell is very rapid relative to diffusion of the solute through the unstirred water/mucus layer adjacent to the cell. The dimensions, and resistance, of this layer can be modified by perturbing hydrodynamics which shift the plateau to a new, limiting permeability. In the case of ionizable solutes, permeability is also pH dependent. The neutral, uncharged species is capable of transcellular, passive diffusion while the charged species is restricted to the paracellular pathway. Thus the observed permeability of such molecules is dependent upon the relative concentrations of charged and neutral species. In the case of a weak acid such as salicylic acid for example, at pH less than about 5.5, rat intestinal permeability is aqueous boundary controlled.

Increasing pH results in progressively lower permeability coefficients until at pH greater than 9, a limiting, small permeability is achieved which is independent of further pH increases. This limiting permeability represents the paracellular diffusion of the charged anion. For such ionizable solutes a correlation with permeability and logD, defined as the apparent octanol-water partition coefficient at a specific pH, frequently 6.5 or 7.4, is used to predict permeability. Both ClogP and

ClogD algorithms are available to help optimize permeability characteristics of lead structures where this property is not found to be in a favorable range to effect the desired *in vivo* performance. However, octanol-water partition coefficients are not always useful for predicting permeability of all solutes.

In the case of highly functionalized solutes such as peptides, permeability has been found to be better correlated with structural and physicochemical measures of hydrogen bond potential. One such system is the ΔlogP model where ΔlogP is the difference between octanol-water partition coefficient and hydrocarbon-water partition coefficient of the solute. At the present time, no reliable computational models are available for predicting ΔlogP, although recent work in this direction has appeared in the literature. Polar surface area (PSA) of a solute, which can be computed from structure, has been proposed as an alternative to octanol-water and/or ClogP for predicting intestinal permeability. Recent examples of successful correlations using this simple metric have been published suggesting this as a useful general tool for optimizing permeability characteristics of a solute. However, failure of PSA to correlate cellular permeability of a series of peptidomimetics has also been reported. In this case, a reasonable correlation could be obtained by taking a weighted average of PSA and NPSA (non-polar surface area) of the solutes. The conclusion from these examples is that, as in the case of computational solubility modeling, application of a specific tool for predicting permeability must be validated by the user as appropriate for the molecular structure of interest.

Solute Clearance

Solute clearance as a PB-PK parameter includes both metabolism and excretion mechanisms. As shown in Figure 6, metabolism and excretion rates are dependent upon biopharmaceutical processes such as active and passive permeability, and physicochemical properties of the solute which influence these processes. With regards to SPR models of these pathways, considerable work has been focused on models for predicting metabolism of potential drug candidates, primarily identifying sites of reactivity in the solute. These include models for most of the major cytochrome P450 (CYP450) isoforms.

Octanol-water partition coefficients have been correlated with CYP450 binding for both substrates and inhibitors and may be considered as a design strategy to decrease association with the enzymes. Further, given that lipophilicity also plays a role in passive diffusion, decreasing logP may both decrease binding to CYP's and solute permeability in the liver to the site of metabolism. Along with passive permeability, a major contribution from active, frequently vectorial transport both to metabolism and excretion pathways is increasingly being recognized. SPR models for these processes is of considerable current interest and some progress has been made towards identifying substrates and/or inhibitors for some of them, most notably P-glycoprotein.

As in the case of metabolism, these models focus on recognition events such as binding and do not generally predict rates of reaction. However, at least in one case, a structure-based predictive model for hepatic clearance rates has been reported, derived from an internal training set of such data. Until such kinetic models are more generally available, *in vitro* data will likely continue to be required for use in PB-PK modeling.

Tissue Distribution

As is the situation with permeability, metabolism and clearance, both biopharmaceutical processes and physicochemical solute properties define tissue distribution. Reasonably accurate predictions of blood-level time profiles can be obtained in PB-PK models considering only plasma-tissue partition coefficients derived from unbound plasma solute concentrations and octanol-water partition coefficients, as shown earlier in this review and in other, similar models. Alternatively, a distribution parameter utilizing octanol-water partition coefficients and fraction unionized at pH 7.4, has been shown to provide good estimates of volume of distribution in humans for neutral and basic compounds.

At the present time, these are experimentally determined inputs in contrast to exclusively computed values. Again, complicating the issue is the role of active transport processes on distribution especially in tissues such as brain and lung. In the case of brain for example, P-glycoprotein plays a significant role in limiting distribution of solutes which are substrates into the brain (Raub, *this volume*). This can be clearly seen in cases where P-glycoprotein inhibitors were administered with the therapeutic agent and significantly greater CNS toxicities were noted which were not observed in the absence of inhibitor. The ability to predict these processes from structure alone presently remains an unsolved challenge.

Computational Model

Fraction unbound is an important determinant of both tissue distribution and clearance in the PB-PK model. Computational SPR models of protein binding are focused almost entirely on serum albumin, primarily human serum albumin. Generally, these models use lipophilicity and charge of the solute to estimate association between the solute and protein. A recent review challenged these models and proposed an alternative approach employing a pharmacophore similarity concept and partial least squares as a more generally accurate strategy. While this and other approaches are useful as a starting point for consideration of protein binding issues, a major weakness is the limitation to albumin as the only relevant potential binding component in blood. Other blood components important for binding certain classes of drugs are lipoprotein particles and α-1-acid glycoprotein.

CONCLUDING REMARKS

Biopharmaceutical properties such as solubility, cellular permeability and metabolic stability, among others are important in the successful discovery and development of new drug candidates. Significant progress has been made in recent years towards development and use of computational, structure-based models of these properties and such models are increasingly being used in the lead optimization process. While these biopharmaceutical properties are important individually, the ultimate *in vivo* performance of a drug candidate is a complex function of all the properties.

Physiology-based pharmacokinetic (PBPK) models using *in vitro* and physicochemical data can be useful in defining these functional relationships and identifying the important biopharmaceutical characteristics of a lead candidate, helping focus the optimization strategy. Presently, such models do not easily include non-ideal drug behavior such as active transport, for example. However, this can be an advantage in some cases. Significant deviations of experimental observations from prediction can be an alert that such processes are important for a specific lead candidate and needs to be more closely examined. Further, these early generation PB-PK models do not easily accomodate interdependencies of physicochemical properties on multiple biopharmaceutical processes. Generally, these must be considered individually in carrying out a sensitivity analysis query. However, as the models are evolved, these limitations are likely to be addressed and the use of integrated structure-based property models with PB-PK models will find more general applications in the lead optimization process.

5

Bioinformatics in Drug Discovery

The Pathway Tools software developed by SRI International is a powerful environment for genomics-based antimicrobial drug discovery. With approximately 70 complete microbial genomes now in the public domain, and many others in the private domain, bioinformatics tools for analysis of microbial genomes at a global level are important at all phases of the drug discovery process. The Pathway Tools use biochemical pathways as an organizing framework for genomic data. The pathway framework is an information-reduction device: It transforms the genome from a list of thousands of gene products to a cognitively more manageable list of hundreds of pathways. More important, it transforms our conception of the genome from a list whose members are arranged in unknown relationships to a network whose members are related through explicit causal links. The Pathway Tools software revolves around the concept of a Pathway/Genome Database (PGDB).

A PGDB describes the genome of an organism [its chromosome(s), plasmid(s), genes, and genome sequence], the product of each gene, the biochemical reaction(s) catalyzed by each gene product, the substrates of each reaction, and the organization of those reactions into pathways. The power of the Pathway Tools is derived from both its database schema and its software components. Both were originally developed for the EcoCyc project. The database schema (ontology) has been carefully designed to encode the complexities of genome and pathway data. Once a genome is encoded within the Pathway Tools schema, new types of complex analyses become possible because so many important semantic relationships have been described in a structured fashion.

The Pathway Tools software components support several different tasks, including querying and visualization, analysis, interactive editing, and Web publishing. The PGDB constitutes an evolving knowledge model of the organism that can be published on an organization's internal Web site using the Pathway Tools. The Pathway Tools provides a comprehensive and unique collection of query and visualization tools that can be used to determine the current state of knowledge regarding the organism: What function is attributed to a given gene product? What role does it play in a given pathway? How are the genes in that pathway distributed in the genome?

To ensure that the PGDB reflects the most up-to-date knowledge about an organism, the Pathway Tools provides a rich set of interactive software tools for curation of a PGDB. This chapter provides an example of how a drug discovery team would use the Pathway Tools in a genomics-based drug discovery project.

COMPONENTS OF PATHWAY TOOLS

The Pathway Tools software components, and their primary functions, are as follows.

PathoLogic

1. Given an annotated genome of a pathogenic micro-organism as its input, PathoLogic predicts the metabolic pathways of the organism, providing new global insights about its biochemistry.
2. PathoLogic creates a new PGDB describing the genome and the pathways of the organism.
3. PathoLogic can generate reports that summarize the evidence for the presence of each predicted metabolic pathway.

Editors

1. The Editors are a set of forms-based editing tools that are specially designed to update the different data types within a PGDB.
2. The Editors are used to alter the content of a PGDB to reflect the evolving knowledge of the biochemical machinery of the organism, such as modifying the functional annotations of genes, the assignment of gene products to pathways, and the connectivity of reactions assigned to a pathway.

Navigator

1. Scientists employ the Navigator to query and visualize the contents of a PGDB.
2. Visualization tools include a chromosome browser, pathway viewer, and displays of operons and regulons.
3. The Navigator can run in a WWW mode to publish a PGDB on the Internet or an intranet.
4. The Navigator can perform comparisons of the entire metabolic networks of multiple organisms.
5. The Navigator can paint gene-expression or protein-expression data sets on a diagram of the full metabolic network of an organism.

We next consider each of the three software components in more detail.

A Complete Genomic Sequence

Use of the Pathway Tools begins with an annotated genome of a pathogenic microorganism, meaning a complete genome sequence (closed or gapped) for which bioinformatics analyses have already identified the locations of likely coding regions and have predicted the functions of these genes. The genome can be provided in the form of a Genbank file or in a PathoLogic-specific format that can accommodate more information than can the Genbank format. If the Genbank format is used, it is important that the input file adhere closely to the Genbank standard, which, unfortunately, many completed genomes fail to do. In brief, PathoLogic imports the genes and proteins described by the input files into a new PGDB that is structured using the Pathway Tools schema and then matches the enzymes listed in the annotated genome against the enzymes required by every pathway in the MetaCyc DB. Those pathways with significant matches are imported into the new PGDB. More precisely, the algorithm followed by PathoLogic is as follows.

1. Initialize the schema of the new PGDB by replicating the MetaCyc schema.
2. Create a DB object for each chromosome or plasmid in the input file(s). Then create a DB object for each gene in the input file(s) and for the gene product of each gene.
3. Attempt to determine the reaction catalyzed (if any) by each gene product in the organism. If an EC number was assigned to the gene product in the input file, then rely on the EC number. Otherwise, match the name of each gene product against the extensive dictionary of enzyme names within MetaCyc.
4. Match the list of reactions now known to be catalyzed by the organism from step against all MetaCyc pathways. For pathways with significant numbers of matches, import the pathway and its associated reactions and substrates from MetaCyc into the new PGDB.

The diversity of pathways in MetaCyc influences the power of the Patho-Logic pathway predictions: The broader the space of pathways in MetaCyc, the larger the range of pathways that PathoLogic can recognize in a new organism. MetaCyc version 7.0 (February 2003) contains 467 pathways from 174 different organisms. Although the majority of MetaCyc pathways were experimentally elucidated in bacteria, approximately 50 MetaCyc pathways were elucidated in plants or animals, and many of the bacterial pathways are likely to be shared by plants and animals.

Table 5.1 shows the number of pathways present in MetaCyc from the most frequently occurring species. Some MetaCyc pathways are labeled as occurring in more than one species. PathoLogic can generate reports that summarize the amount of evidence supporting each pathway predicted to be present in the new PGDB and that list the "*pathway holes*"–the enzymes missing from each predicted pathway.

Navigator Software

The Navigator software provides a scientist with the ability to interrogate a PGDB and visualize the results of a query in an intuitive, graphical fashion. It also provides analysis operations, such as whole-metabolic-map comparisons across multiple organisms. The Navigator functionality is available through both X-windows and the World Wide Web (WWW). The most advanced capability within the Navigator is its ability to generate a diagram of the full metabolic network of the organism, called the Overview Diagram.

The Overview is a device for visualizing global relationships within the metabolic network of the organism. The user can employ an extensive menu of query operations to interrogate the PGDB and visualize the results on the Overview. Queries supported by the Overview include the following:

1. Highlight all reaction steps shared with other organisms for which PGDBs are available or unique to the current organism with respect to other organisms.
2. Highlight all reaction steps for which the enzyme is activated or inhibited by a specified metabolite or for which the enzyme is located in a specified cellular compartment.
3. Highlight all reaction steps for which the reaction is used in multiple metabolic pathways (in which case, knocking out that enzyme will knock out multiple pathways), or highlight those reactions for which the cell has multiple isozymes that catalyze the reaction.

4. Paint a protein-expression or gene-expression data set onto the Overview by painting each reaction step in the diagram with a color that reflects the expression level of the enzyme that catalyzes that reaction or of its gene.
5. Highlight all reaction steps for which the enzymes catalyzing the reaction are all controlled by a user-specified transcription factor.
6. Highlight specific entities in the diagram by name or substring search, such as the name of a pathway, enzyme, or metabolite.

Table 5.1 The number of MetaCyc pathways marked as occurring in specified species

Pathways	*Organism*
173	*E. coli*
35	*Salmonella typhimurium*
31	*H. sapiens*
20	*Sulfolobus solfataricus*
18	*B. subtilis*
18	*Soybean*
17	*Pseudomonas*
15	*Hm. influenzae*
12	*Mycoplasma capricolum*
7	*Pseudomonas putida*
7	*Mycoplasma pneumoniae*
6	*Ascomycotina*
5	*Rhizobiaceae*
5	*S. cerevisiae*
4	*Clostridium*
4	*Thauera aromatica*
4	*Pseudomonas aeruginosa*
4	*Thermotoga maritima*
4	*Thermotoga maritime*
4	*Rhodococcus*
4	*Klebsiella pneumoniae*
3	*Pseudomonadacea*
3	*Pseudomonas sp.*
3	*Neisseriaceae*
3	*Klebsiella aerogenes*
3	*Rattus norvegicus*
3	*Archaebacteria*
3	*Methanosarcina barkeri*
3	*Archaea*
2	*Pseudomonas aureofaciens*

Pathways	*Organism*
2	*Brevibacterium*
2	*Arthrobacter globiformis*
2	*Oryctolagus cuniculus*
2	*Mammalia*
2	*Lactobacillaceae*
2	*Pseudomonas acidovorans*
2	*Pseudomonas putida ATCC 17453*
2	*Sphingomonas sp RW1*
2	*Azotobacter beijerinckii*
2	*Acinetobacter*
2	*Comamonas testosteroni T-2*
2	*Aerobacter aerogenes*
2	*Methanosarcina therm ophila*
2	*Methanogens*
2	*Synechocystis sp. strain PCC 6803*
2	*Sinorhizobium meliloti*
2	*Halobacterium salinarium*
2	*Thiobacillus ferrooxidans*

The user can also select an entity within the Overview for closer inspection, such as to display a given pathway, enzyme, or metabolite in a window of its own. Each of the Navigator display windows for individual entities (such as a pathway) allows the user to display other related entities by clicking on them. For example, the Navigator pathway window displays all of the reactions, enzymes, and metabolites within a pathway, all of which are clickable. The user can alter the display of a metabolic pathway to show the same pathway at multiple levels of detail, from a skeletal view of the pathway that shows only the main substrates at the pathway ends and internal branch points to a detailed view of the pathway that includes enzyme names, gene names, EC numbers, and chemical structures for the substrates.

The Navigator allows users to retrieve metabolic pathways by name search, by substring search, and by querying a pathway-classification system. For example, the classification system allows users to retrieve all biosynthetic pathways or all pathways involved in amino acid biosynthesis. When a user clicks on a gene within a pathway display, a gene-display window shows information about the gene such as its name and synonyms, the name of the gene product, and chromosomal position in nucleotides. When the gene is part of a known operon, the structure of the operon is drawn. Clicking on the operon produces an operon-display window that provides references and commentary about the promoter and the transcription-factor binding sites within the operon. The Navigator includes a chromosome browser that can depict both linear and circular chromosomes and can produce multiple high-resolution views of a user-selected chromosomal region.

Semantic zooming adds new visual features at higher levels of resolution, such as depicting the extents of coding regions and the positions of promoters. Clicking on the name of the gene product within a gene-display window generates a display of the gene product. The exact style of the display that is produced depends on the type of the gene product as follows:

1. For enzymes, the display shows one or more chemical reactions catalyzed by the enzyme and displays information about each reaction, including the cofactors, activators, and inhibitors that modulate the activity of the enzyme
2. For transporters, the display provides a graphic depiction of the transport activity, indicating its energy-coupling mechanism (e.g., ATP-driven transport versus symport) and the transported substrate
3. For transcription factors, the display depicts all of the operons that contain a binding site for that transcription factor (the regulon of the transcription factor)

Interpretation of Data

The accuracy of global genome analyses such as comparisons of the metabolic networks of multiple organisms, and interpretation of expression data, depends intimately on the accuracy of the genome annotation. In a typical microbial genome, 30-40% of the genes will have no predicted function. Of those genes that do have predicted functions, some number of those functional predictions will be incorrect. We postulate that another source of errors in genome annotation is underannotation of multifunctional proteins. Most complete genomes have few proteins that are annotated with more than one predicted function, yet of the 607 enzymes in the EcoCyc DB, 100 are multifunctional. If this frequency of multifunctional proteins is representative of the full genomes of other organisms, *many* protein functions are being missed by genome analysis. The point of this discussion is that the initial annotation of most genomes is incomplete and error prone and that as more knowledge about the genome is gained, it is important to update the PGDB to reflect the current state of knowledge regarding the genome.

A PGDB can serve as a vehicle for recording the current best knowledge of the genome annotation and the rationale for the annotation and can be used to disseminate that knowledge within an organization through the WWW Navigator. The rationale for the genome annotation can be encoded both through PGDB fields that indicate whether a given gene function was determined computationally or experimentally and through *history notes* within a gene that allow users to record time-stamped, name-stamped comments describing why annotations were changed. The purpose of the Pathway/Genome Editors is to allow users to efficiently update the genome and pathway annotations within a PGDB to reflect the evolving understanding of the organism. Each form is designed to support the editing of one of the datatypes within a PGDB. The editing tools include the following:

1. A pathway editor for creating new metabolic pathways and modifying existing pathways
2. A protein editor for creating and modifying enzyme and transporter descriptions
3. A gene editor for creating and modifying genes
4. A compound editor for creating and modifying small molecules within a PGDB
5. A transcription-unit editor for creating and modifying descriptions of transcription units and transcription factors

DATA MANAGEMENT

Data management services for the Pathway Tools are provided by an object- oriented database system developed called Ocelot. Ocelot combines the expressive power of the frame knowledge representation systems developed within the Artificial Intelligence (AI) community with the scalability, multiuser access capabilities, and robust operation of relational database systems. The Ocelot data model structures all data as *frames*, which are of two types: classes and instances. Class frames describe general classes of entities, such as the class of all genes, or the class of all pathways.

Instance frames describe specific biological entities, such as a specific gene or pathway. The object data model allows the complexities of biological data to be modeled far more compactly than does the relational data model. Ocelot databases (such as PGDBs) can be stored persistently in Oracle databases and in disk files. The disk-file approach is advantageous for an organization that does not wish to face the complexity of installing and managing Oracle. However, Oracle storage is required when multiple users want to be able to update a PGDB simultaneously. The Oracle approach also allows faster incremental saving of PGDB updates, in contrast to the disk-file approach, which requires that a PGDB be saved in its entirety when it has undergone any change. The performance of Ocelot has thus far been adequate for managing 10 PGDBs for microbial genomes simultaneously on low-end Sun workstations such as the Ultra-5.

CONCLUDING REMARKS

The Pathway Tools will be particularly useful to organizations that are interested in metabolic enzymes as a class of drug targets. Metabolic pathways provide the energy and manufacturing plant of the cell. Compounds that interfere with those processes will disrupt cell growth. We advocate a two-phased approach to microbial drug design. Phase I is the search for *essential in vivo metabolic pathways*: pathways whose function is essential for microbial growth in the host. Phase II is the search for targets within essential *in vivo* pathways. This approach is advantageous for two reasons. First, by dividing the search for targets into two phases, we drastically reduce the size of the target search space.

A strategy that simply considers each gene product as a potential target faces a target space of thousands of genes. In contrast, the two-phased pathway approach reduces the initial search to consider hundreds of essential pathways. If we assume that 10–20 essential pathways are identified, and that each pathway contains 10 gene products, Phase II will consider 100-200 proteins as targets. The second advantage is that knowledge of pathway topology is extremely valuable in Phase II. For example, targets that occur in multiple essential pathways are preferred over targets in a single essential pathway. Targets without multiple isozymes are preferred over those with multiple isozymes. Targets that occur in a location within the pathway that cannot be circumvented by another branch of the pathway are preferred over targets in parallel branches.

Targets that occupy pathway holes (pathway steps whose enzymes have not yet been determined) are preferred over identified enzymes because the enzyme that fills a pathway hole (if it can be found within the genome) is unlikely to have been patented. As well as providing global insights about the biochemistry of the organism, the pathway-prediction process is a method for validating the genome annotation produced through sequence analysis. Pathway holes indicate what gene functions still remain to be identified within the genome, whereas singleton enzymes in pathways indicate possible false-positive function predictions, since it is unlikely that an organism would contain a single enzyme in the middle of a pathway. The metabolic Overview diagram can be used to analyze gene-expression and protein-expression data in a pathway context. For example, by using this tool to analyze gene-expression data from growth of a microorganism in the presence of a lead compound, the Overview can be used to detect compensating pathways that might become active under those conditions. The Overview can also be used for comparative pathway analysis. A menu-driven query interface allows the user to visualize reactions that are common to one specified set of organisms, but absent from a second set, in order to identify targets present in a preferred group of organisms.

6

Quantitative Proteomics

Quantitative proteomics has become a widely applied analytical tool for protein identification and quantification in complex biological samples. One of the goals of proteomics is to measure protein expression and characterize the protein expression profile in specific tissues and biofluids for potential biomarker discovery. Even though a tremendous effort has been made to improve the proteomics technologies, there are still numerous challenges associated with even the most advanced technologies in analyzing global protein expression due to the inherent complexity of biological samples. These challenges include: (i) sensitivity of the instrument and ability to identify novel proteins; (ii) low to moderate throughput of the system from sample preparation to data analysis; (iii) wide coverage of protein mass and abundance (dynamic range); and (iv) the ability to quantitatively analyze protein expression. At the present time, there is no consensus within the field of proteomics that any one technology can attain a complete and quantitative protein coverage of all proteins in a given tissue or biofluid.

The most commonly used proteomic approach is accomplished by a combination of either two-dimensional gel electrophoresis (2-DE) or liquid chromatography (LC) to separate and visualize proteins/peptides and mass spectrometry (MS) to identify, characterize, and quantify them. Although 2-DE has been the workhorse in proteomics research efforts in the past decade, its lack of ability to widen the protein dynamic range and to analyze hydrophobic proteins or those with very high or low molecular weight is still its biggest disadvantage. One alternative approach to 2-DE is the non-gel-based "bottom up" LC/MS-based shotgun proteomic technology. Although some successes using stable isotopic labeling technology for protein quantification have been reported, it remains technically difficult to comprehensively characterize the global proteome due to the high costs of the labeling reagents and the nature of the methodology (i.e., in isobaric tagging of peptide (iTRAQ) or isotope-coded affinity tag (ICAT), it depends on labeling efficiency). In addition, chromatographic shifts can make quantitative analysis of differentially labeled peptides computationally difficult and error-prone. Furthermore, simultaneous quantification of proteins from multiple samples is problematic due to limitation of available labeling reagents.

Recently, the ion intensity-based label-free quantitative approach has gradually gained more popularity as mass spectrometer performance has improved significantly. It provides a powerful tool to resolve and identify thousands of proteins from a complex biological sample. In this method, proteins are first digested with a protease into a peptide mixture that is then analyzed by LC/MS and LC/tandem mass spectrometry (MS/MS) and subsequently identified by database searching and relative protein abundance determined by chromatographic peak intensity measurement. This approach is rapid and more sensitive, and it increases the protein dynamic range three- to four-fold compared with 2-DE. This method can also be automated and has the ability for large-scale proteome analysis. Along with all these advantages compared with other proteomic technologies, however, significant bioinformatics and biostatistical resources for data analysis are required.

BASIS OF QUANTIFICATION

Label-free and signal intensity-based relative protein quantification shotgun proteomics approaches are promising alternatives to stable isotope labeling approaches and have been applied to quantify differentially expressed proteins from complex biological samples. One approach is to directly compare peptide peak areas between LC/MS runs. This approach is simple and cost effective and has demonstrated high reproducibility and linearity comparing peptide level or protein abundance. Inclusion of statistical analysis with this method allows detection of small significant changes that are biologically meaningful.

Several studies have demonstrated that extracted ion chromatograms (XIC) of selected peptide ions correlate well with protein abundances in complex biological samples. Another label-free method, referred to as spectral counting, compares the number of MS/MS spectra assigned to each protein. One advantage of spectral counting method is its ability to potentially measure relative abundances of different proteins in the same biological system as shown by significant correlations between spectral counts and independent estimates of protein copy number in yeast. However, the application of these methods, especially to mammalian systems, requires more robust computing power and algorithm capable of handling chromatographic peak alignment and peptide ion intensity measurements for analyzing changes in protein abundances in complex biological samples.

METHOD OF ANALYSIS

All label-free LC/MS-based protein quantification methods discussed previously include four fundamental steps: (1) sample preparation (protein extraction, reduction, alkylation, and digestion); (2) sample separation and analysis (LC and MS/MS); (3) data analysis (identification, quantification, and statistical analysis); and (4) data interpretation (protein-protein interaction network and pathway analysis). In this chapter, we will focus on the peak intensity-based method for protein quantification followed by a case study.

Extraction of Proteins

In general, proteins are extracted from tissues, cultured cells, or biofluids in freshly made lysis buffer containing 8 M urea and 10 mM dithiothreitol (DTT). For cells grown in 6-well plates, culture media are first removed by aspiration. Then, 100 μL of lysis buffer is added to each well, and cells are lysed by pipetting. Lysed cells are transferred to microcentrifuge tubes and incubated at room temperature with gentle agitation. Unlysed cells and insoluble particles are removed by centrifugation (12,000 × g for 5 min) prior to protein assay. In order to take the same amount of proteins from each sample, protein concentrations are measured by Bradford assay.

The same lysis buffer should be used as the background reference for protein assay in order to obtain a relatively accurate measurement among all samples (due to the presence of urea in lysis buffer). Resulting protein extracts are subsequently reduced and alkylated with DTT and iodoacetamide to block sulfhydryl groups in proteins. However, in the method described here, the volatile reagents triethylphosphine and iodoethanol are used instead. This volatile reduction and alkylation protocol has been described previously. It allows all sample preparation steps to be carried out in one tube without washing, filtering, or sample transferring, which minimizes sample preparation variations.

Briefly, 100 μg protein is taken from each sample, and pH values are raised by adding 100 μL 100 mM ammonium carbonate (pH 11). Two hundred microliters of reduction/alkylation cocktail (97.5% acetonitrile, 2% iodoethanol, and 0.5% triethylphosphine) is added to protein samples and incubated for 1 h at 37°C. After incubation, protein samples are dried under vacuum overnight to remove residual chemicals. Protein mixtures are then digested by trypsin. Dried pellets are resuspended in 500 μL 100 mM ammonium bicarbonate buffer containing 2 μg trypsin and incubated at 37°C overnight. Protein to trypsin ratio is 50:1, and urea concentration should be below 1.6 M. Tryptic digests are filtered with 0.45-μm spin filters.

Role of Tryptic Peptides

All digested samples should be randomized for injection order to remove systematic bias from data acquisition. Typically, up to 20 μg of the tryptic peptides are required for injection onto a C18 microbore column (i.d. = 1 mm, length = 5 cm). Peptides are eluted with a linear gradient from 5% to 45% acetonitrile developed over 120 min at a flow rate of 50 μL/min, and effluent is electro-sprayed into a LTQ linear ion-trap mass spectrometer. The electron-spr ionization source is operated with 4.8-kV potential, a sheath gas flow of 20 arbitrary units, and a capillary temperature of 225°C. The instrument is tuned using an angiotensin I peptide. Maximum ion time is set to 50 ms for parent ion scan and 500 ms for zoom scan and MS/MS scan. This method requires all the MS data be collected in the data-dependent "Triple-Play" mode.

Parent ion scans and MS/MS scans are collected in centroid mode, and zoom scans are collected in profile mode. Dynamic exclusion is set to a repeat count of one, exclusion duration of 120 s, and rejection widths of –0.75 m/z and +2.0 m/z. The acquired data are filtered and analyzed by an algorithm developed by Higgs et al. and described in detail in the published article. Database searches against the IPI (international protein index) and the Non-Redundant (NCBI) databases are carried out using both the SEQUEST and X!Tandem algorithms. Protein quantification is also carried out using the same algorithm described above. Briefly, once the raw files are acquired from the LTQ, all extracted ion chromatograms (XIC) are aligned by retention time. To be used in the protein quantification procedure, each aligned peak must match parent ion, charge state, daughter ions (MS/MS data), and retention time. After alignment, the area under the curves (AUCs) for individually aligned peaks from identified peptides from each sample are computed; the AUCs are then compared for relative protein abundance.

Taxonomic Study of Proteins

Proteins identified by SEQUEST and X!Tandem are categorized into priority groups based on the quality of the protein identification. The Peptide ID Confidence assigns a protein to a "HIGH" or "LOW" classification based on the peptide with the highest peptide ID Confidence (the best peptide). Proteins whose best peptide has a confidence between 90% and 100% are assigned to the "HIGH" category regardless of whether there are other peptides having low confidence or not. Proteins whose best peptide has a confidence between 75% and 89% are assigned to the "LOW" category. All peptides with confidence less than 75% are filtered out by the software before further analysis. SEQUEST and X!Tandem database search algorithms are used for peptide sequence identification. Each algorithm compares the observed peptide MS/MS spectrum and a theoretically derived spectra from the database to assign quality scores (*XCorr* in SEQUEST and *E-Score* in X!Tandem). These quality scores and other important predictors are combined in the algorithm that assigns an overall score, %ID Confidence, to each peptide.

The assignment is based on a random forest recursive partition supervised learning algorithm. The %ID Confidence score is calibrated so that approximately X% of the peptides with %ID Confidence > X% are correctly identified. The confidence in protein identification is increased with the number of distinct amino acid sequences identified. Therefore, proteins are also categorized depending on whether they have only one or multiple unique sequences of the required confidence. A protein is classified as "YES" in the "Multiple Sequences" column if it has at least two distinct amino acid sequences with the required ID confidence; otherwise it is classified as "NO." Priority assignments reflect the level of confidence in the protein identification. Priority 1 proteins would have the highest likelihood of correct identification and Priority 4 proteins the lowest.

This priority system is based on the quality of the amino acid sequence identification (Peptide ID Confidence) and whether one or more sequences are identified (Multiple Sequences). Many would view any protein identification outside of priority 1 as questionable. All data processing is carried out on a Linux cluster using highly parallel processing and data qualification and filtering software.

Clinical Values of Quantification

One of the key features of the algorithm described here for protein quantification is the chromatographic peak alignment because large biomarker studies can produce chromatographic shifts due to multiple injections of the samples onto the same HPLC column. Unaligned peak comparison will result in larger variability and inaccuracy in peptide quantification. All peak intensities are transformed to a $\log_2$ scale before quantile normalization. Quantile normalization is a method of normalization that essentially ensures that every sample has a peptide intensity histogram of the same scale, location, and shape. This normalization procedure removes trends introduced by sample handling, sample preparation, total protein differences, and changes in instrument sensitivity while running multiple samples. If multiple peptides have the same protein identification, then their quantile normalized $\log_2$ intensities are averaged to obtain $\log_2$ protein intensities.

The average of the normalized log peptide intensities is a weighted average. A peptide is weighted proportionally to the peptide ID Confidence for its protein category and receives a weight of zero if it is outside that category. For example, peptides with < 90% confidence contribute zero quantitative weight to a "HIGH" category protein. The log transformation serves two purposes. First, relative changes in protein expression are best described by ratios. However, ratios are difficult to statistically model, and the log transformation converts a ratio to a difference, which is easier to model. Second, as is frequently the case in biology, the data better approximate the normal distribution on a log scale. This is important because normality is an assumption of the analysis of variance (ANOVA) statistical model used to analyze the data.

The base of the log transform is arbitrary with base 2 the most common with genomic data. Base 2 is popular because a two-fold change (or doubling, or 100% increase) yielding an expression ratio of 2 is transformed to 1 on a log base 2 scale (i.e., a two-fold change is a unit change on the log base 2 scale). The log protein intensity is the final quantity that is fit by a separate ANOVA model for each protein shown below:

[$\log_2$ (Intensity) = overall mean + group effect (fixed) + sample effect (random) + replicate effect (random)]

In this model, group effect refers to the effect caused by the experimental conditions or treatments being evaluated. Sample effect represents the random effects from individual biological samples. It also includes random effects from sample preparation. The replicate effect refers to the random effects from replicate injections of the same sample. All of the injections should be therefore randomized and the instrument be operated by the same operator for a particular study. The inverse log of each sample mean is calculated to determine the fold change between samples.

Management

To assess the stability of the HPLC system and MS instrument, a known purified standard protein is commonly spiked into every sample at a constant amount as an internal reference for assessment of technical variations before tryptic digestion. Several considerations should be given for the selection of the standard: (i) the protein must not come from the same species as the sample of interest; (ii) a series of signature peptides should be easily detected and identified by the instrument; and (iii) the amount of the standard protein spiked into each sample should be comparable with the amount of median abundant proteins in the sample. After global protein identification and quantification, these peptides and their relative quantities should be inspected for quality assurance and quality control purposes.

Mathematical Calculations

The number of significant changes between groups, the fold changes, and the variability (coefficient of variation) for each priority level can be determined from the ANOVA. The threshold for significance is set to control the false-discovery-rate (FDR) for each comparison at an investigator-desired percentile, normally 5%. The FDR is estimated by the *q* value, which is an adjusted *p* value. The FDR is the proportion of significant changes that are false positives. If proteins with a q value ≤ 0.05 are declared significant, it is expected that 5% of the declared changes will be false positives. In this method, the *p* value to *q* value adjustment is done separately for Priority 1, Priority 2, and the "LOW" confidence categories. Fold change (FC) is computed from the means on the AUC scale (antilog) as follows:

When Mean Treated Group ≥ Mean Control Group (up regulation):

FC = Mean Treated Group / Mean Control Group

When Mean Control Group > Mean Treated Group (down regulation):

FC = - Mean Control Group / Mean Treated Group

and

Absolute FC = |FC| = absolute or positive value of the FC

A fold change of 1 means there is no change. Also, the median % coefficient of variation (%CV) for each priority level is determined. The %CV is the standard deviation/mean on a % scale. The %CV is given both for the replicate variation as well as the combined replicate plus sample variation.

Biological Significance of Protein Expression

To understand the biological significance of the protein expression changes, the results including protein IDs and FC from the LC/MS analysis can then be analyzed using protein-protein interaction and/or pathway analysis software. This software allows for creation of protein-protein interaction networks, biological pathways, and gene regulation networks from a data set, which will help better understand specific biological processes that are involved in a particular study.

DESIGNING OF A PROTEOMIC EXPERIMENT

To illustrate the utility of this LC/MS-based method described previously, a practical example is provided that outlines how a proteomic experiment is designed and what steps are involved in data interpretation. Prior to any experiment, study design must be completed with a statistician in order to ensure that the study answers the questions of interest and has sufficient technical and biological replicates to detect small but significant changes using appropriate statistical methods. A technical replicate is a replicate sample from the same biological sample. For example, split a single biological sample into two parts and run both replicates in the experiment. This will allow for assessment of instrument errors.

Biological replicates are samples from independent experimental units (e.g., each of 10 human plasma samples from different individuals). Whereas a technical replicate estimates the precision for the assay itself, biological replicates provide an estimation of biological variation. In general, biological replicates are more informative than technical replicates. Group size determination depends on the size of effect to be detected (FC) and the sample-to-sample biological variation expected (CV), and which error rates to be controlled.

It is best to control the FDR instead of the false-positive-rate (FPR) when hundreds of proteins are analyzed. The FDR can be large (e.g., > 0.05) even if the FPR is small (e.g., <0.05). If control of FDR is chosen, then the proportion of proteins that will change (the prevalence) has to be estimated. With this information, the group size required for given power (probability of determining a true change, i.e., the sensitivity) can be computed. As the percent of proteins expected to change varies, the group size required should be matched accordingly.

Case History of Cell Lines

Two pairs of cell lines, A2780 and 2008 cisplatin-sensitive human ovarian cancer cell lines and their resistant counterparts, A2780/CP and 2008/C13*5.25, were used in this study. These cell lines have been used as model systems for drug-resistance studies. All cell lines were handled under identical conditions and maintained at 37°C in a humidified incubator containing 5% CO_2 in RPMI-1640 supplemented with 15% fetal bovine serum. Proteins were prepared and subjected to LC/MS/MS analysis as previously described.

There were eight groups in this study: two pairs of cell lines each including a sensitive cell line and a resistant cell line with no drug treatment the resistant cell lines (A2780-CP and 2008-C13*5.25) with 10 μM and 20 μM of cisplatin treatment, respectively. There were six samples per group except for group 3, yielding a total of 93 randomized HPLC injections. Samples were run on a Surveyor HPLC system (Thermo-Finnigan) with a C18 microbore column.

Quality Assessment

In this study, all injections were performed using the same C18 microbore column. Chicken lysozyme was spiked into every sample at a constant amount before tryptic digestion. After tryptic digestion, nine chicken lysozyme peptides were quantified (all peptides with ID confidence >75%). The individual protein quantities (as presented by the peak intensities) are displayed for each injection. The overall mean for each group is displayed by the line across the plot. This plot presents a visual quality control assessment of the ability of the instrument to measure a constant amount of protein over sequential injections.

Because a constant amount of chicken lysozyme was spiked into all the samples, it should show no significant change between groups. If there is a significant group effect (i.e., if q value <0.05), then one should interpret significant changes in other proteins with smaller fold changes as possibly due to spurious effects. In this experiment, the largest absolute fold change for chicken lysozyme was 1.16 with a q value of 0.0667. Even though this is not a significant change based on the q-value threshold of <0.05, any significant fold changes of absolute magnitude less than 1.16 in this study should be interpreted with caution.

Quantification of Data

After database search, protein identification and quantification, and statistical analysis, a total of 5282 distinct amino acid sequences corresponding with 2117 proteins were identified and quantified from all samples. Among them, 855 proteins were in the Priority 1 group and 95 of them showed significant changes that are potentially associated with cisplatin resistance. The significance threshold is set to control the FDR at less than 5%. The replicate median %CV (technical variation) for the Priority 1 proteins was 11.71% and the combined replicate plus sample median %CV was 12.87% (technical plus biological variations). There were also 41 proteins that had significant changes among the 1262 proteins

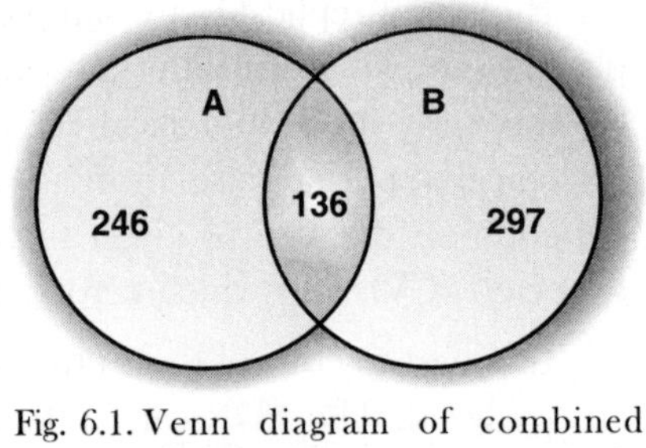

Fig. 6.1. Venn diagram of combined differentially expressed proteins in the two cell lines. All priority group proteins are compared between cell lines (A) A2780/A2780-CP and (B) 2008/2008-C1*5.25.

that were less confidently identified (Priorities 2 to 4). The cisplatin resistance-specific protein changes were extracted when the two types of cell lines were directly compared (A2780/A2780-CP vs. 2008/2008-C13*5.25), suggesting that other protein expression level changes in the two cell lines may be cell-type-specific.

Regulating Networks

The results from the LC/MS were then analyzed by PathwayStudio 4.0. This software allows for creation of biological pathways, gene regulation networks, and protein interaction diagrams from a data set. With a list of gene names derived from the proteomic results, more than 1700 connections were made from the protein list. A "Classification Table" was thus created based on these results. These processes included proliferation, DNA repair, apoptosis, and more. Based on these experimental data and the information obtained from the pathway analysis and other literature searches, a hypothesis can be generated involving several specific pathways that may contribute to cisplatin drug resistance. Although the need for the development of high-throughput, sensitive, and cost-effective MS-based quantitative protein analysis methods for biomarker discovery remains a significant challenge in proteomics, the promise of the proteomic technologies to enhance the biomarker discovery has been shown in this chapter. To reduce false-positive discoveries, significant development in bioinformatics and robust validation methods will be required. As the future of

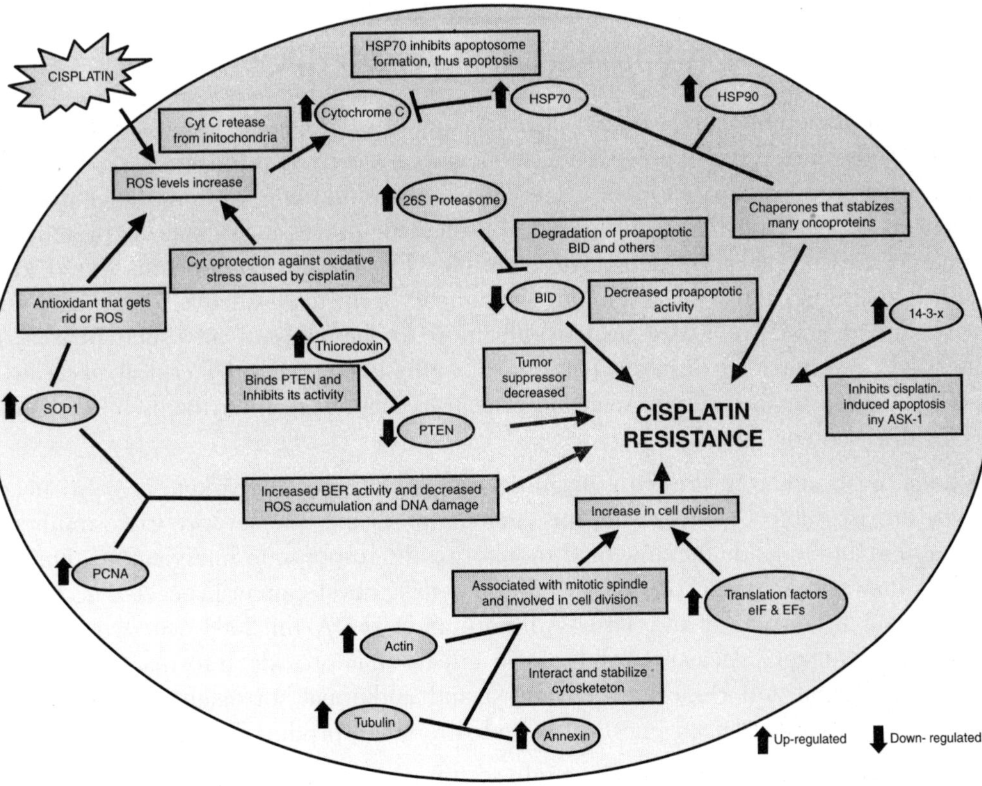

Fig. 6.2. A proposed biological pathway involved in cisplatin resistance in ovarian cancer.

proteomics unfolds, it will certainly create many more opportunities for the development of novel drug targets, improved understanding of proteomic data, and underlying biology. The development in label-free protein quantification technologies will no doubt meet the demands of both proteomics and clinical applications.

Table 6.1 Biological processes in which the differentially expressed proteins are involved

Biological process	*Number of proteins involved*
Amplification	6
Angiogenesis	2
Apoptosis	24
Differentiation	16
Mitosis	9
Regulation of signal transduction	13
Cell survival	16
Damage	6
DNA damage recognition	2
DNA recombination	35
Transformation	8

EXPRESSION PROTEOMICS

Expression proteomics involves the characterization of cell- and tissue-specific proteins. In addition to synthesis of primary polypeptide sequences, a variety of post-translational modifications (e.g., glycosylation, phosphorylation, acetylation, sulfation) contribute to both the charge and mass heterogeneity for a given protein. Furthermore, proteolytic processing of specific proteins can also contribute to the complexity of a biological sample. The protein components found in a given sample are, therefore, the result of multiple synthetic, reductive-oxidative, enzymatic, transport, secretory, and degradative processes. Modification to the rates of any such process can be hypothesized to result in a diseased or abnormal physiologic state. A critical need in clinical proteomics involves the discovery of validated protein biomarkers that may be useful in the diagnosis of specific diseased conditions.

In addition to aiding in the early diagnosis and prediction, biomarkers are also needed for discerning disease subtypes, identifying the etiology, predicting the severity (risk stratification for prognostication and to guide therapy), and monitoring the response to interventions. Furthermore, biomarkers may play a critical role in expediting the drug development process. The Critical Path Initiative issued by the Food and Drug Administration (FDA) in 2004 stated that "Additional biomarkers (quantitative measures of biologic effects that provide informative links between mechanism of action and clinical effectiveness) and additional surrogate markers (quantitative measures that can predict effectiveness) are needed to guide product development".

The concept of developing a new toolbox for earlier diagnosis of disease states is also prominently featured in the National Institutes of Health (NIH) Road Map for biomedical research.

Biomarker discovery efforts require technologies enabling screening mechanisms that are sensitive, reproducible, and have high-throughput capacity. These properties can allow for reliable assessment of large sample numbers, which may be available only in limited volumes. One technology that possesses each of these features is surface-enhanced laser desorption ionization mass spectrometry (SELDI-MS). SELDI, introduced by Hutchens and Yip, combines selective retentate chromatography with time-of-flight mass spectrometry.

SELDI techniques have been the subject of a rapidly growing literature, featuring biomarker studies in a variety fields including oncology, neurology, toxicology, infectious disease, and nephrology. SELDI instrumentation and products were initially brought to market in 1997 by Ciphergen Biosystems, Inc. In November 2006, Bio-Rad Laboratories purchased the rights to produce and market SELDI instrumentation, products, and services. Unique to the SELDI process are specific chip surfaces that can bind a subset of proteins present in a given sample. The ProteinChip surfaces contain either complexed chromatographic groups or preactivated chemistries to which user-defined "bait" molecules can be bound.

PROTEINCHIP ARRAYS

ProteinChip NP20 Arrays

The NP20 chips contain silicon oxide (SiO_2) surface groups capable of electrostatic and dipole-dipole interactions. Hydrophilic and charged amino acid residues on the protein surface can bind, and therefore the NP20 array is relatively nonselective. Buffers with a wide pH range are tolerated. Contaminating buffer salts and detergents can easily be removed by washing with distilled water. These properties allow NP20 arrays to be used in collecting mass calibration spectra using defined peptide or protein standards. In addition, the NP20 array is useful when monitoring larger-scale chromatographic column eluant fractions.

ProteinChip Q10 Arrays

The Q10 chips contain quarternary ammonium groups, providing a cationic surface that can take part in electrostatic interactions with negatively charged aspartic acid and glutamic acid residues. Binding is highly dependent on buffer pH and ionic strength. In the profiling phase of biomarker discovery, permissive binding conditions (i.e., 50 mM Tris-HCl, pH 8 to 9) are recommended. However, more stringent binding buffers (i.e., 50 mM sodium acetate, pH 5) can also be useful in selecting for strongly acidic proteins. If the isoelectric point (pI) of a target protein is known, a buffer pH at least one unit greater than that pI is recommended.

ProteinChip CM10 Arrays

The CM10 chips are derivatized with carboxylate groups, thus providing an anionic surface. Electrostatic interactions with positively charged arginine, lysine, and histidine residues can take place. As with the Q10 arrays, binding is dependent on buffer pH and ionic strength. Permissive binding conditions (i.e., 100 mM sodium acetate, pH 4.0) are again useful in initial biomarker discovery studies. More stringent binding buffers (i.e., 50 mM sodium Hepes, pH 7 to 8) will select for strongly basic proteins. If the pI of a target protein is known, a buffer pH at least one unit lower than that pI is recommended.

ProteinChip IMAC30 Arrays

The IMAC30 chips incorporate nitrilotriacetic acid (NTA) groups that form stable octahedral complexes with polyvalent metal ions (i.e., Cu^{+2}, Ni^{+2}, Zn^{+2}, Fe^{+3}, Ga^{+3}). Once "charged" with a selected metal ion, two free sites on the formed octahedral complex can interact with specific amino acids (e.g., His, Cys, Trp) or phosphorylated amino acids (when using Fe^{+3} or Ga^{+3} as metal). Binding buffers (pH 6 to 8) with high ionic strength (0.5 M NaCl) are commonly used, promoting higher selectivity. Specific proteins can be eluted with increasing concentrations of competitors (such as imidazole) that can displace the protein from the coordinated metal.

ProteinChip H4 and H50 Arrays

These surfaces contain methylene group chains of varying length: H4 (C_{16}), and H50 (C_6-C_{12}). Proteins bind via partitioning of surface hydrophobic residues (Ala, Val, Ile, Leu, Phe, Trp, Tyr) into the lipophilic chip surface. Two methods of binding can be used with the H4 and H50 arrays. In the classic reverse-phase mode, aqueous buffer with low organic content (i.e., 10% acetonitrile) is used for binding. Proteins or peptides can be selectively eluted by raising the organic solvent content of subsequent washes. These surfaces can also be used in a hydrophobic interaction chromatography (HIC) format. Here, high ionic strength (2 M ammonium sulfate) buffers are used for binding. Surface polar groups are occupied by the high salt. Surface nonpolar groups again interact with the lipophilic chip solid phase. By then reducing the buffer ionic strength, proteins can be selectively rehydrated and eluted from the chip surface.

RS100 and PS20 Protein Chip Arrays

ProteinChip RS 100 Arrays are preactivated with carbonyldiimidazole groups, and ProteinChip PS20 Arrays are preactivated with epoxide groups. These two reactive SELDI chip surfaces allow specific, user-defined proteins to be covalently bound via free primary amine groups. The PS20 epoxide group can also link via free sulfhydryl (–SH) groups. Those specific "bait" molecules can then serve as affinity surfaces for selective adsorption of complimentary proteins. Typical applications include immunoassays, receptor-ligand binding studies, and transcription factor analysis. Both surfaces are susceptible to nucleophilic attack, principally by protein amine groups. Therefore, the coupling step must be carried out in amine-free buffers. In addition, free sulfhydryl compounds (i.e., 2-mercaptoethanol, dithiothreitol) should not be present.

ProteinChip Gold, SEND-ID, and PG20 Arrays

In the ProteinChip Gold Array, the gold chip surface is nonselective in binding characteristics. When samples are allowed to dry on the gold surface followed by matrix application, the SELDI system is comparable with a conventional matrix assisted laser desorption/Ionization time of flight (MALDI-TOF) instrument.

In the ProteinChip SEND-ID Array, the SEND-ID chip surface has α-cyano-4-hydroxy cinnamic acid (CHCA) matrix incorporated as well as C_{18} chemistry. This feature greatly reduces the background noise to desorption of matrix and thereby increases the sensitivity in the low mass region. The SEND-ID chip is commonly used to characterize specific proteolytic enzyme digests. In the ProteinChip PG20 Array, protein G is coupled to the PS20 surface. The PG20 array can be used in the development of specific immunoassays.

TECHNIQUE FOR SAMPLING

The SELDI retentive chromatography process has been shown to accommodate a wide variety of sample types including serum, plasma, urine, cerebral spinal fluid, cell and tissue lysates, subcellular fractions, conditioned cell culture media, and laser capture microdissected tissue. Because specific proteins are retained on the chip surface, the sample does not need to be desalted prior to application thus facilitating use of a wide variety of sample extraction and lysis buffers.

Because biological samples often contain a complex assortment of proteins with distinct chemical properties, it can be advantageous to prefractionate samples prior to application to the SELDI chip surface. This approach is especially useful when profiling serum or plasma samples, as highly abundant proteins (albumin, immunoglobulins) can preclude binding of other proteins to the chip surface. The separate fractions are then profiled on different chip chemistries, excluding the Q10 anion exchange chip. An alternative prefractionation technique, which also allows enrichment of lower-abundance proteins, involves Bio-Rad Proteo Miner Beads.

The beads possess synthetic peptides produced in a combinatorial manner resulting in a wide variety of affinity surfaces. Proteo Miner Bead eluates have shown to increase the number of detectable SELDI peaks and sodium dodecyl sulfate polyacrylamide gel electrophoresis (SDS-PAGE) spots in plasma and urine studies.

CHROMATOGRAPHIC FEATURES

A major attribute inherent to the retentate chromatography feature of SELDI is the flexibility of experimental design allowed by the system. This point can be appreciated by an overview of SELDI chip processing.

Chip Surface

As described above, a variety of chemical and user-designed chip surfaces can be used. The chips are sold in an 8 × 12 format. Small sample volumes (5 to 8 μL) can be applied directly onto the chip surfaces. Alternatively, the chips can be assembled within a Bioprocessor assembly, which includes a fitted reservoir for application of larger sample volumes (up to 400 μL).

Initial Step

For a given selected chip type, a variety of binding buffers can be used. Buffer pH, ionic strength, presence of non-ionic detergents (i.e., Triton X-100, CHAPS) or denaturants (urea, GuHCl) can all be varied for experimental optimization. Typically, the chip arrays are preincubated in a selected binding buffer for two 5-min periods prior to addition of sample. Depending on the selected chip surface, additional preequilibration steps may be required, for example charging of the IMAC30 surface with a specific polyvalent metal ion.

Incubating Step

Sample can then be diluted into the same binding buffer and incubated on the chip surface. The length of time for the binding step can be varied. Thirty- to sixty-minute binding steps are common for biomarker discovery profiling studies.

Washing the Chip Surface

The chip surfaces can then be washed with buffer of the same or increased stringency relative to the binding buffer. Three 5-min washes with the same binding buffer is common. However, variable wash buffers can provide valuable information regarding protein-protein interactions as well as in optimization of protein purification strategies. Finally, a low ionic strength wash (distilled water or 5 mM sodium HEPES, pH 7) is recommended to remove high concentrations of salt ions. This will help reduce the presence of sodium adducts that may contribute to the mass heterogeneity of a given peptide/protein.

Desorption and Ionization

Energy-absorbing molecules (EAMs) is a generic name for molecules that assist in desorption and ionization of the analyte. These "matrix" molecules are applied in organic solvent, solubilizing many proteins on the array surface. A standard EAMs solvent contains 50% acetonitrile, 0.5% TFA. As the EAMs solvent evaporates, proteins cocrystallize with the matrix. Commonly used EAMs include CHCA and sinipinic acid (SPA). The processed chip array is then ready to be loaded into a ProteinChip SELDI TOF-MS for spectra acquisition.

ACQUISITION PROTOCOLS

The chip array, containing its complement of bound peptides or proteins, is transferred into the sample post of the ProteinChip SELDI-TOF Reader. The sample is then equilibrated with the evacuated flight tube. The chip surface can then be impacted by a focused nitrogen laser (337 nm). The energy absorbed facilitates proton transfer from the cocrystallized matrix compound and desorption of the protein from the chip surface. The potential difference across the flight tube draws the positively charged ions to the detector. Impact of protein ions on the detector is converted into an ion current displayed in the form of a chromatographic tracing. In designing acquisition protocols, several variables can be adjusted to achieve well-resolved peaks with optimal recovery. The laser intensities required to desorb a given protein efficiently does increase with the mass of the peptide/protein. However, measures must be taken to control for detector saturation, signal decay, and to minimize spectral noise particularly evident in the low mass (<5 kDa) region.

Laser Settings

The SELDI-MS has variable laser settings that must be optimized for a given target mass region. The ProteinChip SELDI System has the advanced feature that its laser settings are defined in nanojoules (nJ), such that doubling the intensity actually doubles the energy imparted on the chip target. Laser intensities from 500 to 15,000 nJ can be used.

Focus Mass Range

The ProteinChip SELDI-MS has detection capabilities extending above 300 kDa. However, spectra with such a wide mass range will have an increasingly large accompanying data file. The mass range should fit the experimental purpose (i.e., peptide mapping can be done from 0 to 5 kDa). Peak resolution within a given mass range can be optimized by the "Focus Mass" setting. This can be especially critical in the case of peptide digest spectra.

Feature of Mass Attenuation

To guard against detector saturation due to desorbed matrix, the ProteinChip SELDI System has an adjustable mass attenuation feature that keeps the detector in the "off" position until a proscribed time (set by the corresponding mass value) has elapsed. This feature is used to diminish the corresponding noise as a consequence of high laser intensity settings.

Method of Spot Partition

The method of EAMs application allows uniform dispersion of protein/matrix cocrystals. Laser pulses can sample multiple spot positions without the need for the operator to find the best signal position or "hot spot." The SELDI System raster laser design has 215 "pixels" to which the laser can be focused and allows for complete spot coverage. The uniform sample distribution also allows multiple protocols to be run on the same spot. For example, both low laser and high laser protocols can be run sequentially, each directed at different spot partitions.

Detecting Pulse Number

The laser repeatedly pulses at an individual spot position for a proscribed number of times before moving to the next position. Typically, 10 to 15 such pulses can allow for optimal peak detection before significant signal decay might ensue. By varying the factors listed above, optimized protocols can be designed to then profile large sample numbers. Typically, reference or pooled samples are used in the protocol design step. Use of automated, optimized protocols is the essence of the high-throughput capability of the SELDI process. The ProteinChip SELDI Data Manager software can facilitate the collection of hundreds of spectra without further operator input.

MATHEMATICAL CALCULATION OF DATA

The raw data obtained in collected spectra is a data average of the multiple pulse transients specified in the acquisition protocol. The data includes time-of-flight increment plotted against ion current, displayed as intensity. Each individual spectrum can include 10^5 to 10^6 data points. Fung et al. provide a detailed review of the biostatistical strategies that can be applied to groups of spectra in search of unique protein biomarkers. A few key points will be emphasized here. When considering a set of SELDI spectra to be screened for potential biomarkers, they all must contain the following characteristics:

1. Sample type (e.g., plasma), including identical steps for sample processing and preparation, including fractionation. Clarke et al. discuss avoiding preanalytical biases.
2. Chip selection.
3. Matrix selection.
4. Instrument settings.

Spectra, so collected, then must be processed in an identical fashion, including baseline fit and subtraction, noise adjustment, and mass calibration. Each spectrum is then normalized relative to the average total ion current across the group. The normalization step provides a correction for possible pipetting errors during processing of the chip arrays and also allows for identification of spectral outliers that may need to be reprofiled or eliminated from the data set. The Bio-Rad

ProteinChip Data Manager software facilitates the discovery of distinct protein biomarkers that may be differentially expressed in a subgroup of spectra. Multiple subgroups can be compared simultaneously.

Spectra are subjected to Expression Difference Mapping (EDM) analysis, which creates peak clusters within a proscribed mass window. A minimum signal/noise (S/N) threshold must be met within a proscribed percentage of all spectra to assign a peak cluster at a particular m/z value. Spectra are rescreened using a second, less-stringent S/N threshold. Finally, peak intensity values are estimated using the average m/z value of the existing members of the cluster. ProteinChip Data Manager then uses the Mann-Whitney test (nonparametric equivalent of Student's *t*-test) to calculate *p* values revealing the comparative levels of each peak across two defined groups (i.e., control vs. diseased). This univariate analysis is an initial step toward discovery of protein biomarkers that potentially could differentiate normal versus abnormal phenotypes.

ProteinChip Data Manager also provides box-and-whisker plot and scatterplot visualization of peak intensity distributions. Such scatterplots reveal group mean differences as well as the wide distribution often characteristic of clinical samples. Even though large sample numbers may yield significant *p* values, the ability for single markers to discriminate the two groups may still be poor. The wide variation in clinical populations points to the need for multivariate analysis of spectra groups. ProteinChip Data Manager also contains two multivariate unsupervised learning analysis methods. Principle component analysis and hierarchical clustering assess the distribution of the spectral data.

These tools allow rapid visualization of spectral outliers, can indicate the presence of subgroups within the data set, and can point to key peak clusters that contribute to the variance across the total sample group. Bio-Rad Pattern Analysis Software is a supervised learning technique, based on a classification and regression tree method. Peak clusters are assessed for the ability to distinguish defined subgroups. A decision tree is constructed using the most discriminating peak clusters as primary splitting factors. Assignment of unknown samples (spectra) to a specific group is based on comparison with multiple threshold peak intensity values.

In the development of potential diagnostic proteomic assays, use of multiple biomarkers has resulted in increased sensitivity and specificity values. The high-throughput capacity of SELDI data acquisition allows a large number of independent samples to be used in the construction and cross-validation of classification trees using Pattern Analysis Software. By testing a robust data set in the biomarker discovery phase, whether using univariate or multivariate analysis programs, markers with a higher degree of validation can then be the focus of identification and assay design.

IMPLEMENTATION OF MS DATA

The selective binding chemistries present on the ProteinChip array surfaces are identical to chemistries present on solid-phase bead (cellulose, dextran, zirconia) supports. By varying the binding and wash conditions on a specific chip array, subsequent mass spectra can give a rapid assay for the retention of a specific biomarker. In addition, the differential binding and elution of neighboring contaminant protein peaks can be monitored. Indeed, monitoring of a specific marker to several chemical surfaces can be done in parallel. This allows for the rapid design of potential purification schemes. The basic strategy for enrichment of a protein biomarker based on SELDI-MS data will implement the following features:

1. The particular chip array to which the biomarker of interest does bind will be further characterized.
2. Alternate binding conditions (i.e., changes in pH) should be tested to optimize biomarker retention on the chip surface.
3. Various wash procedures are then screened to establish conditions for elution of the biomarker from the chip array.
4. The conditions established for selective binding and elution are then implemented at a larger scale on a compatible solid phase. For example, Bio-Rad Q HyperD zirconia resins are compatible with the ProteinChip Q10 surface chemistry. Fractions collected from the solid phase resin can be monitored using the nonselective ProteinChip NP20 array.
5. When the best enrichment step is established, a control sample, in which the same biomarker is known to be deficient (as in the discovery phase of the study), should be processed in parallel.
6. A second enrichment step can then be carried out using an alternate ProteinChip array.

Ultimately, the specific biomarker needs to be identified. Often, the selective enrichment steps are adequate to allow subsequent SDS-PAGE isolation of the proteins.

CONSTRUCTING PEPTIDE MAP

After SDS-PAGE, standard protocols for Coomassie Blue staining can be used. Subsequent protease digestion of selected bands can be achieved using the following approaches: in-gel digestion; passive elution of protein from the gel followed by protease digestion in solution; passive elution of protein followed by protease digestion on-chip.

Hydrolysis of Protein

1. To carry out in-gel protease digestion, SDS must first be removed from the gel. The specific Coomassie-stained gel band (destained in solution containing 50% methanol, 10% acetic acid) should be excised, cut into 1-mm sections, and transferred to an 0.5-mL microcentrifuge tube. The corresponding molecular weight (MW) region of a control lane (containing the processed "deficient" sample) should also be excised. Alternatively, a protein-free sample lane can be processed in parallel.
2. Add 0.4 mL 50% methanol, 10% acetic acid and incubate with agitation for 1 h. Replace the solution with fresh 50% methanol, 10% acetic acid; incubate with agitation for an additional 30 min. Remove solution.
3. Incubate in 0.4 mL 50% acetonitrile, 0.1 M ammonium bicarbonate, with agitation for 1 h. Remove solution.
4. Add 50 µL 100% acetonitrile. Incubate with agitation for 15 min. Remove solution.
5. Dry the gel pieces in a SpeedVac for 15 min or until completely dry. Prepare trypsin solution. Sequencing grade trypsin is dissolved in 10 mM HCl (0.2 µg/µL) = 10X stock stored at –20°C

or –80°C. A 1X trypsin solution is made by diluting with 9 volumes of 25 mM ammonium bicarbonate, pH 8.0.

6. Add 10 μL 1X trypsin solution to the dried gel piece(s) in an 0.5-mL microcentrifuge tube. Incubate at room temperature for 15 min to rehydrate the gel. Process the "no protein" control tube in the same manner.
7. If necessary, add up to 40 μL 25 mM ammonium bicarbonate, pH 8.0, to cover the gel. Only add as much as is necessary to cover the pieces so that the resultant digest solution is as concentrated as possible. Seal the tube tightly and incubate at 37 °C for up to 16 h.

Mechanism of Digestion in Solution

1. After destaining, transfer the gel piece to a fresh microcentrifuge tube and crush the gel piece at the bottom of the tube with a dissection needle.
2. Passive elution of protein from polyacrylamide gels can be accomplished using either of two extraction solutions. Add 40 μL elution solution for an entire protein band or 20 μL for half of a band. The solution should completely cover the gel pieces.
3. Sonicate for 30 min at room temperature. Vortex for 4 h at room temperature. The samples can be stored overnight at 4°C.
4. Collect the eluant in a fresh microcentrifuge tube; rinse the gel pieces with an equal volume of fresh elution solution. Combine the solutions.
5. Evaporate the eluant in a SpeedVac. Redissolve the eluted protein in 5 to 10 μL water (for low-molecular-weight proteins) or 0.5% *n*-octyl glucopyranoside in water (for high-molecular-weight proteins). Vortex for 15 min at room temperature.
6. The samples are ready for analysis. If you are not able to recover the protein after the SpeedVac drying, remove the liquid, wash the vial with a small volume of 20% acetonitrile, and examine. If a specific ProteinChip array was used to detect the protein of interest, the same type of array should be used to analyze the retrieved protein. Binding and washing conditions should be similar to those used originally. Use 1 or 2 μL of recovered protein solution to confirm that the protein of interest was eluted from the gel.
7. The eluted protein can then be incubated with protease (i.e., add an equal volume of 50 mM ammonium bicarbonate, containing 2X trypsin) at 37° C.
8. The proteolytic digests resulting from these two protocols can then be profiled on several ProteinChip arrays. For NP20: Apply 1 to 2 μL of the digested sample. Let dry. Apply 2 × 0.5 μL aliquots CHCA (20% saturation) allowing 5 min for matrix to dry between each application. For H4 or H50: Apply 1 to 2 μL sample as above.
9. The hydrophobic chains allow for a brief wash step with 2 μL water to reduce salt contaminants. Apply 2 × 0.5 μL (20%) CHCA. For SEND-ID: Dilute the digested sample with an equal volume 50% acetonitrile, 0.2% TFA. Apply 1 to 2 μL onto the SEND array and let dry. As CHCA is already incorporated onto the surface, the sample is ready to be inserted into the ProteinChip reader.

Mechanism of Digestion On-Chip

1. Protein that has been passively eluted from a gel band or purified by alternate methods can be digested directly on the chip surface. From the initial profiling study, a specific ProteinChip array as well as the best binding-washing conditions can be adopted.
2. Add the eluted protein to the array surface. Rinse with water for 1 min. Repeat two times for a total of three washes.
3. Prepare a spot with elution solution alone and no protein as a control to identify background signal from protease autolysis. Dry the spots for 5 min at 60°C (heat denaturation of protein before digestion).
4. Add 4 to 4.5 μL of buffer for digestion (i.e., 25 mM ammonium bicarbonate). Add 0.5 to 1 μL of protease (1 to 10 ng/μL).
5. Incubate for 2 h at optimal temperature. Dry the spot for 2 min at 60°C.
6. Add 1 μL of saturated CHCA solution in 50% acetonitrile/0.05% TFA. Allow the array to dry.
7. Read the array using the ProteinChip Reader.
8. The unique proteolytic fragment m/z values are can then searched against protein databases for the identification of the protein biomarker.

SEQUENCE ANALYSIS

Sequence analysis of peptides bound to Bio-Rad ProteinChip arrays can be obtained using a ProteinChip Tandem MS Interface as a front-end SELDI ion source for the Applied Biosystems/ MDS Sciex QStar Hybrid LC/MS/MS System. The ProteinChip Interface uses a 337-nm nitrogen laser with a lensed fiber optic, delivering 150 mJ of energy per pulse at 30 pulses/s. Peptides up to 5 kDa can be selected in Q1 and transmitted to Q2 for collisional induced dissociation, using an applied collisional energy of 50 eV/kDa. Sample spot scanning is controlled via ProteinChip Interface Control Software, and data analysis and acquisition are carried out using the QStar System's Analyst software. This system is capable of high attomole-low femtomole MS and tandem mass spectrometry (MS/MS) sensitivities. The resulting sequence obtained from spectral data can be submitted to the database mining tools Mascot or ProFound for identification.

SPECIFIC PROTEIN ASSAYS

As mentioned in the discussion of the ProteinChip RS 100 and PS20 arrays, user-defined ligands can be coupled to the chip creating an affinity surface for specific proteins or protein complexes. This strategy can be used via antibody capture to probe for specific antigens. Similarly, affinity surfaces have been used for the detection of autoantibodies, DNA transcription factors, and a variety of other protein-ligand interactions. Several key points must be considered in designing optimal affinity chip surfaces and assays:

1. Retention of binding capacity for the "bait" molecule:
   (a) When coupling antibodies to a PS20 or RS100 chip surface, it is useful to test several pH conditions. Although coupling with lysine amino groups can occur at basic pH (0.1 M

sodium bicarbonate, pH 9), preferential involvement of the alpha-amino group can be favored in lower pH solutions. Coupling in PBS, pH 7.0, or 50 mM sodium acetate, pH 5.5, should be tested to assess the most productive binding of antigen.

(b) Avoid stearic hindrance of ligand adjacent to the chip surface.

2. Blocking of unreacted epoxy or carbonyldiimidazole groups can also be achieved using nonspecific (BSA) protein solutions (0.1 mg/mL in Tris buffer).
3. Binding conditions (time and buffer components) must be optimized.
4. Nonspecific binding may be reduced by inclusion of high salt and 1 M urea washes.
5. Relevant controls (i.e., nonspecific or preimmune antibodies or scrambled DNA sequences) should be incorporated to help define the specifically bound m/z factors.

Annotation

1. As the SELDI MS system involves positive ion flight, it is recommended that SDS be removed from cell lysis buffers. Non-ionic detergents (i.e. Triton X-100, CHAPS) can be substituted.
2. IMAC chips are charged with selected metals (e.g., 50 μL 0.1 M copper sulfate) for 5 min. When using copper metal, an additional neutralization wash with 0.1 M sodium acetate, pH 4, is included prior to preincubation with binding buffer.
3. The hydrophobic carbon chains on the H4 and H50 chips are activated by incubation with 50% acetonitrile (or methanol) for 5 min prior to preincubation with binding buffer.
4. When using the preactivated RS100 and PS20 arrays to which specific "bait" molecules have been covalently bound, the coupling step is followed by a wash step to remove unbound ligand and then incubation in amine buffer (50 mM Tris-HCl, pH 8, or 0.5 M ethanolamine). This serves to block unreacted sites on the chip surface. Finally, the chips can be preequilibrated in the selected binding buffer.
5. Appropriate binding times will be expected to vary when using ligand-coupled RS100 or PS20 arrays. For example, binding of specific antigens by antibody- coupled arrays may require overnight incubation at 4°C.
6. CHCA is useful in peptide profiling as matrix ions are generally <600 Da. However, production of multiply protonated protein ions can be seen. The matrix ions for SPA can be >1000 Da, with a lesser tendency to form multiply protonated ions. SPA is more commonly used over a wide mass range (2 to 200 kDa) biomarker discovery study.
7. Bio-Rad Laboratories markets the PCS4000 edition ProteinChip Reader. An earlier version, PBSIIc, accounts for a larger number of SELDI instruments in the field and is also supported by Bio-Rad.
8. The PBSIIc instrument uses laterally positioned (–) charged deflector plates that can be activated, transiently, suppressing ion entry into the drift tube. This strategy diminishes the impact of matrix ions onto the detector, also limiting detector saturation.
9. The PBSIIc laser makes a focused lateral impact on the spot. The lateral spot positions are on a 0 to 100 scale. Positions between 15 and 85 are commonly used for quantitative analysis.

10. Spectra can be externally calibrated by application of known m/z peptide or protein standards on an NP20 chip. Standards should be chosen to span the mass range of interest. The standard spectrum should be collected using matrix and instrument settings matching the corresponding sample set. A second polynomial equation relating TOF to the standard m/z values can then be saved and applied to the spectral group being analyzed.
11. Shorter incubation times may be adopted. Remove 1-μL aliquots for analysis throughout the digestion period to determine the optimum/minimum digestion time.
12. FAPH solution (50% formic acid, 25% acetonitrile, 15% isopropanol, 10% water, v/v/v/v) is recommended for proteins greater than 10 kDa. ACN/TFA solution (70% acetonitrile, 30% 0.2% trifluoroacetic acid, v/v) is effective for proteins less than 10 kDa.
13. Protein to protease molar ratio should be at least 50:1. Significant excess of protease over the protein will produce ion suppression and nonspecific cleavage.
14. Accurate m/z estimates can be achieved using peptide standards, most effectively when mixed with the sample for internal calibration.
15. Binding of biotin-DNA molecules to a streptavidin-coated chip surface can benefit from the use of biotin adducts possessing long-chain linkers and the use of concatamers of the specific DNA sequences.
16. Use of the ProteinChip PG20 array, containing bound protein G, can provide the correct orientation for adsorbed immunoglobulins used to screen for specific antigens.

7

Adverse Drug Reactions

Adverse drug reactions continue to pose a major impediment to drug development, and the clinical management of marketed products. These are typically divided into acute, dose dependent reactions (Type A), and reactions that may occur in only a small percentage of patients, where the frequency of occurrence in the population is not dependent on dose (Type B), although more complex classifications have also been proposed. According to a recent PhRMA review, the most frequently encountered toxicities in pre clinical drug development are *hepatotoxicity* and *dermal* reactions. Hepatotoxicity may take many forms, so implying drug metabolism generally would be presumptive, but as the major organ responsible for the metabolism of drugs it seems clear that metabolism is important in many cases. Type A reactions generally may be mediated through parent compound or metabolites, and the role of chemically reactive metabolites has been well recognized. Type B reactions, also referred to as idiosyncratic or hypersensitivity reactions, have been the subject of extensive reviews in recent years.

These are generally believed to be immune mediated, and are not predictable from pre clinical animal studies, thus they may not be identified until late clinical stages, or post marketing. Type B reactions have been reported to constitute 25% of all clinical adverse events. While direct T cell stimulation has been proposed as a possible mechanism, bioactivation to reactive metabolites that covalently bind to proteins is still believed to be a key event in the origin of most Type B reactions, and is the basis of the Hapten hypothesis. Much has been written on the role of covalent binding for both type A and Type B reactions, and clearly protein binding can be dissociated from toxicity. The dilemma of the significance of protein binding in type B reactions can be addressed in part by the Danger Hypothesis, where by the immune system requires a second signal to trigger a response. This may be cellular damage or an infection. Thus in general covalent protein binding is still considered a key factor in most Type B reactions. Drug metabolism therefore, is implicated in the most problematical toxicities in drug development and clinical practice. This raises the question whether such properties can be rationally designed out at an early stage of development, based on knowledge of metabolic pathways or simply structure, and such consideration have been widely discussed. The ability to do this effectively remains elusive, but it seems clear that an understanding of how functional groups can be metabolized to reactive metabolites is likely to reduce the prevalence of both type A and B reactions. This review will survey the more extensively studied reactive metabolites which have been encountered in drug molecules.

METABOLIC ACTIVATION

Up to 30 different types of functional groups have been shown or implicated to undergo some type of metabolic activation. However, those most frequently encountered in drug development may be only about one third of this. Reactive metabolites can be broadly grouped into either electrophiles or radicals. Carbenes would not fall into either group, and while well recognized as a species involved in CYP (Cytochrome P450) inactivation they have not been significantly implicated in toxicities. To some degree, the different biological effects of electrophiles and radicals can be rationalized based on their chemistry. Thus the closed shell electrophiles will react readily with biological nucleophiles, such as proteins to give covalent adducts, where as radicals are more likely to abstract a hydrogen radical from the protein. However, oxygen existing in the triplet state, is ideally set up to react with unpaired electron species, and so this tends to be the more prevalent reaction course for many radicals. Consequently, free radical formation is more frequently associated with oxidative stress, and electrophiles with covalent binding.

While radicals may be implicated in some idiosyncratic reactions, for most drugs showing this form of toxicity where a bioactivation mechanism has been proposed, most appear to involve electrophiles. This statement is something of a generalization however, and as more work is done specifically looking for radical formation more examples may be found. Reactive metabolites may be derived from a wide range of metabolic reactions including oxidation (CYP, peroxidases, MAO (Monoamine oxidase)), reduction (CYP, CYP reductase), and conjugation (glucuronyl transferases, sulfotransferases, glutathione transferase, Acyl Co-A synthase). While oxidative CYP mediated reactions predominate, peroxidases present in blood cells may play a major role in agranulocytosis and other blood toxicities. There are a number of recent reviews of metabolic activation covering the ranges of structures which have been shown to have bioactivation potential. This review will focus more on the commonality of reactivity patterns for reactive metabolites derived from these widely different structural features. This may help to simplify the mechanistic aspects of this undesirable drug development property.

METABOLICALLY DERIVED ELECTROPHILES

Regardless of the structure from which they are metabolically derived, electrophiles may be grouped as either: 1. Acylators, 2. Activated double bonds, 3. Other electrophilc carbon centers, or 4. Electrophiles based on nitrogen or sulfur, or derived from sulfur oxidation. An important and quite common observation is that metabolism of a particular functional group may give rise to more than one type of electrophile, as shown in some of the examples below.

Acylators

These may be defined as electrophiles which result in the biological nucleophile being bonded to an sp2 hybridized carbon, usually double bonded to oxygen. Examples of functional groups which may produce acylating electrophiles are: carboxylic acids, acetylenes, formamides, halogenated hydrocarbons, and thiazolidinediones. Carboxylic acids may act as acylators either via glucuronidation or acyl CoA ester formation. Acyl glucuronides are perhaps the most widely studied of reactive metabolites, and there are many good reviews on this subject. A large number of carboxylic acid containing drugs known to be metabolized by acyl glucuronidation, have been marketed. This level of clinical experience makes this functionality some what unique as regards the potential for an empirical risk assessment. Of note are the number of drugs metabolized via acyl glucuronidation which have been withdrawn from the market for safety reasons.

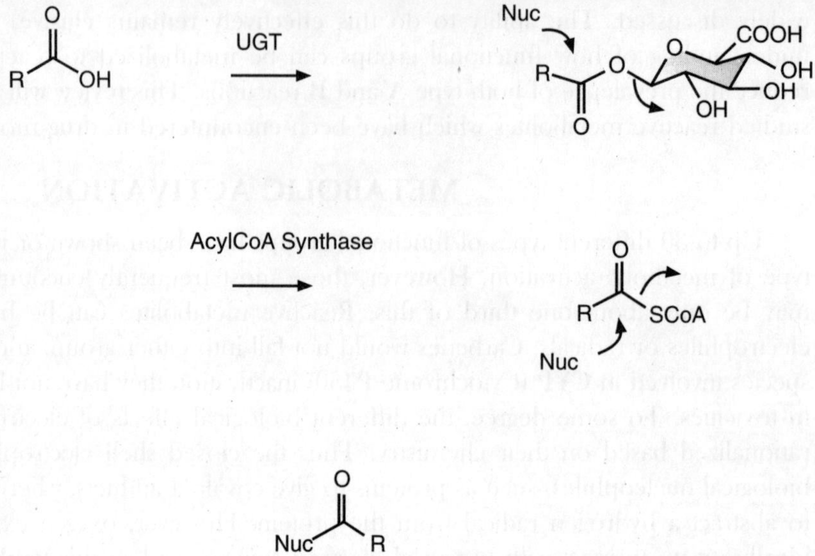

Fig. 7.1. Acylating metabolites derived from conjugation of carboxylic acids: acylglucuronide and acyl CoA esters.

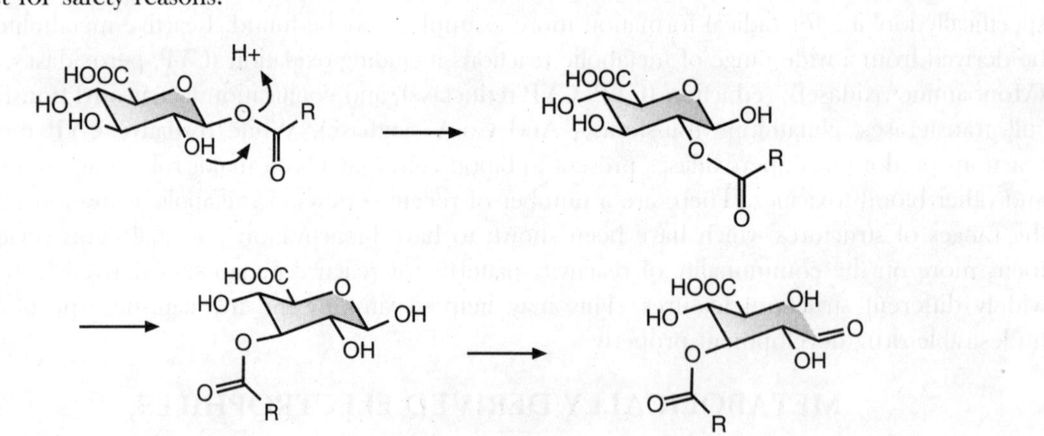

Fig. 7.2. Migration of acylglucuronides to generate an electrophilic carbon center (aldehyde), reaction with nitrogen nucleophile of proteins, and rearrangement to a stable product.

Between 1974-1993, there were a total of 29 drug withdrawals for safety reasons from the U.S., U.K and Spanish markets, and seven of these are reported to be metabolized by acyl glucuronidation. While no causative link has been found between acyl glucuronidation and the limiting toxicities for these compounds, it is a pathway of metabolism observed in a significant percentage of compounds which have been withdrawn from the market due to safety factors. In addition to acting as acylating agents, acyl glucuronides can undergo an intramolecular rearrangement to generate an aldehyde which can bind to proteins. This mechanism is distinct from the acylating mechanism, thus the acyl glucuronidation pathway can give rise to two distinct types of reactive species.

Carboxylic acids may also be metabolized to acylating agents via formation of acyl CoA esters, which can lead to covalent binding, and this constitutes an alternative bioactivation pathway which has been proposed to lead to toxicity. Acetylenes may form ketenes during CYP450 mediated oxidation, and these also function as an acylators, most particularly as mechanism based inhibitors of CYP. Similarly, formamides on oxidation have been proposed to form isocyanates which may act as acylators, but examples of drugs containing acetylene or formamide functions are less common.

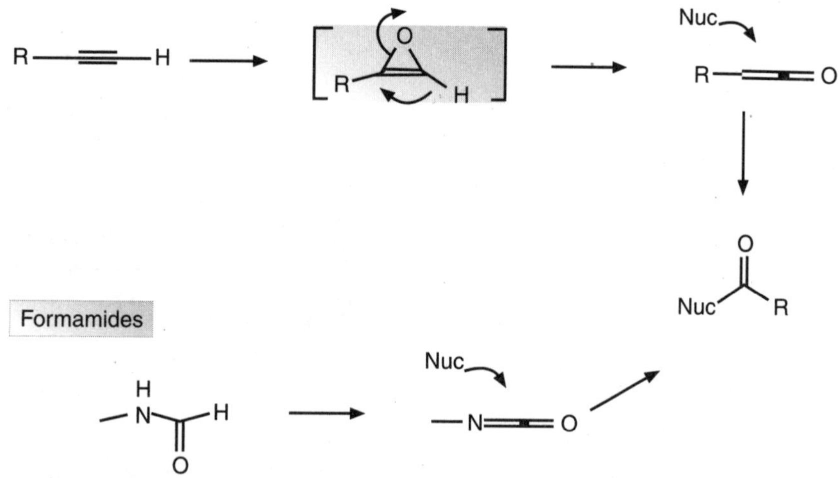

Fig. 7.3. Oxidation of acetylenes and formamides to give acylating metabolites.

Halogenated hydrocarbons (in contrast to halogenated aromatics) are a structural feature not widely encountered in drugs, but certain anesthetics agents in this class of compounds are among the most mechanistically well studied in regard to metabolic activation and idiosyncratic reactions. The best example is halothane which undergoes oxidation to generate an acyl halide, a very effective acylating agent. The protein targets for the acyl halides, and their implication in the immune response have also been studied. Halogenated hydrocarbons can also be bioactivated by glutathione conjugation, followed by a series of steps to generate thioacyl halides.

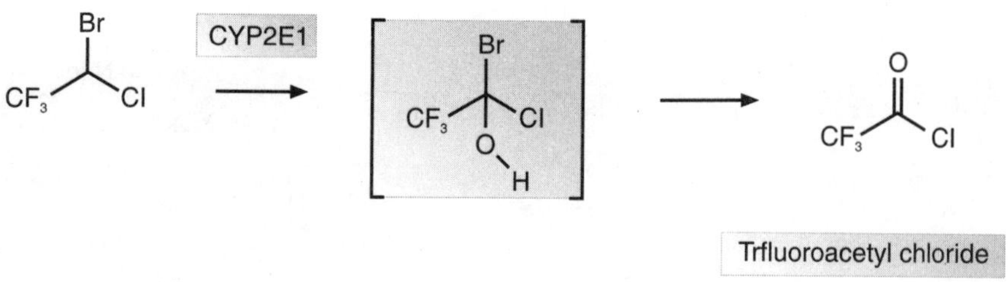

Fig. 7.4. Oxidation of halothane to an acyl chloride.

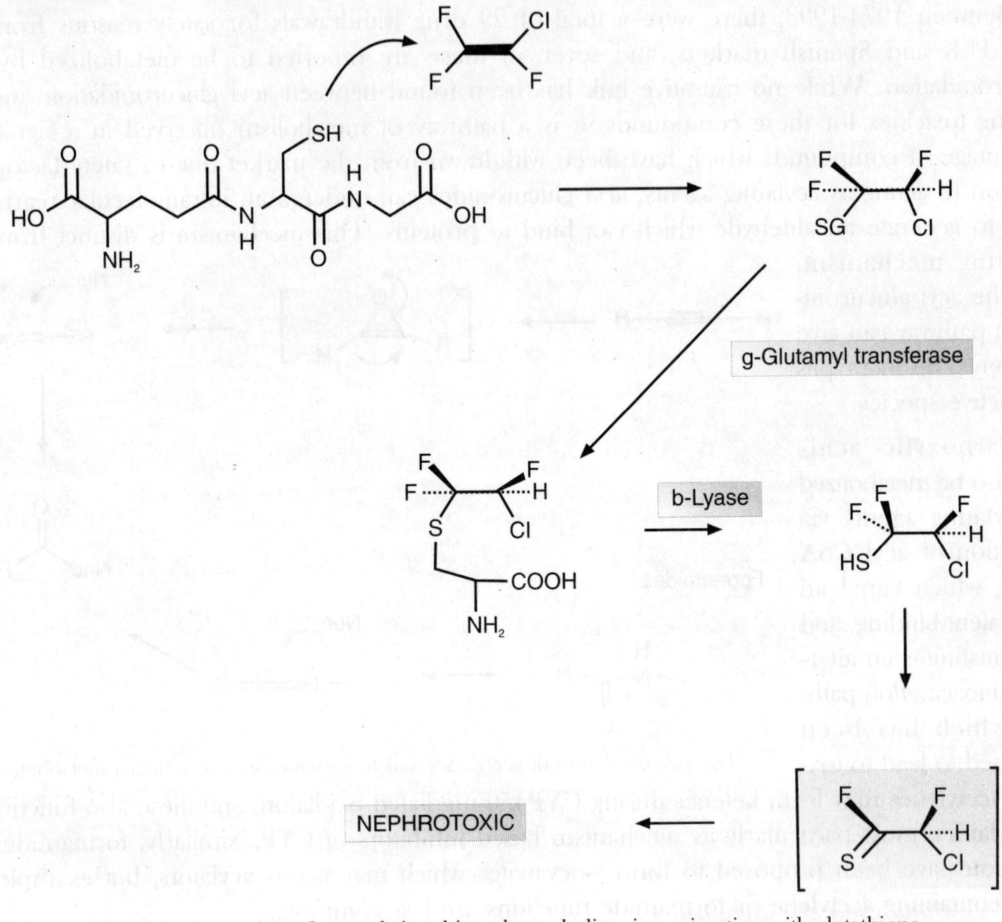

Fig. 7.5. Generation of an acylating thioacyl metabolite via conjugation with glutathione.

Nephrotoxicity has been a common observation for compounds showing this pathway, but impairment of mitochondrial function and other toxicities have also been noted. A third bioactivation

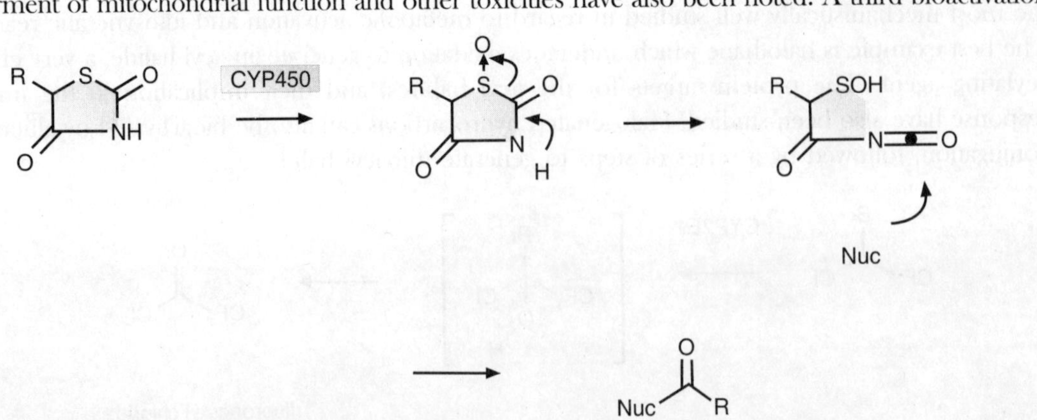

Fig. 7.6. Oxidation of the thiazolidinedione ring system to give thiocyanate which may act as an acylator, and a sulfenic acid.

pathway for halogenated hydrocarbons involves reductive metabolism to form radicals which will be discussed below. Thiazolidinediones may undergo oxidation to produce both an isocyanate (acylator) and a sulfenic acid, a sulfur based electrophile. This was first noted for troglitazone, which has been withdrawn from the market due to idiosyncratic hepatic toxicity. Other thiazolidinedione containing drugs such as pioglitazone and rosiglitazone, are also likely to have this oxidative pathway, however these drugs are considered to be essentially non hepatotoxic. This is most likely due to the much lower doses at which these drugs are used.

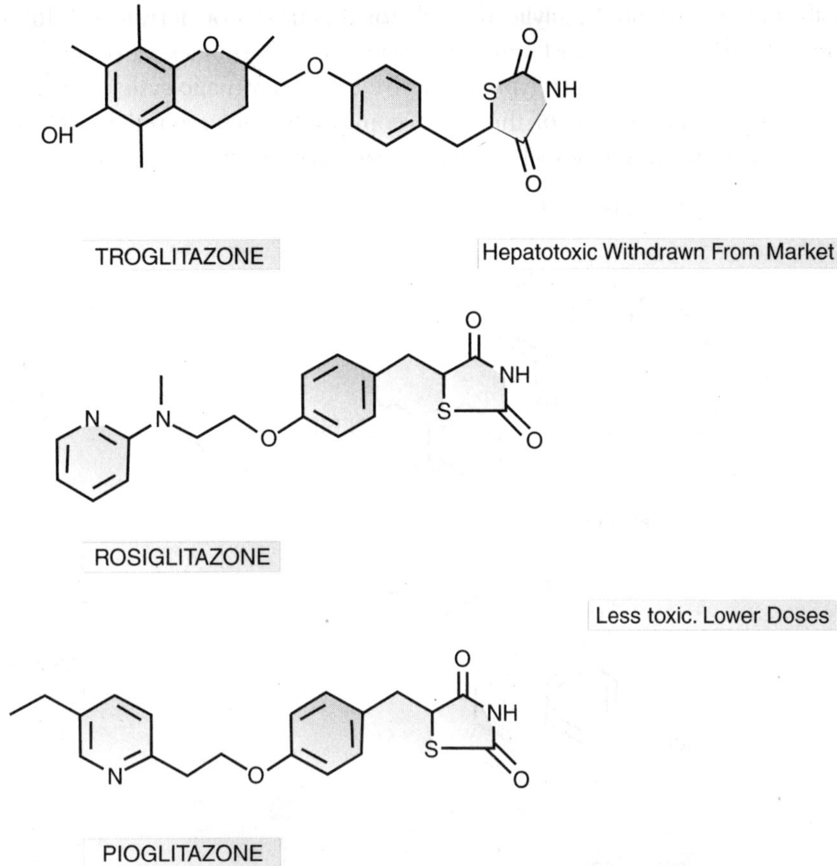

Fig. 7.7. Examples of thiazolidindione containing drugs.

Functionalities

A wide range of functionalities can be metabolized to electrophiles which are capable of reacting with nucleophiles by a 1,4 addition process. These functionalities include: nitrogen or oxygen

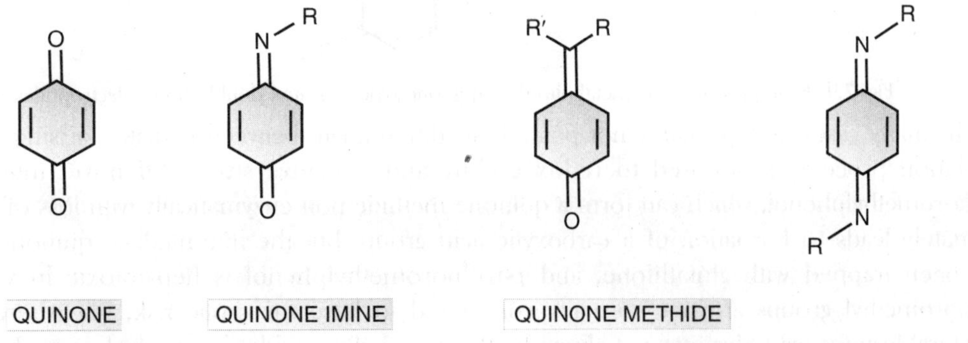

Fig. 7.8. Quinone type structures which may be formed metabolically, which may act as Michael type acceptors.

substituted benzenoids, allylic alcohols (or precursors or derivatives), furans, thiophenes, and some indoles. Among the most important type of electrophiles associated with toxicities are quinone type structures. These are typically derived from aromatic systems activated by nitrogen or oxygen substituents, and is one of the more commonly encountered bioactivation pathways, due to the prevalence of these types of aromatic systems in drugs.

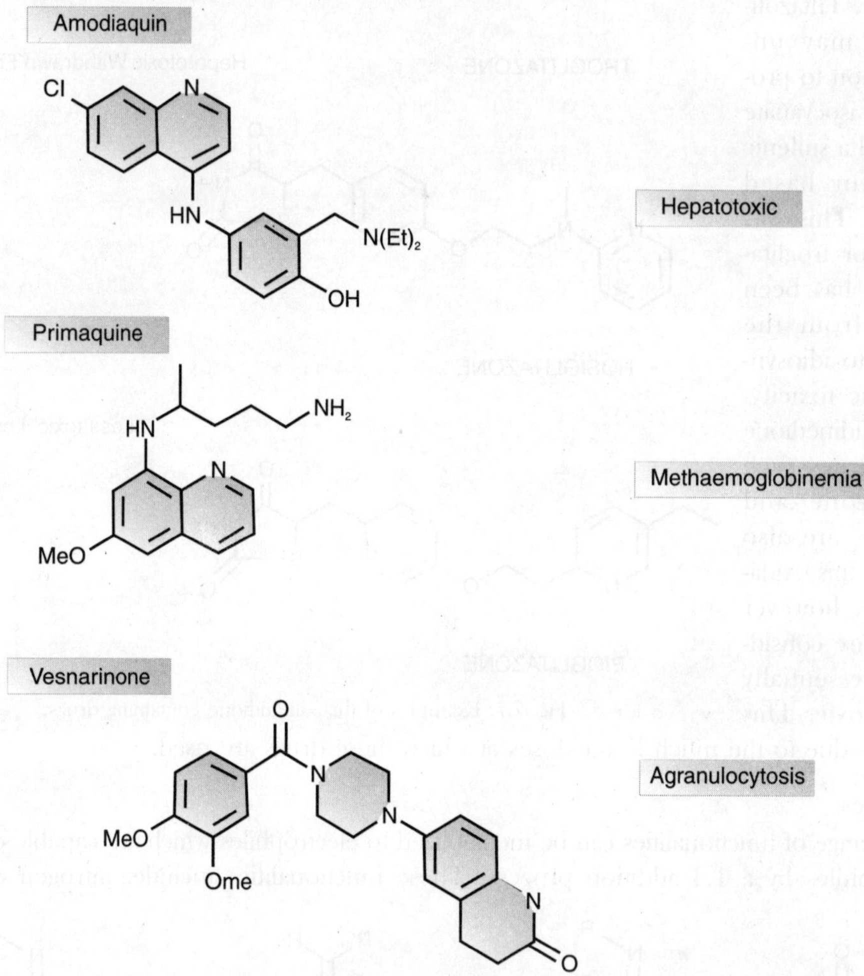

Fig. 7.9. Examples of drugs metabolized to quinone type activated double bond electrophiles.

In many cases, it is probably not possible to differentiate between toxicities arising due to an alkylation process as opposed to redox cycling and oxidative stress. Of novel interest is *p*-trifluoromethylphenol, which can form a quinone methide non enzymatically with loss of HF. This ultimately leads to formation of a carboxylic acid group, but the intermediate quinone methide has been trapped with glutathione, and *p*-trifluoromethylphenol is hepatotoxic in vitro. Aryl trifluoromethyl groups are not normally considered to have metabolic risk, but where relevant structural features exist, this may not always be the case. Leflunomide shows a high enough incidence

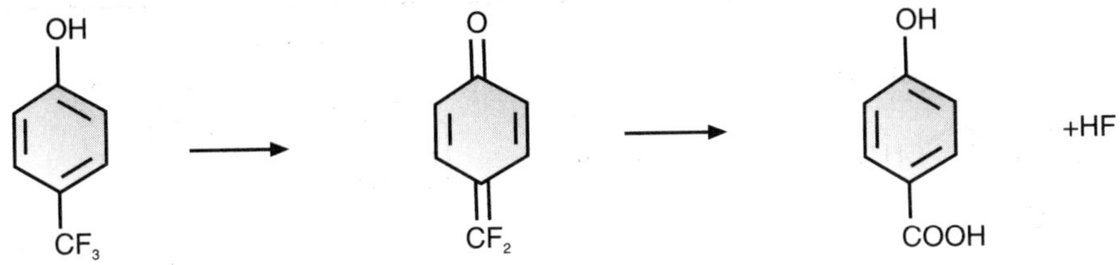

Fig. 7.10. Hydrolysis of p-trifluoromethylphenol to form an electrophilic quinone methide.

of hepatotoxicity that public interest groups have lobbied for it's withdrawal. The FDA has subsequently ruled in it's favor, and more recent studies suggest that the hepatotoxicity is manageable. Imino methide formation would seem to be feasible here, either directly or after amide hydrolysis. However, no such mechanism has yet been demonstrated.

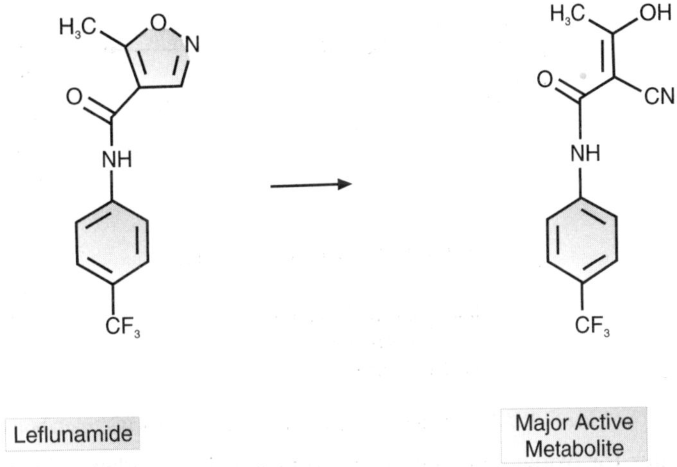

Fig. 7.11. The metabolism of leflunomide.

The major metabolite observed is the ring opened structure shown which is a malononitrile, isomeric with the isooxazole ring. Mechanistically distinct from the above aromatic systems is the case of the antiepileptic agent felbamate, which also produces a metabolite with an activated double bond. While not toxic in rats, Felbamate showed idiosyncratic toxicities in humans, chiefly hepatotoxicity and aplastic anemia which resulted in a Black Box warning for the drug. This may be rationalized based on differences in metabolism. Felbamate is a bis-carbamate of a propane 1,3 diol, and following hydrolysis and oxidation, carbamic acid can be eliminated to give an $\alpha\beta$ unsaturated aldehyde (designated as ATPAL by the authors). This will form glutathione conjugates in vivo confirming it's electrophilicity, and this pathway appears to be more prominent in human.

Rats metabolize the drug predominantly by direct oxidation of parent. The reverse transcriptase inhibitor Abacavir marketed for the treatment of HIV infection shows hypersensitivity reactions in a small percentage of patients. The drug is metabolized by alcohol dehydrogenase isozymes to produce an aldehyde intermediate that binds to proteins and this is likely to involve a 1,4 addition mechanism. Drugs may also indirectly produce activated double bond electrophiles from endogenous compounds via lipid peroxidation. This is associated primarily with drugs that form free radicals. For example, 4-hydroxynonenal, a known product derived from radical mediated lipid oxidation, is believed to be toxic due to its inherent electrophilicity and protein binding potential. Thus for idiosyncratic immune based reactions, this provides an alternative mechanism by which haptenized proteins may be generated, involving an electrophilic process but, initiated by free radicals. This mechanism however, remains hypothetical. The π excessive heterocycles furan and thiophene are present in a wide array of drugs and their electron rich state makes them prime

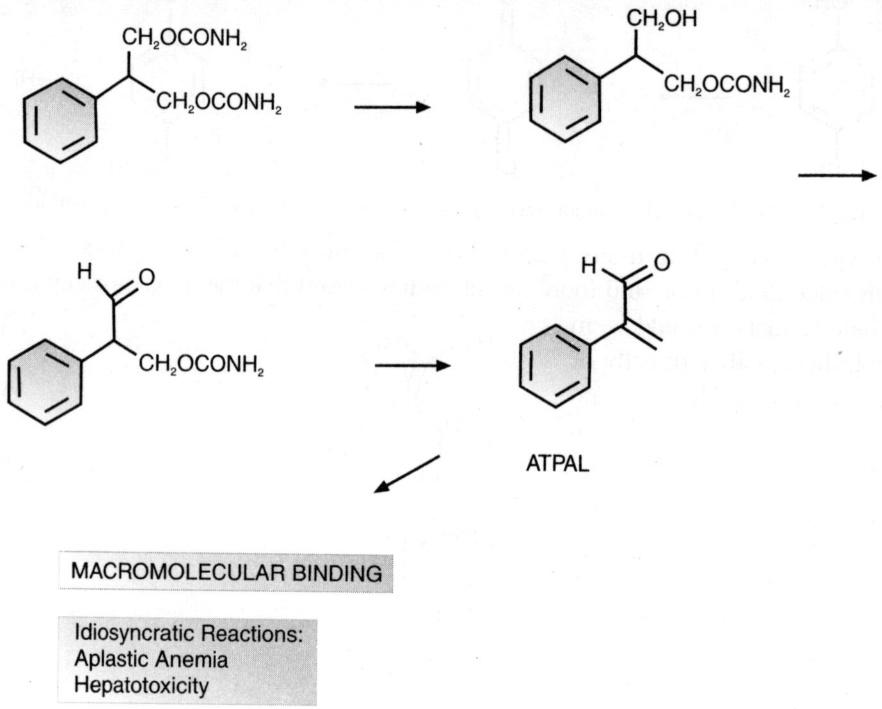

Fig. 7.12. The metabolism of felbamate to a reactive metabolite containing an activated double bond.

sites for oxidative metabolism. In both cases, oxidation leads to structures whose chemical reactivity is basically that of activated double bonds.

Menthofuran is a hepatotoxic terpenoid present in mint plants, which contains a furan that is bioactivated via oxidation to a ring opened γ ketoenal, a reactive Michael type acceptor. Thiophene can be oxidized to S-oxides which have activated double bond character, but in this case the ring system stays intact. Tienilic acid, a thiophene containing diuretic, was withdrawn from the US market in 1980 due to immunoallergic hepatitis.

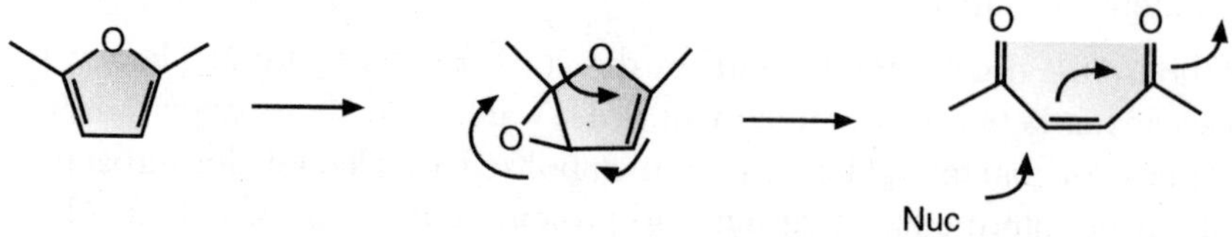

Fig. 7.13. The oxidation of furan to an activated double bond type metabolite.

The metabolism of this drug has been extensively studied, and toxicity is believed to be due to this type of S-oxidation pathway (although an arene oxide type mechanism has also been

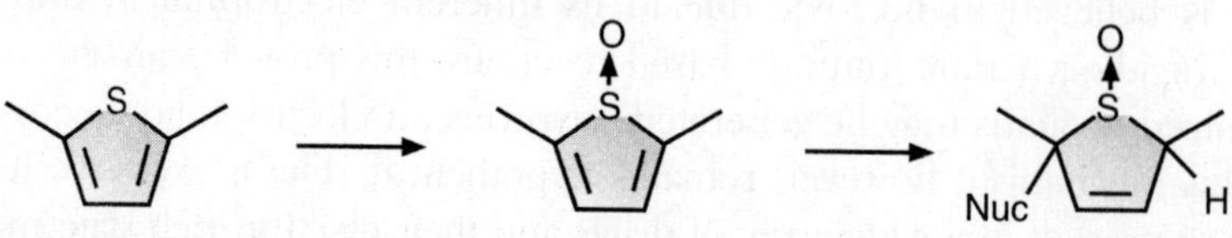

Fig. 7.14. The oxidation of thiophene to a thiophene -S- oxide, reactive by vitrue of an activated double bond character.

proposed). Mechanistic aspects leading to the immune mediated hepatitis, have been reported. Another heterocyclic ring system capable of bioactivation is the indole ring which has appropriate 3-substitution. 3-Methyl indole is a pulmonary toxin that undergoes dehydrogenation of the methyl group by CYP2F1 to produce a methylene imine which can covalently bind to proteins.

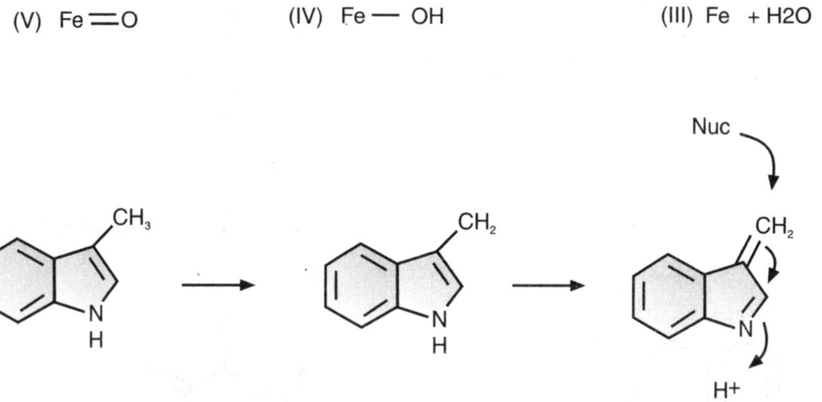

Fig. 7.15. Activated double bond metabolite derived from oxidation of 3-methylindole.

The above indicated reactions for these five membered heterocycles constitute only a portion of the types of transformations possible for these structures, yet represent the most probable bioactivation pathways. The metabolism of five membered heterocycles has been recently reviewed.

Carbon Centered Electrophiles

There are a number of other carbon centered electrophiles that can be distinguished from acylators or activated double bonds. These include epoxides, arene oxides, imines and some carbonyl compounds. The most important example of the latter is the aldehyde/Amadori mechanism for covalent binding of acyl glucuronides to proteins mentioned above. It is important to note that a large number enzymes have evolved specifically for the metabolism of the carbonyl group. This is likely due to two reasons: 1. The low reactivity towards one electron oxidative or reductive processes, making them poor substrates for CYP mediated transformations, and 2. Their inherent electrophilicity. Thus the diversity of enzyme systems and isozymes that have evolved to process aldehydes and ketones speaks to their biological significance.

Aldehydes are initial products of O-and N-dealkylation reactions, although these are not frequently associated with adverse events. The aldose reductase inhibitor, Sorbinil, shows hypersensitivity reactions manifested by rash and fever, and in rare cases more severe dermal reactions. Bioactivation has been proposed to be due to an O-dealkylation process with subsequent covalent binding of the aldehyde to form a Schiff base. One mechanism for ethanol induced hepatitis has been proposed to involve oxidation to acetaldehyde and Schiff base formation with N-terminal amino groups on proteins. These can be subsequently stabilized by a cyclization process to form imidazolidindiones. Aliphatic epoxides (as distinct from arene oxides), are probably of more toxicological importance for industrial chemicals than for drugs. The toxicity of mycotoxins, such as aflatoxin is believed to be mediated through epoxide formation. Arene oxides are more widely implicated in drug toxicities, even though much earlier work on these electrophiles was

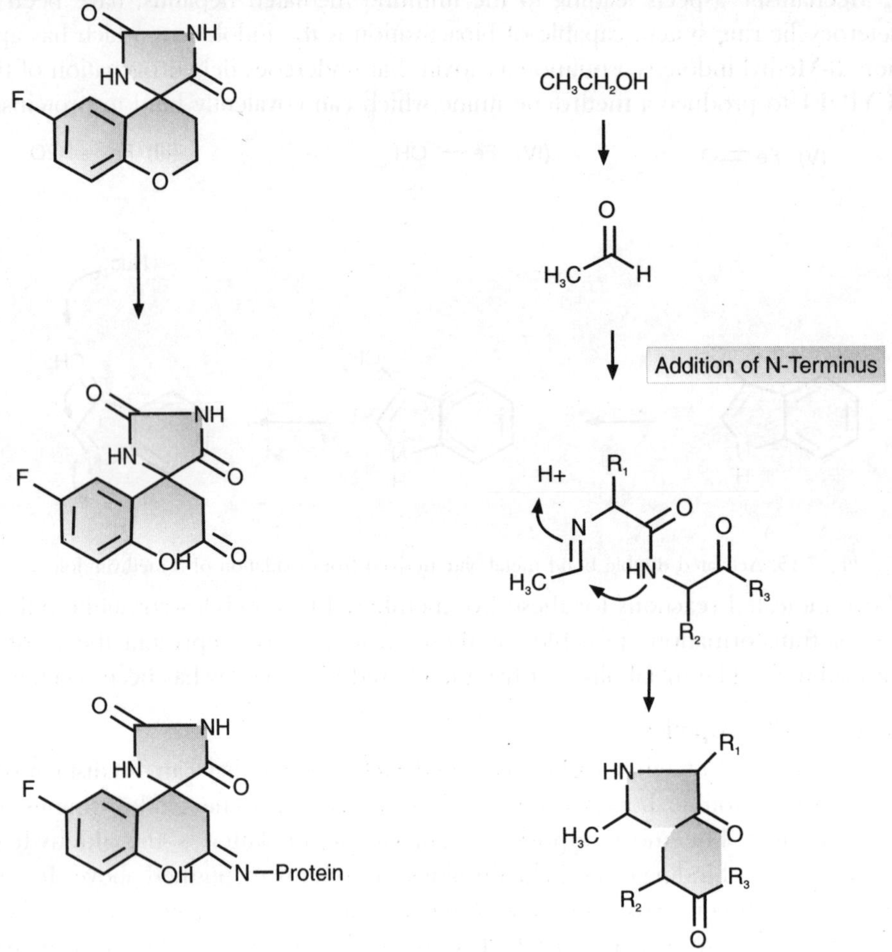

Fig. 7.16. Sorbinil and ethanol. Examples of compounds where toxicity has been proposed derived from aldehyde formation, showing their proposed reaction with nitrogen protein nucleophiles.

done to elucidate the mechanism of carcinogenicity for polycyclic aromatic hydrocarbons. Cabamazepine is an anticonvulsant associated with various hypersensitivity reactions. Metabolism is complex, and an arene oxide has been suggested as a metabolite that might be associated with the idiosyncratic reaction, but an epoxide, and quinone imines have also been proposed.

Lamotrigine is used as adjunct therapy in epilepsy, but shows a rare incidence of a severe dermal reaction. Presumed to be immunologically based, bioactivation pathways have been investigated, and evidence for arene oxide formation in rats has been demonstrated. While this pathway has not yet been demonstrated in humans, it does constitute the most likely metabolic activation pathway associated with the idiosyncratic reaction. Imines form another class of electrophiles in this group, and are formed from oxidation of alicyclic amines. These functionalities are very common in drugs, but to date there appear to be few cases where their formation has been associated with toxicity.

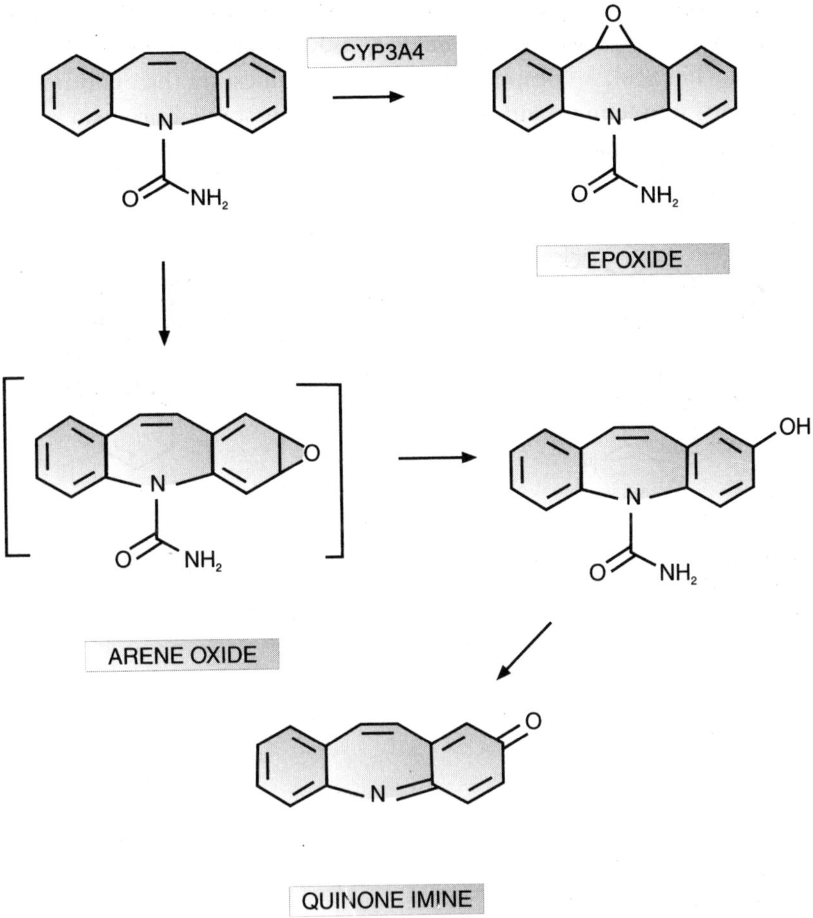

Fig. 7.17. The metabolism of carbamazepine to reactive carbon centered electrophiles: epoxide and arena oxide.

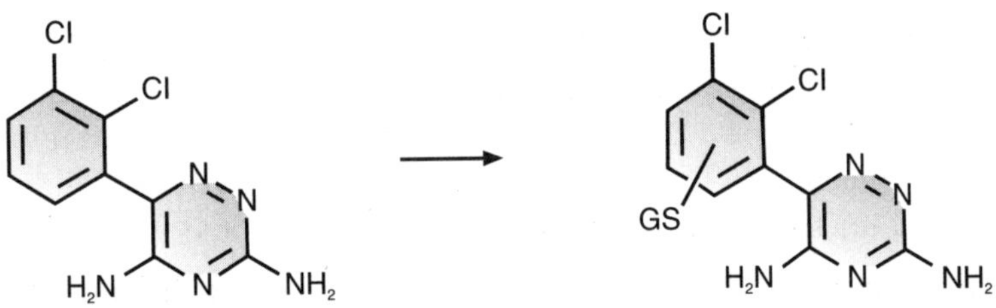

Fig. 7.18. The metabolism of lamotrigine to a glutathione adduct, consistent with an arene oxide intermediate.

Other Types of Electrophiles

Functional groups that are capable of generating this type of electrophile include : nitroaromatics, aromatic amines and hydroxylamines, thioureas, and thiazolidinediones. Nitoaromatics may undergo reduction usually mediated by CYP or CYP reductase. The initial one electron reduced product is the nitro radical anion, which may undergo redox cycling generating active oxygen species. Further reduction generates nitroso and hydroxylamine species, both of which posses electrophilic nitrogen centers. While nitro aromatic groups are generally avoided in drug development for this reason, this metabolic pathway may be encountered inadvertently in natural products. An example is the procarcinogen aristolochic acid, a component of certain Chinese herbs. The herbal product had been taken as a dietary supplement and weight loss agent in some European countries, and has been associated with carcinogenicity and renal nephropathy.

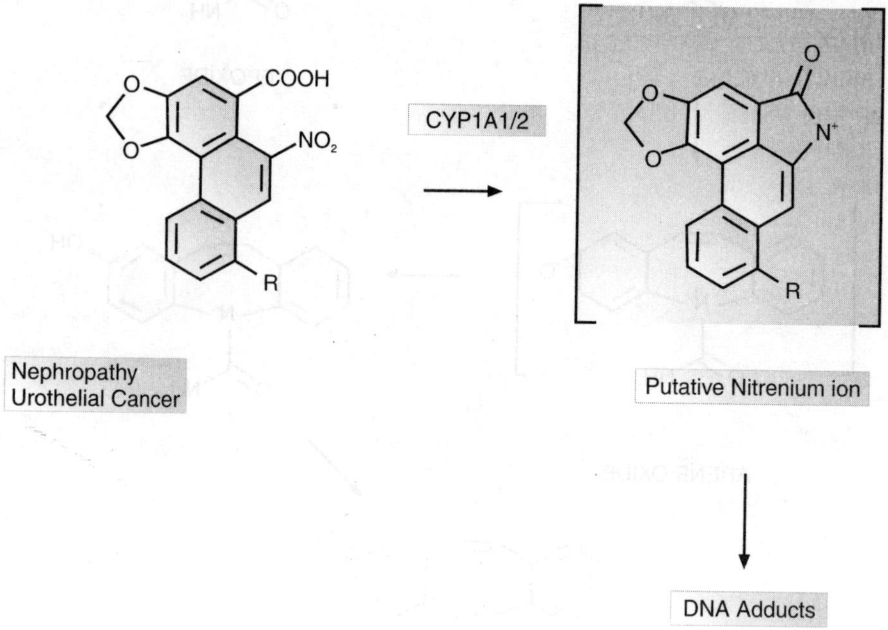

Fig. 7.19. The reduction of the nitro group of aristolochic acid to generate an electrophilic nitrogen center, and its reaction with nucleosides.

Aristolochic acid is an aromatic nitro derivative, and has been shown to undergo CYP mediated reduction to an electrophilic nitroso derivative which subsequently forms DNA adducts, the presumed mechanism for the observed toxicity. Aromatic amines may be oxidized to these same intermediates, and Sulphamethoxazole is probably one of the most studied compounds in this respect. This aromatic amine containing sulphonamide is associated with a range of idiosyncratic reactions (e.g. fever and rash) in 2-3% of the population. Metabolism is known to involve oxidation of the aromatic amine function to hydroxylamine and nitroso derivatives. These have been shown to react with biological nucleophiles, and haptenize lymphocytes and neutrophils.

Other aromatic amines such as acetylaminofluorene and 4-dimethyl-aminobenzene, have well recognized carcinogenic properties. This appears to derive from the formation of hydroxylamines which subsequently undergo sulfation followed by nitrogen/oxygen bond cleavage to form electrophilic nitrogen species, most likely nitrenes. Simple thiourea containing compounds may show hepatic or pulmonary toxicities believed to be associated with bioactivation, and covalent binding of sulfur, although the mechanism has not been clearly defined. Both CYP and FMO mediated metabolism has been implicated. Thioamides will also inhibit thyroid functions, and have been used in the treatment of hyperthyroidism. The major adverse reactions for these drugs are

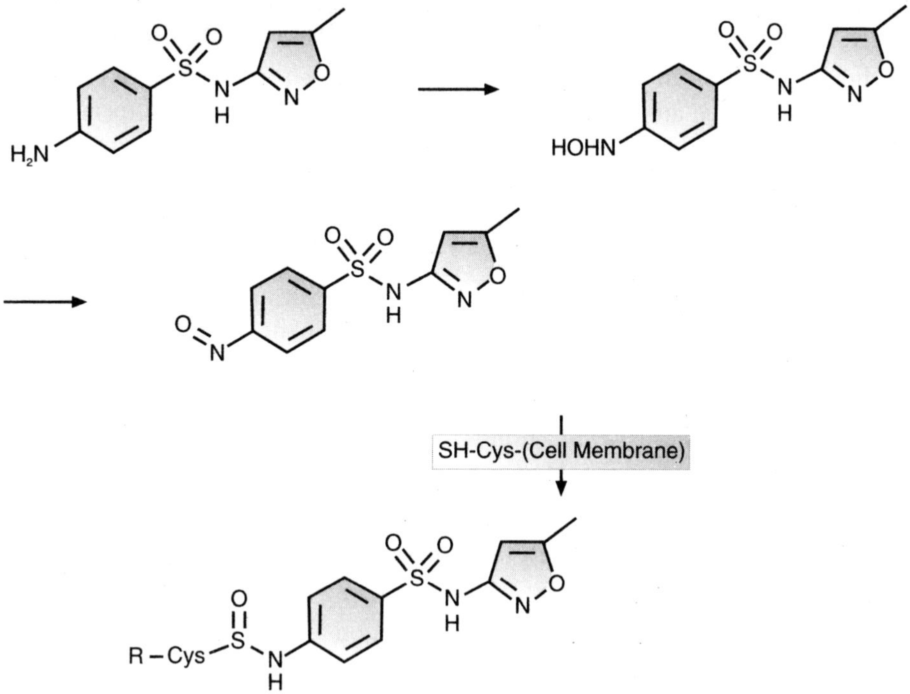

Fig. 7.20. The oxidation of sulfamethoxazole to an electrophilic nitroso metabolite and its reaction with cell surface thiols.

hematological dysfunctions and immunosuppression. The thyrostatic agent propyl thiouracil is associated with idiosyncratic agranulocytosis, and metabolism has been shown to involve oxidation of sulfur to a sulfonate derivative. This is likely to involve the formation of the intermediate oxidation states corresponding to sulfenates and sulfinates.

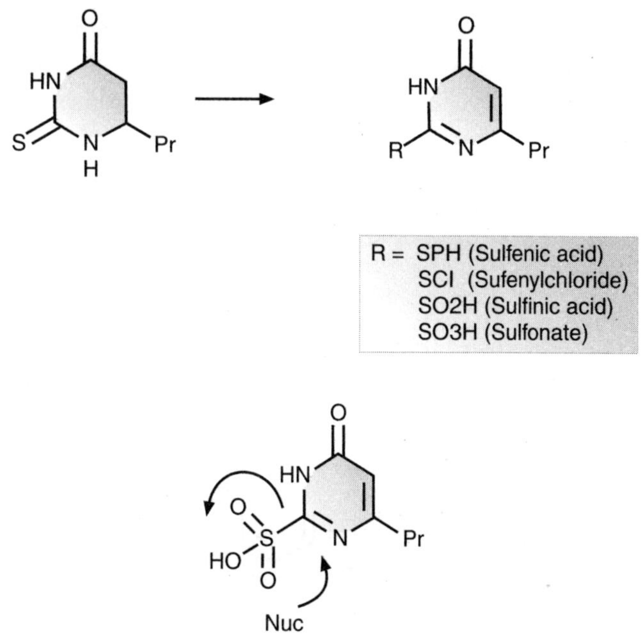

Fig. 7.21. The sulfur oxidation of propylthiouracil to generate electrophilic metabolites.

Sulfenyl chloride formation via myeloperoxidases and hypochlorite was also proposed, and both this, and the sulfenate possess electrophilic sulfur centers. However, the sulfonate metabolite of propylthiouracil itself also appears to be chemically reactive, and has been shown to cause T-cell sensitization. Sulfonates are not usually considered good leaving groups, but in the case of those derived from thioamides their displacement by nucleophiles is precedented. Therefore, reaction of propylthiouracil sulfonate with sulfhydryl groups of proteins, is likely to be the means by which this drug covalently binds to proteins. Thiazolidinediones as discussed above, may undergo oxidative ring opening to form sulfenic acid derivatives, and a glutathione adduct, consistent with the formation of such a metabolite, has been isolated. More recently a different glutathione adduct of troglitazone has been identified, which has also been proposed to derive from sulfur oxidation, but via sulfonium ion formation.

RADICALS AND ELECTROPHILES

While radicals have been widely implicated in many toxicities, the chemical details of the mechanisms associated with these toxicities are less well defined. This is complicated in part by the fact that many radical forming drugs also generate electrophiles, and due to the difficulty in studying radicals in general. Radicals contain one or more unpaired electrons, and may participate in nucleophilc or electrophilic type reactions, but these are mechanistically distinct from the electrophilic processes discussed above. Reactivity can be markedly influenced by orbital symmetry, which in turn is effected by substituents. The reactions of relevance to biological systems involve:

1. Direct reaction with cellular constituents to form a covalent adduct (and a resulting more stable radical).
2. Hydrogen radical abstraction to form a radical based on an endogenous cellular constituent or
3. Direct reaction with a second unpaired electron species, most importantly, molecular oxygen, generating reactive oxygen species and oxidative stress.

It is likely that this last process has the most important toxicological consequences. It should be noted that the first two processes result in the formation of new radicals which potentially could also react with oxygen, and it is perhaps for this reason that the literature frequently does not address this level of mechanistic detail when oxidative stress is observed. Examples of free radicals directly forming covalent adducts with biological molecules include certain compounds noted as mechanism based inhibitors of CYP. Thus the free radical formed during oxidation reacts with the heme group to form a covalent bonds. The particular functionalities capable of this type of

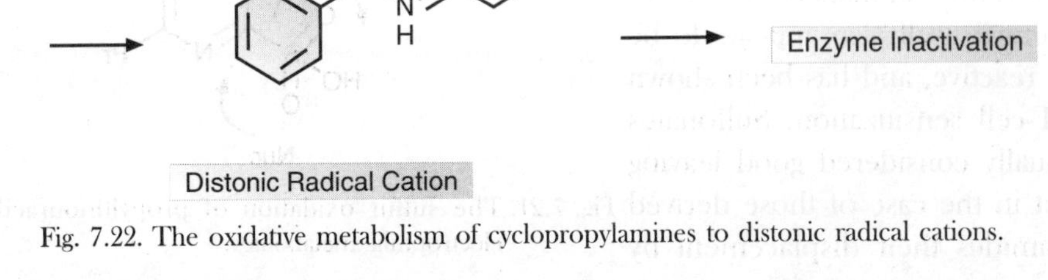

Fig. 7.22. The oxidative metabolism of cyclopropylamines to distonic radical cations.

mechanism based inhibition is in part a reflection the combination of their capacity for one electron oxidation chemistry mediated by CYP, and the highly delocalized nature of the heme prosthetic group, where a more stable radical localized on the heme can be formed, (although a number of different reactions are possible). Functionalities which can cause mechanism based inactivation of CYP by this mechanism include halocarbons, hydrazines, olefins, sydnones, cylopropylamines, and some dihydropyridines and dihydroquinolines.

The cyclopropylamine group is encountered in a number of drugs, and may be bioactivated to form a distonic radical cation. The direct relevance of cyclopropylamines oxidation to toxicity seems to be predominantly associated with a higher risk for drug interactions, since to date there seems to be little evidence that this property is the causative event for other forms of toxicity, idiosyncratic or otherwise. Radicals may react with proteins to abstract a hydrogen radical, and sulfhydryl and tyrosyl groups may be likely sites for this to occur. Subsequent molecular events associated with this reaction and their possible toxicological significance have not been significantly studied, but it seems a likely initial event which may lead to covalent adduct formation, in that reaction with a second drug radical could form a closed shell species.

Thus the mechanism by which radicals may form a covalent adduct to proteins is likely to be a two step process, (in contrast to electrophiles). Perhaps for this reason covalent binding of radicals to proteins with subsequent toxicological significance has been significantly less well demonstrated than for electrophiles. Among the few examples are ethanol, halogenated hydrocarbons, and possibly hydrazines. α Hydroxyl ethyl radicals have been shown to be formed during microsomal incubations with ethanol, and to subsequently covalently bind to proteins. These adducts have also been shown to be immunogenic, and therefore provide an alternative mechanism from acetaldehyde covalent binding, for ethanol induced immune mediated hepatitis. In addition to the oxidative pathways discussed earlier, low molecular weight halocarbons can undergo reductive metabolism to form carbon or chlorine centered radicals and these have been

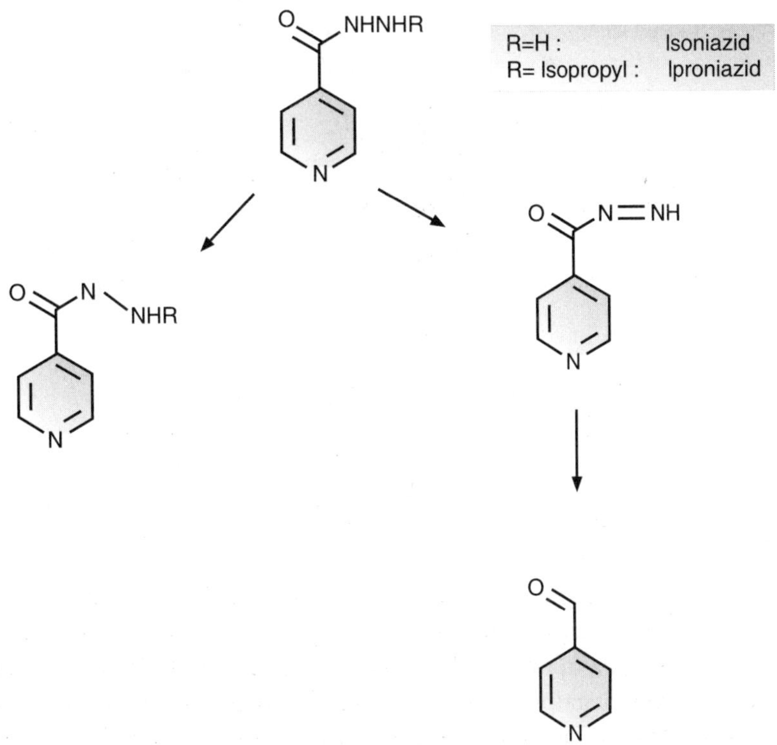

Fig. 7.23. Proposed radical produced by metabolic oxidation of isoniazid and iproniazid.

considered as possible alternative mediators of observed hepatotoxicity. For chloroform, the reductive pathway leading to dichloromethyl radical formation is now thought to have less toxicological relevance than oxidative pathways, since the former pathway requires very high concentrations and anaerobic conditions.

Isoniazid is a drug used to treat tuberculosis, and is associated with idiosyncratic toxicities such as Lupus. It is a hydrazine derivative, and hydrazine itself has been shown to form acetyl radicals following metabolic acetylation in perfused rat livers. Iproniazid was an MAO inhibitor used for depression and was withdrawn from the market due to hepatotoxicity. Isoniazid and iproniazid have been shown to generate free radicals in vitro. Both macromolecular binding and oxidative stress have been proposed as biological consequences. Isoniazid acetyl radicals also appear to form covalent adducts with NAD(H). The reaction of radicals with oxygen to produce reactive oxygen species and their subsequent biological significance has been well reviewed. The formation of superoxide anion and hydrogen peroxide by this process may lead to glutathione depletion and oxidation of cellular components such as lipids (oxidative stress).

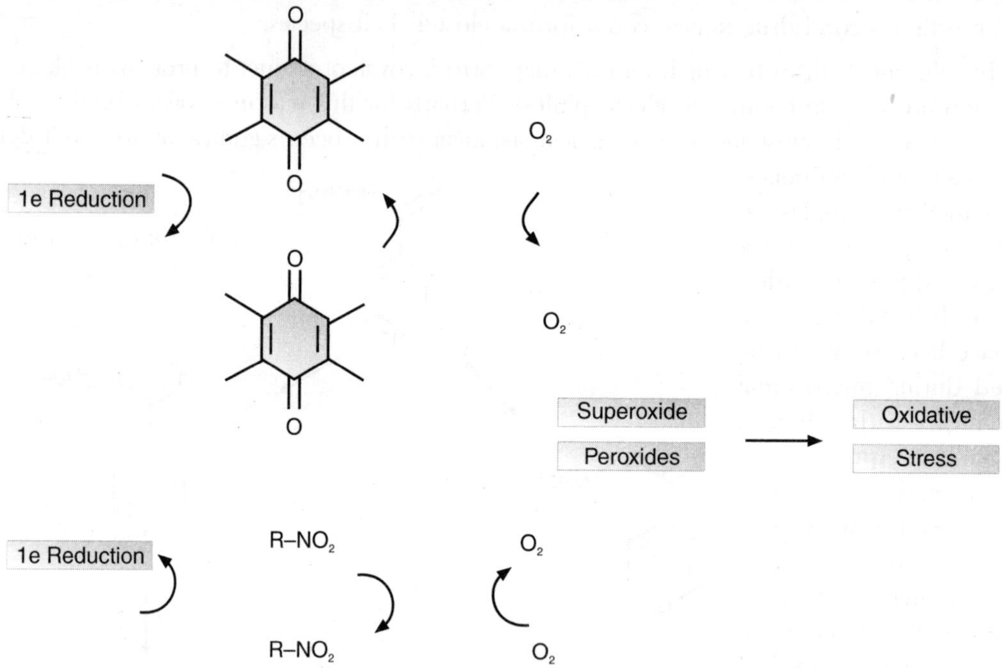

Fig. 7.24. Redox cycling by one electron oxidation of quinones and nitro aromatics, to produce reduced, active oxygen species.

While any radical in principle may potentially react with molecular oxygen, compounds that can reversibly form radical anions are likely to cause greater effect, since the formation of the reduced oxygen species will be in excess of the stoichiometric amount of drug. From a drug development standpoint however, this is mitigated by the fact that a rather limited number of functionalities so far have been noted that are capable of redox cycling by this mechanism: quinones, nitroaromatics, some quaternary ammonium compounds and transition metal complexes. Quinone redox cycling has been most extensively studied with the anthraquinone antineoplastic agents such as Doxorubicin. Long term use of Doxorubicin is associated with cardiomyopathy due to mitochondrial toxicity, and despite continued investigations, reversible quinone reduction leading to oxidative stress remains the most likely mechanism.

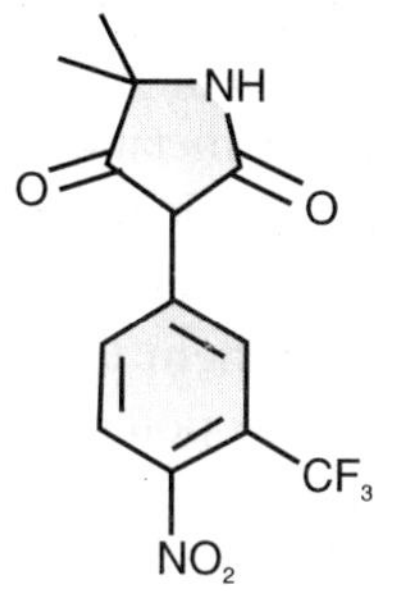

Nilutamide

Therapeutic use: Prostate Cancer
Toxicity: Acute Hepatitis
Mechanism: CYP450 Reductase mediated Nitro reduction

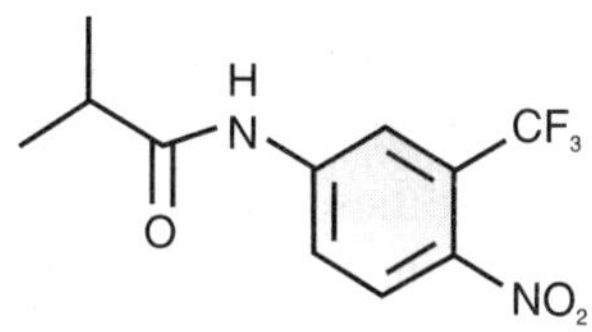

Flutanmide

Therapeutic use: Prostate Cancer
Toxicity: Hepatoxicity
Mechanism: Nitro reduction
Oxidative stress

Fig. 7.25. The structures of nilutamide and flutamide.

The reduction of aromatic nitro compounds may lead to nitrogen centered electrophiles as discussed above, but the initial one electron reduction product is the nitroradical anion which may undergo redox cycling. The antiandrogen Nilutamide is used to treat prostate cancer, is hepatotoxic, and to a lesser extent also a pulmonary toxin. This appears to be due to redox cycling, but the free radical can also undergo further reduction to nitroso and hydroxylamine intermediates. Flutamide is another nitro aromatic antiandrogen associated with hepatotoxicty, and redox cycling via nitro reduction is implicated.

CONCLUDING REMARKS

Reactive intermediates may be grouped according to their reactivity patterns, and reactive metabolites with similar reaction properties may derive from quite different functional groups. Reactive metabolites may be broadly classified as either electrophiles or radicals. At present, electrophiles constitute the more important group due to the wide diversity of structures that can produce these, and are hence more likely to be encountered in drug development. However, as more work focuses on searching for radical intermediates, their implication in toxicity mechanisms is likely to increase.

In all but a few cases, proposed reactive metabolites have not been proven to be the cause of the toxicities to which they are associated, and in many cases a particular compound, or even a particular functional group, is capable of forming more than one type of reactive metabolite. This makes application of this type of information to drug development problematical. Never the less, approaches have been proposed by which this type of information might be applied to minimize safety risk in drug development. The emphasis on bioactivation or covalent binding frequently remains a point of contention in selecting candidates for drug development.

Bioactivation must be put in context with other properties of the compound, (most particularly anticipated dose level), to gauge the level of concern, and in the future is more likely to find value in combination with other types of data. The use of genomic and proteomic information to gauge safety risk in preclinical toxicology will increase significantly in coming years. Thus if changes in the expression of genes associated with the processing of electrophilic metabolites or oxidative stress are observed, knowing if such mechanisms are operating for a particular compound will help significantly in interpreting these findings. Also, if such properties are identified in a lead candidate, it is only by an understanding of the underlying chemistry behind them that such properties can be rationally designed out of follow up compounds. While risk can potentially be minimized by these approaches, managing the consequences of idiosyncratic toxicities will remain a clinical issue for the foreseeable future. Here, genomics may also have significant potential. The feasibility of identifying genetic risk factors for individuals towards a particular idiosyncratic reaction has now been demonstrated, and if this approach is found to be general, may have significant impact in managing these important metabolically mediated toxicities.

8

Drugs for Chronic Heart Diseases

CARDIAC HYPERTROPHY

Aortic Banding in Rats

Blood flow restriction of the aorta in rats induces not only hypertension but also cardiac hypertrophy within several weeks. Angiotensin converting enzyme inhibitors, even at subantihypertensive doses, but not other antihypertensive drugs, inhibit cardiac hypertrophy.

Method

Male Sprague Dawley rats weighing 270-280g are fasted 12h before surgery. Anesthesia is induced by i.p. injection of 200 mg/kg hexobarbital. The abdomen is shaved, moistened with a disinfectant and opened by a cut parallel to the linea alba. The intestine is moistened with saline and placed in a plastic cover to prevent desiccation. The aorta is prepared free from connective tissue above the left renal artery and underlaid with a silk thread. Then, a cannula no. 1 (0.9 × 40 mm) is placed longitudinally to the aorta and both aorta and cannula are tied. The cannula is removed, leaving the aortic lumen determined by the diameter of the cannula. The intestine is placed back into the abdominal cavity with the application of 5.0mg rolitetracycline (Reverin). In sham-operated controls no banding is performed. The skin is closed by clipping. The animals are treated once daily over a period of 6 weeks with doses of the ACE-inhibitor or other anti-hypertensive drugs found previously effective to lower blood pressure in rats. At the end of the experiment blood pressure is measured under hexobarbital anesthesia (200 mg/kg i.p.) via indwelling catheters in the left carotid artery.

Blood pressure measurement in conscious rats with the convential tail-cuff method is not possible due to the large pressure difference across the ligature. Therefore, only one measurement at the end of the study is possible. The hearts are removed, rinsed in saline until free of blood and gently blotted to dryness. Total cardiac mass is determined by weighing on an electronic balance to the nearest 0.1 mg. The atria and all adjacent tissues are trimmed off and the weight of the left ventricle including the septum as well as the remaining cardiac tissue representing the right ventricle are determined separately. Weights are calculated per 100 g body weight.

Assessment

Total cardiac mass, weight of left and right ventricle of treated rats are compared with operated controls and sham-operated controls.

Latest technology

Uetmasu et al. (1989) described a simple method for producing graded aortic insufficiencies in rats and subsequent development of cardiac hypertrophy. Selective perforation of the right cup of the aortic valve or in combination with that of the left valve cup was performed using a plastic rod inserted from the right common carotid artery. Hypertrophy of the heart, but no hypertension or cardiac insufficiency, was observed. Similar methods were used by Yamazaki et al. (1989) to study the alterations of cardiac adrenoceptors and calcium channels subsequent to aortic insufficiency, by Umemura et al. (1992) to study baroreflex and β-adrenoceptor function and by Ishiye et al. (1995) to study the effects of an angiotensin II antagonist on the development of cardiac hypertrophy due to volume overload. Hyperplastic growth response of vascular smooth muscle cells in the thoracic aorta was found following induction of acute hypertension in rats by aortic coarctation by Owens and Reidy (1985).

Changes in cardiac gene expression during compensated hypertrophy and the transition to cardiac decompensation in rats with aortic banding were studied by Feldman et al. (1993). Muders et al. (1995) produced aortic stenosis in rats by placing a silver clip (inner diameter 0.6 mm) on the ascending aorta. Schunkert et al. (1995) studied alteration of growth responses in established cardiac pressure overload hypertrophy in rats with aortic banding. Prevention of cardiac hypertrophy after aortic banding by ACE inhibitors probably mediated by bradykinin could be shown. Weinberg et al. (1997) studied the effect of angiotensin AT_1 receptor inhibition on hypertrophic remodelling and ACE expression in rats with pressure- overload hypertrophy due to ascending aortic stenosis.

Chronic Heart Failure in Rats

Rat models of heart failure were reviewed by Muders and Elsner (2000). Chronic heart failure can be induced in rats by occlusion of coronary arteries. One of the first reports was by Selye et al. (1960). More recent reports are by Pfeffer et al. (1979), Hodsman et al. (1988), Van Veldhuisen et al. (1994, 1995), Kajstura et al. (1994), Gomez et al. (1997), Liu et al. (1997a, 1997b), and Jadavo et al. (2005). Itter et al. (2004) described a model of chronic heart failure (CHF) in spontaneously hypertensive rats.

Method

Adult male 4-month-old SHR/NHsd and WKY/NHsd rats weighing 250-300 g were used. Cardiovascular failure was induced by permanent (8 weeks) occlusion of the left coronary artery 2 mm distal to the origin from the aorta resulting in a large infarction of the free left ventricular wall. Eight weeks after surgery, parameters indicating CHF were measured. Cardiac hypertrophy, function, and geometric properties were determined by the "*working heart*" mode and *in vivo* determinations by MRI and heart weight. Hydroxyproline/proline ratio was measured as an indicator of heart fibrosis.

Operative technique

The rats were anesthetized with a mixture of ketamine/xylazine (35/2 mg/kg) i.p. The left ventrolateral thorax was shaved and prepared to create a disinfected surgical access area. When a stable anesthesia was achieved the animals were placed on a small animal operation table, intubated and ventilated with room air using a small animal ventilator (KTR-4). The level of anesthesia was deemed as adequate following loss of the pedal withdrawal reflex and absence of the palpebral reflex. The tidal volume was adjusted at 3-5 ml and the ventilation rate was 40 breaths/min. Left thoracotomy was performed via the third intercostal space. The heart was exposed and the pericardium opened. The left main coronary artery was ligated with Perma-Hand silk 4-0 USP near its origin at the aorta (2 mm distal to the edge of the left atrium).

Ligation resulted in infarction of the free left ventricular wall. Ligation was deemed successful when the anterior wall of the left ventricle turned pale. At this point the lungs were hyperinflated by increasing the positive end expiratory pressure, and the chest was closed. The rats were placed on a heating pad. They were continuously monitored until they start moving in their cages. To avoid ventricular arrhythmias, lidocaine (2 mg/kg i.m.) was given before surgery. The sham procedure consisted of opening the pericardium and placing a superficial suture in the epicardium of the LV. To prevent acute lung edema, the rats received furosemide 2 mg/kg twice daily for 3 days via the drinking water.

Interpretation

Before killing the animals 8 weeks after MI, non-invasive sequential nuclear magnetic resonance (NMR) measurements of heart geometric properties were done. Thereafter the animals were anesthetized with pentobarbitone (180 mg/kg i.p.) and subsequently heparinized (Heparin sodium 500 IU/100 g body weight i.p.). Once stable anesthesia was achieved (stage III 3, reflexes absent), the animals were connected to an artificial respirator via a PE tube inserted into the trachea and ventilated with room air. The right carotid artery was cannulated with a polyethylene catheter to monitor mean blood pressure, systolic blood pressure, diastolic blood pressure and heart rate over a stable time course of 10 min. A transverse laparotomy and a right anterolateral thoracotomy were performed, and the heart was rapidly removed for the evaluation of its function in the working heart mode. Thereafter the heart weight, and the left and right ventricular weights were determined. For infarct size determination the left ventricle was sectioned transversely into four slices from the apex to the base. The infarct size was determined by planimetry and expressed as a percentage of LV mass. Lung weight and further lung histology sections were evaluated. Hydroxyproline/proline ratio was determined in paraffin-embedded slices of the left ventricle.

Monitoring device

The animals were monitored by MRI at day 7 and day 42 post-MI. The rats were anesthetized with a mixture of 1% halothane and 30/70N_2O/oxygen with a specially manufactured rat mask. The fully anesthetized rats (phase III 3) were placed on a cradle made of Plexiglas in a supine position. Respiration and ECG were monitored continuously. MRI experiments were performed according to Rudin et al. (1991). The images were acquired by a spinecho sequence SE (500/20), the field of view was 50 mm, the image resolution was 256 × 256 pixels with a dimension of 0.2 × 0.2

mm. Four adjacent transverse slices were recorded; slice thickness was 1.5 mm. Before the acquisition of data, a coronary pilot scan was measured for adequate positioning of the transverse slices. MRI data acquisition was gated to the cardiac cycle by a Physiograd SM 785 MR monitoring system. Two sets of transverse images were acquired, one at end-systole and another at end- diastole. End-diastole was defined as the image obtained 8 ms after the onset of the R wave of the ECG, corresponding to the largest cavity area.

End-systole was defined as the image with the smallest LV cavity area. The image analysis was done using Bruker software. The parameters of left ventricular end-diastolic volume (LVEDV), left ventricular end-systolic volume (LVESV), septum size, infarct size, ejection fraction (EF), left ventricular chamber diameter (*r*) and circumference were measured. EF was estimated in percentage terms by the subtraction LVEDV-LVESV. After the procedure the rats were ventilated with oxygen, the mask was replaced, and they were brought back into their cages. They were monitored until they started moving in the cage.

Cardiac drugs

The animals were anesthetized with pentobarbitone (180 mg/kg i.p.) and subsequently heparinized (Heparin sodium 500 IU/100 g body weight i.p.). Once stable anesthesia was achieved, the animals were connected to an artificial respirator via a PE tube inserted into the trachea and ventilated with room air. The right carotid artery was cannulated with a polyethylene catheter. The catheter was connected to a PLUGSYS measuring system to monitor mean blood pressure, systolic blood pressure, diastolic blood pressure and heart rate over a stable time course of 10 min.

Cardiac cycle

For the final investigations, the heart of the anesthetized rat was rapidly removed and immersed in physiological buffer chilled to 4°C. The aorta was dissected free and mounted onto a cannula (internal diameter: 1.4 mm) attached to a perfusion apparatus. The hearts were perfused according to the method of Langendorff with an oxygenated (95% O_2/5% CO_2) non-circulating Krebs-Henseleit solution of the following compositions (mM): NaCl, 118; KCl, 4.7; $CaCl_2$, 2.52; $MgSO_4$, 1.64; $NaHCO_3$, 24.88; KH_2PO_4, 1.18; glucose, 5.55; and Na-pyruvate, 2.0 at a perfusion pressure of 60 mmHg. Any connective tissue, thymus or lung was carefully removed. A catheter placed into the pulmonary artery drained the coronary effluent perfusate that was collected for the determination of coronary flow and venous pO_2 measurements. The left atrium was cannulated via an incision of the left auricle. All pulmonary veins were ligated close to the surface of the atria. When a tight seal with no leaks had been established and after a 15-min equilibration period, the hearts were switched into the working mode, using a filling pressure (preload) of 12 mm Hg in WKY/NHsd and 18 mm Hg in SH rats.

The after load pressure was 60 mm Hg in WKY/NHsd and 80 mm Hg in SH rats. After validation of the basis parameters the afterload pressure was enhanced in a cumulative manner from an additional 20 mm Hg to 140 mm Hg. Thereafter the isovolumetric maxima were determined by enhancing the preload pressure in steps of 5 mm Hg to 30 mm Hg. Flow and pressure signals for computation were obtained from the PLUGSYS-measuring system. Computation of data was performed with a sampling rate of 500 Hz, averaged every 2 s, using the software Aquire Plus V1.21f.

Evaluating external heart work

After the evaluation of the external heart work, the total heart weight, the left and right ventricular weights were determined. The left ventricle was then sectioned transversely into four slices from the apex to the base. Eight pictures were taken of each rat heart, two from each slice. Total infarct size was determined by planimetry of the projected and magnified slices. The area of infarcted tissue as well as the intact myocardium of each slice were added together and averaged. The infarcted fraction of the left ventricle was calculated from these measurements and expressed as a percentage of the LV mass. The left ventricular perimeter, diameter, infarct scar length, as well as wall thickness and infarct wall thinning were determined as well. According to Pfeffer et al. (1985) and Pfeffer and Pfeffer (1987), rats with infarct sizes <20% and >40% were excluded from the study.

Pulmonary histomorphology

After lung weight determination, the organ was immersed in 4% formalin (pH 7.0-7.5; 0.1 M). The lung was cut into small pieces, dehydrated and embedded in paraffin. Hematoxylin and eosin (HE) sections were evaluated by light microscopy.

Hydroxyproline/proline ratio

After embedding, the rest of the fixed left ventricular tissue was freeze-dried. Proline and hydroxyproline was then analyzed according to the method of Lopez de Leon and Rojkind (1985) and the ratio of both were calculated.

Latest technique

Jain et al. (2000) studied the effects of angiotensin II receptor blockade after coronary ligation and exercise training on treadmill in rats. Medvedev and Gorodetskaya (1993) induced heart failure in rats by micro-embolization of coronary vessels with 15-μm plastic microspheres. Katona et al. (2004) found that selective sensory denervation by capsaicin aggravates adriamycin-induced cardiomyopathy in rats. A simple and rapid method of developing high output heart failure and cardiac hypertrophy in rats by producing aortacaval shunts was described by Garcia and Diebold (1990). Rats weighing 180-200 g were anesthetized with 30 mg/kg i.p. pentobarbitone. The vena cava and the abdominal aorta were exposed by opening the abdominal cavity via a midline incision.

The aorta was punctured at the union of the segment two-thirds caudal to the renal artery and one-third cephalic to the aortic bifurcation with an 18-gauge disposable needle. The needle was advanced into the aorta, perforating its adjacent wall and penetrating in the vena cava. A bulldog vascular clamp was placed across the aorta caudal to the left renal artery. Once the aorta was clamped, the needle was fully withdrawn and a drop of cyanoacrylate glue was used to seal the aorta-punctured point. The clamp was removed 30s later. The patency of the shunt was verified visually by swelling vena cava and the mixing of arterial and venous blood. The peritoneal cavity was closed with silk thread stitches and the skin with metallic clips.

Rats with aorta-caval shunts developed cardiac hypertrophy with significantly higher absolute and relative heart weights. Other studies with aortocaval shunts in rats were published by Flaim et al. (1979) and Liu et al. (1991). Isoyama et al. (1988) studied myocardial hypertrophy after creating aortic insufficiency in rats. Terlink et al. (1998) studied ventricular dysfunction in rats with diffuse isoproterenol-induced myocardial necrosis. Sabbah et al. (1991) described a model of chronic heart failure produced by multiple sequential coronary microembolizations in dogs.

INVESTIGATING CARDIAC HYPERTROPHY AND INSUFFICIENCY IN MICE

Cardiac Hypertrophy in Mice

Rockman et al. (1991, 1993) developed a model of ventricular hypertrophy in the intact mouse by use of microsurgical techniques.

Method

Eight-week-old adult mice weighing 18-22 g are anesthetized by intraperitoneal injection of a mixture of 100 mg/kg ketamine, 5 mg/kg xylazine, and 2.5 mg/kg morphine. Animals are placed under a dissecting microscope in the supine position and a midline cervical incision is made to expose the trachea and carotid arteries. After endotracheal intubation, the cannula is connected to a volume cycled rodent ventilator on supplemental oxygen with a tidal volume of 0.2 ml and a respiratory rate of 110 per min. Both left and right carotid arteries are cannulated with flame stretched PE50 tubing. Catheters are connected to modified P50 Statham transducers. The chest cavity is entered in the second intercostal space at the left upper sternal border through a small incision and the thymus is gently deflected out of the field of view to expose the aortic arch.

After the transverse aorta is isolated between the carotid arteries, it is constricted by a 7.0 nylon suture ligature against a 27-gauge needle, the latter being promptly removed to yield a constriction of 0.4mm diameter and provide a reproducible transverse aortic constriction of 65-75%. The hemodynamic effects of acute and chronic constriction are followed by monitoring the pressure gradient between the two carotid arteries in anesthetized animals. Systolic and mean arterial pressure at baseline, during total occlusion when the ligature is tied, and early (15 min) and late (7 days) after transverse aortic constriction are recorded.

The increase in systolic pressure provides an adequate mechanical stimulus for the development of cardial hypertrophy. To confirm myocardial hypertrophy, both sham- operated and aortic-constricted hearts are examined 7 days after operation. Hearts examined for *cell size* are perfused with 4% paraformaldehyde/1% glutaraldehyde through the apex, immersed in osmium tetroxide, dehydrated in graded alcohols, and embedded in araldite.

Tissue blocks are sectioned at a thickness of 1 μm, mounted on slides, and stained with toluidine blue. Cell areas are measured by manually tracing the cell outline on an imaging system connected to a computer. At the end of the experiment, mice were sacrificed in anesthesia, heart excised and weighed, the atria and ventricles separately frozen in liquid nitrogen for Northern blot analysis. Total RNA is extracted by a single step extraction with guanidinium thiocyanate. The RNA is size fractionated by agarose gel electrophoresis, transferred to nylon membranes by vacuum blotting, and hybridized with the appropriate complementary DNA probes labeled with ^{32}P by random priming to a specific activity of $0.95\text{-}1.2 \times 10^6$ cpm/ng.

Assessment

Variables measured are expressed as mean ± SD. Statistical significance of differences between sham-operated and thoracic aortic-constricted animals is assessed by Student's *t*-test.

Latest technology

Dom et al. (1994) studied myosin heavy chain regulation and myocytes' contractile depression after LV hypertrophy in aortic-banded mice. Okada et al. (2004) subjected mice to transverse aortic constriction. Echocardiographic analysis demonstrated cardiac hypertrophy and failure 1 and 4 weeks after surgery. Cardiac expression of endoplasmatic reticulum chaperones was significantly increased indicating that pressure overload by transverse aortic constriction induced prolonged endoplasmatic reticulum stress.

Chronic Heart Failure in Mice

Several authors reported the development of murine models of cardiac failure. Xu et al. (2004) studied cardioprotection in mice with heart failure by dual inhibition of angiotensin converting enzyme (ACE) and neutral endopeptidase (NEP).

Method

Mice with a targeted deletion of the B_2 kinin receptor gene or C57BL/6J mice at an age of 10-12 weeks were anesthetized with 50 mg/kg sodium pentobarbital i.p., intubated and ventilated with room air using a positive-pressure respirator. A left thoracotomy was performed via the fourth intercostal space; the lungs were retracted to expose the heart, and the pericardium was opened. The left anterior descending coronary artery was ligated with an 8-0 nylon suture near its origin between the pulmonary outflow tract and the edge of the left atrium.

Acute myocardial ischemia was considered successful when the anterior wall of the left ventricle turned pale and an obvious ST segment elevation was observed. The lungs were inflated by increasing positive end-expiratory pressure and the thoracotomy site was closed. Sham-operated mice were subjected to the same procedure except that the suture around the left anterior coronary artery was not tied. Systolic blood pressure was measured in conscious mice using a non-invasive computerized tail-cuff system. Cardiac geometry and function were evaluated with a Doppler echocardiographic system. LV diastolic dimension was measured and ejection fraction was calculated from

$$[(LVAd - LVAs)/LVAd] \times 100$$

where LVAd is LV diastolic area and LVAs is LV systolic area.

Four weeks after surgery, each strain was separated into one group treated with an ACE inhibitor, one group treated with a NEP inhibitor, one group treated with both inhibitors and one control group. All drugs were administered in drinking water for 20 weeks. At the end of the study, all mice were anesthetized with pentobarbital and the heart stopped at diastole by intraventricular injection of 15% KCl. The heart, lungs, and liver were weighed to assess hypertrophy and congestion. Infarct size was determined by Gomori trichrome staining and expressed as the ration of the infarcted portion to total LV circumference. Sections (6 μm) from each slice were double-stained with fluorescein-labeled peanut agglutinin to delineate the myocyte cross-sectional area and interstitial space and rhodamine-labeled *Griffonia simplicifolia* lectin I to show the capillaries. To calculate interstitial collagen fraction, the total surface area (microscopic field), interstitial space (collagen plus capillaries), and area occupied by capillaries alone were measured by computer-assisted videodensometry. After 20 weeks of treatment, plasma renin was measured.

Assessment

Data were expressed as mean ± SE. Mortality rates were compared using χ^2 tests. For the echo, blood pressure, heart weight, lung weight, infarct size, plasma renin concentration, and histology data, paired or two-sample tests using non-parametric methods were used to perform all comparisons of interest.

Heart Failure in Transgenic Mice

Purpose and rationale

Several hundreds of papers on transgenic mice and heart failure are published. Only a few can be mentioned here. Chien (1995) described cardiac muscle diseases in genetically engineered mice. Edwards et al. (1996) described severe cardiomyopathy in transgenic mice overexpressing the skeletal muscle myogenic regulator *myf5*. Arber et al. (1997) found that MLP-deficient mice exhibit a disruption of cardiac cytoarchitectural organization, dilated cardiomyopathy, and heart failure. Graham et al. (1997) described a mouse model for mitochondrial myopathy and cardiomyopathy resulting from a deficiency in the heart/muscle isoforms of the adenine nucleotide translocator. Iwase et al. (1997) studied cardiomyopathy in transgenic mice induced by overexpression of the cardiac stimulatory G protein α subunit. Knollmannn et al. (2000) reported remodeling of ionic currents in hypertrophied and failing hearts of transgenic mice overexpressing calsequestrin. Beggah et al. (2002) described reversible cardiac fibrosis and heart failure induced by conditional expression of an antisense mRNA of the mineralocorticoid receptor in cardiomyocytes.

Verheule et al. (2004) found increased vulnerability to atrial fibrillation in transgenic mice with selective atrial fibrosis caused by overexpression of TGF-β1. Duncan et al. (2005) found that chronic xanthine oxidase inhibition prevents myofibrillar protein oxidation and preserves cardiac function in a transgenic mouse model of cardiomyopathy. Hartil and Charron (2005) reviewed mouse models where transgenic technology has been utilized to alter expression of genes involved in cardiac uptake and metabolism of either lipid or carbohydrate. Hilfiker-Kleiner et al. (2005) reported that STAT3-knockout mice harboring a cardiomyocyte-restricted deletion of STAT3 showed enhanced susceptibility to cardiac injury caused by myocardial ischemia, systemic inflammation, or drug toxicity. Sanbe et al. (2005) studied reversal of amyloid-induced heart disease in desmin-related cardiomyopathy.

CARDIAC HYPERTROPHY IN GUINEA PIGS

Congestive heart failure in man is characterized by cardiac hypertrophy, peripheral edema, lung and liver congestion, dyspnea, hydrothorax and ascites. Effective treatment is achieved by cardiac glycosides. Based on techniques reported by Selye et al. (1960) a method was developed to induce congestive heart failure in guinea pigs with symptoms very close to human pathology.

Method

Male guinea pigs weighing 250-400 g are used. The fur at the ventral thorax is shaved and the skin disinfected. The animal is anesthetized with ether. The skin is cut with scissors on the left side at a length of 4 cm. The left musculus pectoralis is cut at the costal insertion and elevated. The forth intercostal space is opened with two blunted forceps. The heart is pressed against the opening with the left hand. The pericardium is opened with a fine forceps and pulled back to the basis of the heart. The beating heart is extruded from the thorax wound by pressure with the left hand on the right thorax wall. A ring-shaped clamp covered with a thin rubber tube is placed around the basis of the heart keeping the heart outside of the thorax without closing off the blood circulation. A thread soaked with diluted disinfectant solution is placed as a loop around the apex of the heart and tightened so that the apical third of both ventricles is tied off. The degree of tightening of the loop is essential.

Complete interruption of blood supply to the apical third resulting in necrosis has to be avoided as well as the loop's slipping off. Technical skill is necessary to place the loop around the beating heart into the correct position. After removal of the clamp the heart is placed back, the incision between the forth and fifth costal rib closed and the musculus pectoralis placed over the wound. Intrathoracal air forming a pneumothorax is removed by pressure on both sides of the thorax. After application of an antibiotic emulsion the skin wound is closed. The surgical procedure has to be finished within a short period of time. The animals develop symptoms of severe congestive heart failure with a death rate of 80% within 14 days. Lung weight and relative heart weight are significantly increased. Exudate in the thorax cavity and ascites amount between 3.5 and 7.5 ml with extreme values of 17.5 ml. Lung edema and liver congestion are found histologically. Peripheral edema and preterminal dyspnea and tachypnea are observed. When treated with various doses (0.1 to 100 μg/kg) of cardiac glycosides s.c. or i.m. over a period of 14 days the symptoms of cardiac insufficiency, e.g., volumes of transudate as well as death rate, are dose-dependent diminished.

Assessment

From survival rate, ED_{50} values of cardiac glycosides can be calculated which are in the same dosage range as therapeutic doses in man. The experimental model in guinea pigs reflects very closely the symptoms of cardiac insufficiency in man, e. g., lung congestion, hydrothorax, liver congestion, ascites, peripheral edema and cardiac hypertrophy. The therapeutic potency of cardiac glycosides can be evaluated with this method. Additional factors being known to enhance the symptoms of congestive heart failure in man, like salt load and diphtheria toxin, further increase mortality and hydropic symptoms. The method can be used for special purposes, however, it needs considerable training and technical skill.

Latest technology

Siri et al. (1989, 1991) produced left ventricular hypertrophy in the guinea pig by gradually increasing ventricular afterload. A mildly constricting band was placed around the ascending aorta of very young guinea pigs (225-275 g). With growth to 500-1000 g, left ventricular systolic pressure increased and ventricular hypertrophy developed. Only some of the animals developed dyspnea and severe ventricular dysfunction. Kiss et al. (1995) studied the effects on Ca^{2+} transport and mechanics in compensated pressure-overload hypertrophy and congestive heart failure in guinea pigs. The descending aorta was banded for 4 and 8 weeks in adult guinea pigs. Tweedle et al. (1995) assessed subrenal banding of the abdominal aorta as a method of inducing cardiac hypertrophy in the guinea pig. Pfeffer et al. (1987) induced myocardial infarction in rats by ligation of the left coronary artery and found hemodynamic benefits and prolonged survival with long-term captopril therapy. Acute ischemic left ventricular failure can be induced in anesthetized dogs by repeated injections of plastic microspheres into the left coronary artery. Huang et al. (1997) created congestive heart failure in sheep by selective sequential intracoronary injection of 90 μm microspheres under 1.5% isofluorane injection.

INVESTIGATING CARDIOMYOPATHIC IN SYRIAN HAMSTER

Cardiomyopathy in Syrian hamsters has been described by Bajusz et al. (1966), Bajusz and Lossnitzer (1968), Bajusz (1969), Bajusz et al. (1969a, b), Homburger and Bajusz (1970), Gertz (1972). The disease originates from an autosomal, recessively transmissible disorder, which leads to degenerative lesions in all striated muscles and in particular in the myocardium. Histopathological changes consist of myocytolytic necrosis followed by fibrosis and calcification. The evolution of the cardiomyopathic disease can be characterized by five distinct phases: A prenecrotic stage, in which no pathology is evident, a time of active myocytolysis and cellular necrosis, a phase of fibrosis and calcium deposition, an overlapping period of reactive hypertrophy of the remaining viable myocytes, and a final stage of depressed myocardial performance and failure.

Method

The model of cardiomyopathy in Syrian hamsters has been used by several authors. One has to note, that several strains of cardiomyopathic hamsters have been used: strain Bio 53:58 by Capasso et al. (1989, 1990) and by Chemla et al. (1992, 1993), strain BIO 14.6 by Tapp et al. (1989) and by Sen et al. (1990), strain CHF 146CM by van Meel et al. (1989) and by Haleen et al. (1991), strain BIO82.62by ver Donck et al. (1991), strain J-2-N by Kato et al. (1992), strain CHF 147 by Desjardins et al. (1989), Hanton et al. (1993). Various experimental protocols have been described. Most authors use survival rate and heart weight as end point. Generally, the experiments are started with animals at an age of 120 to 200 days.

Capasso et al. (1989, 1990) studied the mechanical and electrical properties of cardiomyopathic hearts of Syrian hamsters using isolated left ventricular posterior papillary muscles. Tapp et al. (1989) tested stress-induced mortality in cardiomyopathic hamsters by five consecutive daily 2-h periods supine immobilizations at 4°C. Sen et al. (1990) tested the inotropic and calcium kinetic effects of calcium channel agonists and antagonists in primary cultures of isolated cardiac myocytes. Haleen et al. (1991) tested the effects of an angiotensin converting enzyme inhibitor not only on survival, but also on left ventricular failure in the isolated Langendorff heart by measurement of left ventricular end-diastolic pressure, dP/dt_{max} and mean coronary flow. Dixon et al. (1997) tested the effect of an AT_1 receptor antagonist on cardiac collagen remodelling in the cardiomyopathic Syrian hamster.

In addition to the effects on left ventricular papillary muscles strips, Chemla et al. (1992) tested the effects on diaphragm contractility in the cardiomyopathic Syrian hamster. Whitmer et al. (1988) and Kuo et al. (1992) tested sarcolemmal and sarcoplasmatic reticulum calcium transport in the cardiomyopathic Syrian hamster. Nigro et al. (1997) identified the Syrian hamster cardiomyopathy gene. Tanguay et al. (1997) tested the coronary and cardiac sensitivity to a vasoselective benzothiazepine-like calcium antagonist in isolated, perfused failing hearts of Syrian hamsters. Bilate et al. (2003) recommended the Syrian hamster as a model for the dilated cardiomyopathy of Chagas' disease. Female hamsters were infected via the intraperitoneal route with *Trypanosoma cruzi* Y strain blood trypomastigotes. Survival was monitored, echocardiography was performed after 4 and 12 months, and histopathological examinations were carried out at the end of the study period.

Assessment

Positive effects of various drugs have been found in the cardiomyopathic hamster, such as cardiac glycosides, inotropic compounds, beta-blockers, calcium antagonists, and ACE-inhibitors. The specificity of the effects has to be challenged.

Latest technology

The tight-skin (TSK) mouse is a genetic model of pulmonary emphysema connected with right ventricular hypertrophy.

INVESTIGATING CARDIAC FAILURE IN RABBITS

Rabbit models of heart failure were reviewed by Muders and Elsner (2000). Rapid pacing was used by Masaki et al. (1994), Porsa et al. (1994), Eble et al. (1998), Li et al. (2003), and Rose et al. (2005); coronary artery ligation by Pennock et al. (1997), Currie and Smith (1999), Romanic et al. (2001), and Miller et al. (2004); combined pressure and volume overload by Ezzaher et al. (1991), Mohammadi et al. (1997), Dekker et al. (1998), and Baartscheer et al. (2003a, 2003b); aortic insufficiency and aortic constriction by Bouanani et al. (1991) and Pogwizd et al. (1999); regurgitation after damage of the mitral valve by Gunawardena et al. (1999); regurgitation after aortic valve destruction by Magid et al. (1988, 1994), Yoshikawa et al. (1993), King et al. (1997), and Liu et al. (1998); Luchner et al. (2001) used a rabbit model of progressive left ventricular dys-function to investigate differential expression of cardiac atrial natriuretic peptide and brain natriuretic peptide. Ventricular pacing-induced heart failure could be induced with a transvenously implanted pacemaker system.

Method

Male rabbits (chinchilla bastard) underwent implantation of a programmable cardiac pacemaker. Under anesthesia (ketamine 60 mg/kg xylazine 5 mg/kg i.m.), the right internal jugular vein was dissected and cannulated with a single-lumen central venous catheter. The catheter was then advanced into the right ventricle under pressure guidance. A transvenous screw-in pacemaker lead (Medtronic) was advanced through the catheter into the ventricular apex and implanted endocardially. The pacemaker was implanted subcutaneously into the right abdominal wall and the pacemaker lead was connected subcutaneously with the pacemaker. Rapid ventricular pacing-induced heart failure could be induced with a transvenously implanted pacemaker system.

All rabbits were allowed to recover for at least 10 days after surgery before the pacemaker was started for the induction of heart failure. Proper pacemaker function was checked intraoperatively, at the time of programming, and subsequently all 10 days. Rabbits (CHF group) underwent pacing with a stepwise increase of stimulation frequencies over 30 days. During the first 10 days, animals were paced at 330 beats/min (bpm). This protocol results in ELVD, as defined by significant LV systolic dysfunction with cardiac enlargement and decreased perfusion pressure but no clinical signs failure.

The pacing rate was then increased to 360 bpm for 10 days and 380 bpm for another 10 days and ELVD evolved to CHF with further cardiac enlargement and further decreased perfusion pressure together with clinical signs of fluid retention (ascites). At baseline (control), after being paced at 330 bpm for 10 days (ELVD) and at the end of the protocol (CHF), conscious arterial pressure was measured invasively via the medial ear artery and a 2-D guided M-mode echocardiogram was obtained. At the end of the pacing protocol, rabbits were killed by i.v. euthanasia and tissue was rapidly harvested. Hearts were trimmed on ice, snap frozen in liquid nitrogen and stored at -80°C until further processing.

Echocardiography

A long and short-axis echocardiogram (HP Sonos 5500, 12 MHz probe) was performed under light sedation (5 mg midazolam i.m.) in a supine position from the left parasternal window. LV end-diastolic (LVEDd) and end-systolic (LVESd) dimensions and diastolic and systolic thickness of the left ventricular anterior wall (AEDth and AESth) and posterior wall (PEDth and PESth) as well as left atrial diameter (LAd) were determined from three repeated 2-D guided M-mode tracings using the ASE convention. From those measurements, fractional shortening (FS) was calculated as: FS = (LVEDd - LVESd)/LVEDd.

Analytical methods

For analysis of cardiac natriuretic peptide expression, mRNA was extracted from all atrial and left ventricular samples utilizing a commercial kit. As a probe for brain natriuretic peptide (BNP), a 750-bp *Eco*R1/*Hind*III DNA restriction fragment containing the gene for rabbit BNP was used.

Assessment

Results of the quantitative studies were expressed as mean ± SEM. Comparisons between the control, ELVD and CHF groups were performed by analysis of variance (ANOVA) followed by Fisher's least significant difference test. Comparison between the atrial and LV tissues as well as between atrial natriuretic peptide (ANP) and BNP were performed by paired Student's *t*-test. Statistical significance was defined as $P < 0.05$.

Latest technology

Arnolda et al. (1985) studied adriamycin cardiomyopathy in the rabbit. Klimtova et al. (2002) performed a comparative study of chronic toxic effects of daunorubicin and doxorubicin in rabbits. Alexander et al. (1993) studied electrographic changes following corona-virus-induced myocarditis and dilated cardiomyopathy in rabbits. Sanbe et al. (2005) described a transgenic model for human troponin I-based hypertrophic cardiomyopathy in the rabbit.

INVESTIGATING CONGESTIVE HEART FAILURE IN DOGS

Several methods are described, to induce congestive heart failure in dogs, such as rapid ventricular pacing. Luchner et al. (1996) assessed circulating, renal, cardiac, and vascular angiotensin II in a canine model of rapid ventricular pacing-induced heart failure that evolves from early left ventricular dysfunction to overt congestive heart failure.

Method

Male mongrel dogs underwent implantation of a programmable cardiac pacemaker (Medtronic). Under pentobarbital sodium anesthesia and artificial respiration, the heart was exposed via a small left lateral thoracotomy and pericardiotomy, and a screw-in epicardial pacemaker lead was implanted into the right ventricle. The pacemaker was implanted subcutaneously into the left chest wall and connected to the pacemaker lead. The dogs were allowed to recover for at least 10 days after surgery before the pacemaker was started. During the first 10 days dogs were paced at 180 beats/min (bpm), resulting in early left ventricular dysfunction as defined by significant systolic dysfunction with decreased cardiac output, cardiac enlargement, and increased filling pressures but maintained systemic perfusion pressure and renal sodium excretion and no clinical signs of heart failure.

The pacing rate was then increased weekly to 200, 210, 220 and 240 bpm, and early left ventricular dysfunction evolved to overt congestive heart failure with avid sodium retention and clinical signs of congestion. At baseline (control), after dogs had been paced at 180 bpm for 10 days and at the end of the protocol (overt CHF), urine was collected for measurement of sodium excretion; conscious mean arterial pressure was measured via a port catheter; a 2-D guided M-mode echogram was obtained; and arterial blood was drawn. Cardiac filling pressures and cardiac output were measured by the thermodilution method at baseline and at the end of the protocol. Arterial blood was collected in EDTA tubes for measurement of ANP, BNP, cGMP, PRA, aldosterone, and Ang II. After euthanasia, hearts were rapidly trimmed and left ventricles weighted for calculation of the index LV weight to body weight.

Assessment

Results were expressed as mean ± SE. Comparison between the control, early LV dysfunction, and overt CHF were performed by ANOVA followed by Fisher's least significant difference test.

Latest technology

Kleaveland et al. (1988) and Nagatsu et al. (1994) used the technique of experimental *mitral regurgitation* in dogs to induce left ventricular dysfunction. A 30-cm, 7F sheath was introduced across the aortic valve through the carotid artery. A urologic calculus retrieval forceps was advanced through the sheath to the mitral valve apparatus and was used to sever chordae tendineae. When pulmonary capillary wedge pressure rose to 20 mmHg and forward stroke volume was reduced to 50% of its baseline, a ventriculogram was performed to confirm angiographically that severe mitral regurgitation had been created. Dell'Italia et al. (1995) and Su et al. (1999) induced mitral regurgitation by percutaneous chordal rupture in dogs.

Kinney et al. (1991) published a method to induce acute, reversible tricuspid insufficiency in anesthetized dogs. A wire spiral is advanced through the atrioventricular canal from the right atrium.

The spiral causes regurgitation by preventing complete apposition of the valve leaflets while permitting retrograde flow to occur through the spiral lumen. The degree of regurgitation can be controlled by the use of spirals of different size. Creation of tricuspid insufficiency was demonstrated by onset of right atrial pressure V waves, a ballooning of the right atrium during ventricular systole, palpation of an atrial thrill, or by color Doppler echocardiography. The model is reversible and allows repeated trials of various grades of regurgitation. Carlyle and Cohn (1983) described a non-chirurgical model of chronic left ventricular dysfunction. The method is accomplished by repetitive DC shock with a guidewire introduced percutaneously and positioned in the left ventricle along the intraventricular septum and an external paddle at the left ventricular apex.

McDonald et al. (1992) produced localized left ventricular necrosis without obstruction of the coronary blood flow in dogs by transmyocardial direct-current shock. Sabbah et al. (1991, 1993, 1994) and Gengo et al. (1992) produced chronic heart failure in dogs by multiple sequential intracoronary embolizations with microspheres. The dogs underwent three to nine intracoronary embolizations with polystyrene latex micro- spheres (70–102 μm in diameter) performed 1–3 weeks apart. Embolizations were discontinued when left ventricular ejection fraction was less than 35%. Vanoli et al. (2004) used multiple coronary microembolizations in dogs, whereby three to nine embolizations were performed 1 week apart. The first three embolizations consisted of 2 ml of microsphere suspension injected subselectively into either the left anterior descending or left circumflex coronary artery in an alternating fashion.

Subsequent embolizations consisted of 3-6 ml of microspheres divided equally between the left anterior descending or left circumflex coronary artery until LV ejection fraction was <35%. Magovern et al. (1992) described a canine model of left ventricular dysfunction caused by five weekly intracoronary infusions of *adriamycin*. Koide (1997) described premorbid determinants of left ventricular dysfunction in a model of gradually induced pressure overload in dogs. Mongrel dogs were studied through 8 weeks of gradually imposed ascending aortic constriction with the use of a *novel banding technique*. During banding, an initial gradient of 30 mm Hg was created. Before banding, at 2, 4, and 6 weeks after banding, hemodynamics and left ventricular mechanics were examined at cardiac catheterization; then the pressure overload was increased by tightening the band. Valentine et al. (1988) and Devaux et al. (1993) described *X-linked muscular dystrophy in dogs* with cardiac insufficiency similar to Duchenne muscular dystrophy in men and recommended this as an animal model for cardiac insufficiency.

INVESTIGATING CARDIAC FAILURE IN PIGS

Cardiac failure was studied in pigs using several experimental procedures. Chow et al. (1990) recommended rapid ventricular pacing in pigs as an experimental model of congestive heart failure. Farrar et al. (1993) studied pacing-induced dilated cardiomyopathy in pigs. Congestive heart failure was produced by rapid ventricular pacing at 230 bpm for 1 week. Spinale et al. (1990a, 1990b, 1991, 1992) examined the consequences of chronic supraventricular tachycardia on various parameters of ventricular dysfunction and subendocardial changes in pigs. Carroll et al. (1995) investigated gene expression in a swine model of ventricular hypertrophy during pressure overload. Multani et al. (2001) studied long-term angiotensin converting enzyme and angiotensin I-receptor inhibition in pacing-induced heart failure in pigs. Heart failure was induced by rapid atrial pacing (240 bpm for 3 weeks). Kassab et al. (1993, 2000) investigated remodeling of right ventricular branches after hypertrophy in pigs. Krombach et al. (1999) studied the effects of amlodipine in congestive heart failure in pigs at rest and after treadmill exercise.

Method

Left thoracotomy was performed in Yorkshire pigs in anesthesia. Catheters connected to a vascular access port were placed in the thoracic aorta, the pulmonary artery, and the left atrium. The access ports were then placed in a subcutaneous pocket. A 20-mm flow probe was placed around the pulmonary artery immediately distal to the pulmonary catheter and the electrical connection exteriorized through the thoracolumbar fascia. A shielded stimulating electrode was sutured onto the left atrium, connected to a programmable pacemaker and buried in a subcutaneous pocket. The thoracotomy was closed in layers and the pleural space evacuated of air. After a 14- to 21-day recovery, measurements were performed under normal resting conditions and after exercise.

The pacemakers were activated to 240 bpm for a period of 21 days. During the last 3 days, one group was treated with drug, the other served as control. At the day of the study, electrocardiograms were performed, and the pacemakers deactivated. After a 30-min stabilization period, 2-D and M-mode echocardiographic studies were used to image the left ventricle from the parasternal approach. Left ventricular fractional shortening was calculated as (end-diastolic dimension - end-systolic dimension)/diastolic dimension, and was expressed as a percentage. The access ports were entered and pressures obtained using externally calibrated transducers. The flow probe was connected to a digital flowmeter. From the digitized flow signal, stroke volume was computed on a beat-to-beat basis and averaged for a minimum of 25 ejections.

Pulmonary and systemic vascular resistances were computed as the mean pressure divided by cardiac output multiplied by the constant 80 to convert to resistance units of dyne $\cdot$ s $\cdot$ cm^{-5}. Samples were drawn from the pulmonary artery and atrial catheters for measurement of oxygen saturation and hemoglobin content. The plasma samples were assayed for renin activity, endothelin concentration, and catecholamine levels.

Assessment

Results were presented as mean ± SEM. Pairwise tests of individual group means were compared using Bonferroni probabilities.

Latest technology

Zhang et al. (1996) studied functional and bioenergetic consequences of post-infarction left remodeling in a porcine model. Proximal left coronary artery occlusion was used to generate a myocardial infarction in young pigs. The animals were then followed over several months while remodeling of the left ventricle developed. Left ventricular wall thickness, ejection fraction, and wall stress were measured by MRI. Myocardial ATP, creatine phosphate, and inorganic phosphate levels were measured by spatially localized ^{31}P-NMR spectroscopy, and regional myocardial blood flow was measured with radioactive microspheres.

Method

All MRI studies were performed on the standard Siemens Medical System VISION operating at 1.5 T. The animals were anesthetized with sodium pentobarbital. A catheter was placed into the femoral artery and advanced into the LV chamber for LV pressure recording. Animals then were placed on their left side in a Helmholtz coil with a diameter of 18cm, which was used to improve signal to noise. To compute LV wall stress, the image acquisition was triggered by the LV pressure through the fluid-filled LV catheter. All of the imaging sequences were synchronized to the LV pressure trace. The electronic LV pressure signal was recorded and fed to a comparator set to a threshold level of 10% of the upslope of the LV pressure curve at the beginning of systole.

The signal from the comparator was sent to a pulse former and then fed to the ECG port of the magnetic resonance system, where it was treated like the standard electrographic input to run the pulse sequences. Scout images were taken in the axial plane with a single-shot, ultrafast gradient echo sequence. From the axial image, both horizontal and vertical long-axis images were obtained. By alternating back and forth several times, a true vertical long axis of the left ventricle was obtained. From the long-axis scout image, short-axis segmented cine turboflash slices were prescribed to cover the myocardium from apex to base. The double- oblique, short-axis turboflash images cover the heart from apex to base with a slice thickness of 10 mm, with no interslice gap.

MRI technique

The parameters of the segmented cine sequence were TR/TE/flip angle = 33 ms/6.1 ms/25 degrees with an FOV = 17.5 cm and a matrix of 87 × 128 (pixel size, 2 mm × 1.4 mm) and slice thickness of 7-10 mm. The sequence used segmented k-space acquisition such that three phase-encoded lines were gathered per cardiac phase per heartbeat. Total image acquisition required approximately 52 heartbeats for each slice location. The temporal image resolution (data acquisition window) of this sequence was 33 ms per cardiac image. Each myocardial level took <1.5 min to acquire, since two acquisitions were used and the average heart rate of the animals was 120 bpm. The average number of short-axis slices needed to image the entire myocardium from apex to base was six to eight. This 10-min protocol provided high signal-to-noise cine sequences covering the entire heart.

Spin-echo images

To obtain high-resolution anatomic heart images, multislice, single-phase spin-echo images trigged in the systolic phase were acquired to cover the entire heart. These images permitted the precise delineation of the extent of the scar region of the heart. Images were taken with a slice thickness of 5mm and a FOV of 17.5 cm, resulting in a true spatial resolution of 2 mm × 1.4 mm pixel size. The TR for this sequence equals the RR interval (500 ms) and the echo time TE was set to 30 ms. Total measurement time for an average of 10-14 slices was 5 min.

Interpreting image analysis

The imaging data were archived to optical disk and copied to a SUN SPARC 10 workstation for evaluation with the use of an automatic segmentation program. The program is based on robust deformable models of endocardial and epicardial border segmentation of ventricular boundaries in cardiac magnetic resonance images. This segmentation technique has been combined with a user interface that allows one to load, sort, visualize, and analyze a cardiac study in <20 min. The segmentation algorithm is based on the steepest descent as well as dynamic programming strategies integrated via multiscale analysis for minimizing the energy function of the resulting contour. The ventricular boundaries are used to construct a three-dimensional model for visualization and to compute hemodynamic parameters.

Automatic segmentation of endocardial and epicardial boundaries was performed for calculation of ventricular volumes, EF, LV diastolic and systolic volumes, and absolute myocardial mass from multislice, multiphase magnetic resonance cine images. Starting with a user-specified approximate boundary or an interior point of the ventricle for one starting image in one slice, the algorithm generated automatic contours corresponding to the epicardium and the endocardium and automatically propagated them to other slices in the cardiac phase (spatial propagation) and to other phases for a given slice location (temporal propagation) of the cardiac study. The observer then could make some manual corrections to the six or seven pairs of contours in the first column of the temporal-spatial matrix. Manual modifications generally were made on the apex and base levels.

Assessment

Mean LV wall thickness for each short-axis ring was averaged from three measurements of the remote zone (anterior wall and septum wall). The thickness of the scar was averaged from three measurements of the scar area. LVSA measurement in each slice was computed by subtracting the total area enclosed by the endocardium from that enclosed by the subepicardium; the resultant area was multiplied by the slice thickness to obtain the volume of each slice; the total LV mass volume was calculated by adding up the volumes of all the short-axis slices. The total LVSA was obtained by dividing the total LV wall mass volume by the mean of LV wall thickness of each slice. Similarly, the LVSSA was obtained by dividing the total scar volume, which was the sum of the scar volume of each short axis, by the mean of the scar thickness of each short axis.

LV mass was computed by the total LV wall mass volume multiplied by 1.05 (specific gravity of myocardium) to calculate the LV mass. The LV end-diastolic volume (V_d) and end-systolic volume (V_s) of each slice were represented by the area enclosed by the endocardium. The total LV volume was computed by adding the volumes of all slices. LVEF was calculated by $100 \times (V-V_s)/V_d\%$. Interobserver and intraobserver errors for the calculations of LV mass and LV volumes have been shown to be <3 mg and 3 ml, respectively. Meridional wall stress was computed from the LV pressure and simultaneously obtained short-axis view of LV MRI (LV cavity diameter and average thickness the remote LV wall) as described by Grossman et al. (1975).

^{31}P-NMR spectroscopic technique

Measurements were performed in a 40-cm-bore, 4.7-T magnet interfaced with a SISCO console. The LV pressure signal was used to gate NMR data acquisition to the cardiac cycle, while respiratory gating was achieved by triggering the ventilator to the cardiac cycle between data acquisitions. ^{31}P- and ^{1}H-NMR frequencies were 81 and 200.1 MHz, respectively. Spectra were recorded in late diastole with a pulse repetition time of 6-7 s. This repetition time allowed full relaxation for ATP and P_i resonances and ≈ 90% relaxation for the CP resonance. CP resonance intensities were corrected for this minor saturation; the correction factor was determined for each heart from two spectra recorded consecutively without transmural differentiation, one with 15-s repetition time to allow full relaxation and the other with the 6- to 7-s repetition time used in all the other measurements.

Radiofrequency transmission and signal detection were performed with a 25-mm-diameter surface coil. The coil was cemented to a sheet of silicone rubber 0.7mm in thickness and ≈50% larger in diameter than the coil itself. A capillary containing 15 μl of 3M phosphonoacetic acid was placed at the coil center to serve as a reference. The proton signal from water detected with the surface coil was used to homogenize the magnetic field and to adjust the position of the animal in the magnet so that the coil was at or near the magnet and gradient isocenters. This was accomplished with a spin-echo experiment and a readout gradient. The information gathered in this step also was used to determine the spatial coordinates for spectroscopic localization.

Chemical shifts were measured relative to CP, which was assigned a chemical shift of –2.55 ppm relative to 85% phosphoric acid at 0 ppm. Spatial localization across the LV wall was performed with the RAPP-ISIS/FSW method. Signal origin was restricted with the use of B_0 gradients and adiabatic inversion pulses to a column coaxial with the surface coil perpendicular to the LV wall. The column dimensions were 17 × 17 mm. Within this column, the signal was further localized using the B_1 gradient to five voxels centered about 45°, 60°, 90°, 120°, and 135° spin rotation increments. FSW localization used a nine-term Fourier series expansion.

The Fourier coefficients, the number of free induction decays acquired for each term in the Fourier expansion, and the multiplication factors used to construct the voxels have been reported previously. The position of the voxels relative to the coil was set using the B_1 magnitude at the coil center, which was experimentally determined in each case by measurement of the 90° pulse length for the phosphonoacetic acid reference located in the coil center. Each set of spatially localized transmural spectra were acquired in 10 min. A total of 96 scans was accumulated within each 10- min block.

Assessment

Resonance intensities were quantified with the use of integration routines provided by the SISCO software. ATPγ resonance was used for ATP determination. Since data were acquired with the transmitter frequency being positioned between the ATPγ and CP resonance, the off-resonance effects on these peaks were negligible. The numeric values for CP and ATP in each voxel were expressed as ratios of CP/ATP. P_i levels were measured as changes from baseline values (ΔP_i) with the use of integrals obtained in the region covering the P_i resonance.

INVESTIGATING CARDIAC FAILURE IN SHEEP

Various methods have been used to induce cardiac failure in sheep: pressure overload after aortic banding, volume loading after myocardial infarction, rapid ventricular pacing, coronary microembolization. Rademaker et al. (2002a) studied combined angiotensin converting enzyme inhibition and adrenomedullin in an ovine model of heart failure induced by rapid ventricular pacing.

Method

Coopworth ewes (38 ± 47 kg) were instrumented via a left lateral thoracotomy. Under general anesthesia (induced by 17 mg/kg thiopentone; maintained with halothane/nitrous oxide), two polyvinyl chloride catheters were inserted in the left atrium for blood sampling and left atrial pressure (LAP) determination; a Konigsberg pressure-tip transducer was inserted in the aorta to record mean arterial pressure (MAP); an electromagnetic flow probe was placed around the ascending aorta to measure cardiac output (CO); a 7 French Swan-Ganz catheter was inserted in the pulmonary artery for infusions; and a 7 French His-bundle electrode was stitched subepicardially to the wall of the left ventricle for left ventricular pacing. All leads were externalized through incisions in the back. A bladder catheter was inserted per urethra for urine collections. The animals were allowed to recover for 14 days before commencing the study protocol.

During the experiments, the animals were held in metabolic cages, had free access to water and ate a diet of chaff and sheep pellets (containing 40 mmol/day sodium and 200 mmol/day potassium). A further 40 mmol of sodium was administered orally daily as NaCl tablets using an applicator. Heart failure was induced by 7 days of rapid left ventricular pacing (225 bpm) and maintained by continuous pacing for the duration of the study. On four separate days with a rest day between each, the sheep received, in random order, a vehicle control (Haemaccel), human adrenomedullin alone (50 ng/min per kg infusion for 3 h), an ACE inhibitor alone (captopril: 25 mg bolus +2 mg/h infusion for 3 h), and both agents combined.

Infusions were administered in a total volume of 60 ml via the pulmonary artery catheter, commencing at 10:00 hours. Mean arterial pressure, left atrial pressure, cardiac output and calculated total peripheral resistance (CTPR = mean arterial pressure/cardiac output) were recorded at 15-min intervals in the 1-h prior to infusion (baseline), and at 15, 30, 45, 60, 90, 120 and 180 min during both the 3-h infusion and post-infusion periods. Hemodynamic measurements were determined by on-line computer-assisted analysis.

Blood samples were drawn from the left atrium at 30 min and immediately pre-infusion (baseline), and at 30, 60, 120 and 180 min during the 3-h infusion and post-infusion periods. Samples were taken into tubes on ice, centrifuged at 3939 *g* for 10 min at 4°C and stored at either -20°C or -80°C before assay for immunoreactive (ir-) adrenomedullin, cAMP, plasma renin activity, angiotensin II, aldosterone, atrial natriuretic peptide, brain natriuretic peptide, endothelin-1, catecholamines and cortisol. All samples from individual animals were measured in the same assay to avoid inter-assay variability. Plasma electrolytes and hematocrit were measured in every sample taken. Urine volume and samples for the measurement of urine cAMP, sodium, potassium and creatinine excretion were collected every 1 h. Creatinine clearance was calculated as urine creatinine/ plasma creatinine.

Assessment

Results are expressed as mean ± SEM. Baseline hemodynamic and hormone values represent the means of the four and two measurements respectively made in the 1 h immediately pre-infusion. Statistical analysis was performed by repeated-measures ANOVA. Baseline data from all treatments were compared. Treatment- and time-related differences between all four study limbs were determined using a two-way ANOVA (treatment-time interactions are quoted in the text). Statistical significance was assumed when $P < 0.05$

INVESTIGATING CARDIAC FAILURE IN MONKEYS

Several authors used monkeys for studies of cardiac failure. Hollander et al. (1977) investigated the role of hypertension in ischemic heart disease in the *cynomolgus monkey* with coarctation of the aorta. Sieber et al. (1980) studied cardiotoxic effects of adriamycin in *macaques*. Various studies were performed by the group of Hoit and Walsh in *baboons*. Hoit et al. (1997a) studied the effects of thyroid hormone on cardiac β-adrenergic responsiveness in conscious baboons.

Method

Adult male baboons (*Papio anubis*) weighing 21-30 kg were pre-instrumented for physiological monitoring in a lightly anesthetized, sedated state. Animals were pre-instrumented with a Konigsburg micro- manometer and a polyvinyl catheter in the LV apex, miniaturized sonomicrometer pairs (3 MHz, 6 mm) across the LV anteroposterior minor axis, a polyvinyl catheter in the right atrium for central venous access, and pacing wires on the right atrial appendage. Wires and tubes were tunneled subcutaneously into the interscapular area for later use. Postoperative pain was reduced by the use of Buprenex (0.01 mg/kg i.m., q 6 h), and postoperative antibiotic (Monocid 25 mg/kg) was administered for 5 days to reduce the risk of infection. Baseline hemodynamic studies were performed after a minimum of 1 week for postoperative recovery.

Hemodynamic data acquisition

The micromanometers and fluid-filled catheters were calibrated with a mercury manometer. Zero drift of the micromanometer was corrected by matching the LV end- diastolic pressure measured simultaneously through the LV catheter. The fluid-filled LV catheter was connected to a pre-calibrated Statham 23 dB transducer with zero pressure at the level of the mid right atrium. The transit time of ultrasound between the ultrasonic dimension crystals was measured with a multichannel sonomicrometer (Triton Technology) and converted to distance assuming a constant velocity of sound in blood of 1.55 mm/ms. The analog LV dP/dI signal was obtained online by electronic differentiation of the high-fidelity LV pressure signal. τ was derived from the high-fidelity LV pressure tracing by the method of Weiss et al. (1976), which assumes a monoexponential decay of LV pressure to a zero asymptote and has been shown to be directionally equivalent to other mathematical approaches for quantification of isovolumic pressure decay. τ is equal to the time in milliseconds for LV pressure to decay to 1/*e*; thus, decreases in τ reflect improved isovolumic ventricular relaxation.

Fractional shortening of the LV minor axis was calculated as (EDD-ESD)/EDD, where EDD is LV end-diastolic dimension and ESD is LV end-systolic dimension. LV end-diastole was defined as the time in which LV dP/dt_{max} increased by ≥ 150 mm Hg/s for 50 ms, and LV end-systole was defined as the time of the maximum ratio of LV pressure to LV minor-axis dimension. LV volumes were derived from minor-axis diameter (*D*) measurements: LV volume = $\pi/6(D)^3$. V_{cf} was calculated as LV fractional shortening divided by LV ejection time; LV ejection time was defined as the time from peak positive to peak negative dP/dt. Analog signals for high-fidelity and fluid-filled LV pressures, LV short-axis dimension, LV dP/dt, and the ECG were recorded on-line on a Gould multichannel recorder at 25 and 100 mm/s paper speed and digitized through an analog-to-digital board interfaced to an IBM AT computer at 500 Hz and stored on a floppy disk. Data were analyzed using an algorithm and software developed in our laboratory. Steady-state data were acquired over 5-10 s during spontaneous respiration and averaged.

Methodology

Hemodynamic studies were performed a minimum of 1 week after instrumentation and were repeated after 22-30 (26.8 ± 2.7) days of thyroid T_4 administration. Animals were tranquilized with Valium (1-5 mg) and ketamine (100 mg), and cholinergic blockade was achieved with atropine (0.4-0.8 mg i.v.); additional ketamine was administered as necessary, to a maximum cumulative dose of 40 mg/kg. Animals were atrially paced at a rate 40% to 50% greater than the control heart rate in order to obtain data at matched heart rates after thyrotoxicosis was produced.

Observing dobutamine group

After hemodynamic stability was ensured and baseline data were recorded, intravenous dobutamine was infused at 5-min intervals at upwardly titrated rates of 2.5, 5.0, 7.5, and 10.0 $\mu g \cdot kg^{-1} \cdot min^{-1}$ to examine the effects of β_1-adrenergic stimulation. The dose range of catecholamine for these studies was chosen to alter inotropic and lusitropic states without causing an untoward increase in heart rate. Steady-state hemodynamic measurements were made during minutes 4 and 5 of each infusion period. At each level, the pacemaker was briefly turned off to determine the effect of dobutamine on the heart rate. Four of the animals in this group were studied with incremental pacing both before and after β-adrenergic blockade with esmolol (0.3 $mg \cdot kg^{-1} \cdot min^{-1}$ i.v.V).

The pacing protocol and the results from a larger group of animals studied before β-adrenergic blockade were detailed in a previous report. Briefly, atrial pacing was instituted at a rate above the intrinsic heart rate to avoid competing rhythms and was increased at 0.2-Hz increments until the critical heart rate was achieved. The critical heart rate was defined as the rate at which dP/dt_{max} and τ reached a maximum and minimum, respectively, during progressive increases in heart rate. We showed previously that hyperthyroidism significantly increases the critical heart rates for both dP/dt_{max} and τ. The EC_{50} of dobutamine for LV dP/dt_{max} was determined by fitting log(dose)-transformed data to a sigmoidal relation with software from Graph Pad.

Observing terbutaline group

Additional animals were chronically instrumented so that we could examine the effects of β_2-adrenergic stimulation. One animal died suddenly after receiving thyroid hormone for 20 days. In the remaining three animals, the β_2-adrenergic agonist terbutaline was infused both before and after production of the hyperthyroid state. Incremental doses of terbutaline (15 min/dose) were infused over a dosing range of 25 to 300 $ng \cdot kg^{-1} \cdot min^{-1}$.

Analyzing thyroid function

Thyroid function tests were performed before the baseline experiment in the euthyroid state and before the terminal experiment (within 24 h of the last dose of T_4) in the hyperthyroid state. T_3 radioimmunoassay, T_4, and free T_4 levels were measured at each state.

Assessment

Paired mean data were compared by Student's *t*-test. The effects of thyroid status, catecholamine dose, and β-blockade on hemodynamic and dimension variables were examined with repeated-measures ANOVA. When significant differences were found, group means were compared with contrasts. A value of $P < 0.05$ was considered significant. Unless specified, data are expressed as mean ± SD.

INVESTIGATING CARDIAC FAILURE IN OTHER SPECIES

Various species have been used to study experimental cardiac failure. Breisch et al. (1984) studied the effects of pressure-overload hypertrophy in the left myocardium of young adult cats. Hypertrophy was induced by a 90% constriction of the ascending aorta. Genao et al. (1996) recommended dilated cardiomyopathy in turkeys as an animal model for the study of human heart failure. Do et al. (1997) studied energy metabolism in normal and hypertrophied right ventricle of the ferret heart. Wang et al. (1994) studied Ca^{2+} handling and myofibrillar Ca^{2+} sensitivity in ferret cardiac myocytes with pressure-overload hypertrophy. Bovine hereditary cardiomyopathy was recommended as an animal model of human dilated cardiomyopathy by Eschenhagen et al. (1995).

STEM CELL TECHNOLOGY

Kojima et al. (1994), Komuro et al. (1990, 1991, 1993), Yamazaki et al. (1993, 1994, 1996) described a method to induce hypertrophy of cardiomyocytes by mechanical stress *in vitro*.

Method

Primary cultures from cardiomyocytes are prepared from ventricles of 1-day-old neonatal Wistar Kyoto rats. According to the method of Simpson and Savion (1982), the cultures are treated for 3 days with 0.1 mM bromodeoxyuridine to suppress proliferation of non- myocardial cells. Elastic culture dishes (2 × 4 × 1 cm) are made by vulcanizing liquid silicone rubber consisting of methylvinyl polysiloxane and dimethyl hydrogen silicone resin using platinum as a catalyst. The bottom of the disc is 1-mm thick, and it is highly transparent because of no inorganic filler in either component. Cells are plated in a field density of 1×10^5 cells/cm^2 in culture medium consisting of Dulbecco's modified Eagle's medium with 10% fetal bovine serum.

Mechanical stress on cardiac cells is applied by gently pulling and hanging the dish on pegs. A 10% change in length of the dish results in an almost identical change in the length of the cell along a single axis. Cardiocytes are stretched by 5%, 10%, or 20%. Drugs, e. g., an angiotensin-II receptor antagonist, are added 30 min before stretch. *For protein analysis*, the silicone dishes are stretched for 24 h after 2 days of serum starvation and [^{3}H]phenylalanine (1 µCi/ml) is added for 60 min. At the end of each stress, the cells are rapidly rinsed four times with ice-cold phosphate-buffered saline and incubated for 20 min on ice with 1 ml of 5% trichloroacetic acid. The total trichloroacetic acid-insoluble radioactivity in each dish is determined by liquid scintillation counting.

For determination of mitogen-activated protein kinase, cardiomyocytes are lysed on ice and centrifuged. Aliquots of the supernatants of myocyte extracts are incubated in kinase buffer (25 mM/l Tris-HCl, pH7.4, 10 mM/l $MgCl_2$, 1 mM/l dithiothreitol, 40 µM/l APT, 2 µCi [γ-^{32}P]ATP, 2 μM/l protein kinase inhibitor peptide, and 0.5 mM/l EGTA) and substrates (25 μg myelin basic protein). The reaction is stopped by adding stopping solution containing 0.6% HCl, 1 mM/l ATP, and 1% bovine serum albumin. Aliquots of the supernatant are spotted on P81 paper (Whatman), washed in 0.5% phosphoric acid, dried and counted. For determination of *c-fos mRNA*, Northern blot analysis is performed.

Assessment

Values are expressed as mean ± SEM. Comparisons between groups are made by one-way ANOVA followed by Dunnett's modified *t*-test. The interesting approach to induce hypertrophy of cardiac cells *in vitro* has been used predominantly by one research group. Confirmation by other research groups including modifications of the mechanical procedures seems to be necessary.

9

Efficacy Models

Early in the drug development process, *in vivo* studies are often run to help select molecules with appropriate ADME characteristics, generate data in efficacy models, and evaluate toxicity. Poor exposure can lead to highly variable data, and/or ambiguous study results. Use of resource and scarce API for uninterpretable study outcomes is highly inefficient, and based on poor data, a compound may be given an undeservedly low priority. Because exposure is such a critical factor for compound evaluation during lead optimization, the formulation of the compound can be a key component of the drug selection process. Biopharmaceutical modeling is often used to estimate the fraction absorbed of a given dose delivered (F_{abs}). Models range from a simple algebraic calculation of maximum absorbable dose, or MAD, through empirically parameterized models like IDEA to the highly sophisticated GastroPlus.

Each biopharmaceutics model has utility within the drug development process and in capable hands with good inputs, can yield reasonably accurate estimates of bioavailability. While these models are most often used to evaluate the impact of physical and chemical properties of drugs on fraction absorbed, some of these models may also be used to help guide formulation approaches to achieve better bioavailability. At the lead optimization stage, one needs a simple assessment of where a drug sits in the "*biopharmaceutical landscape*" so that strategies for formulation options can be assessed. A model developed at the University of Michigan called the microscopic mass balance approach meets this criterion. Understanding the model output can help answer questions such as: Why am I getting poor exposure? Will a better formulation help me? What formulation technologies should be tried?

DRUG PARTICLES

Conceptually, MiMBA involves placing a dispersed suspension of drug particles at the beginning of a cylindrical tube representing the intestinal tract. As the particles traverse the intestine, dissolution of the particles occurs, increasing the concentration of drug within the tube. Simultaneously, absorption takes place, effectively decreasing the drug concentration within the tube. The trade off between dissolution and absorption dictates the time course of the drug concentration in the intestine. At the end of the intestinal tube, the total fraction absorbed is calculated by subtracting from the initial dose everything remaining inside the intestine, i.e., the fraction remaining as solid particles and the fraction that dissolved but did not get absorbed. Mathematically, MiMBA utilizes three dimensionless variables that encompass the physical and chemical properties of the drug as well as the physiological system in which dissolution and absorption takes place.

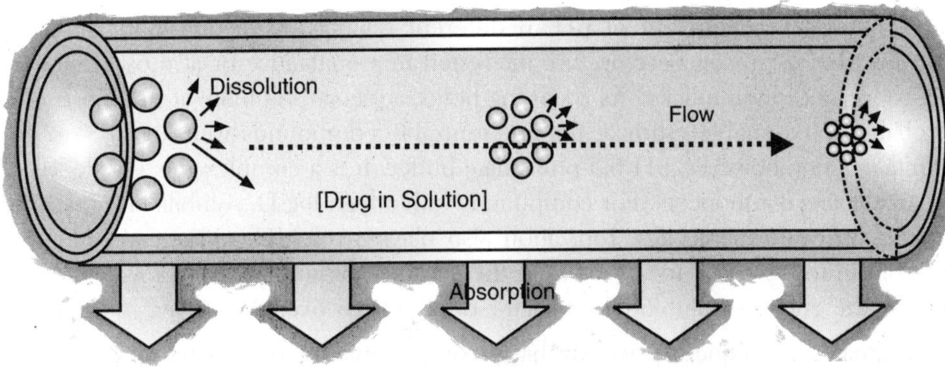

Fig. 9.1. Conceptual model upon which MiMBA is based showing the dissolution and absorption processes as particles transit the intestine.

These variables are then incorporated into a system of two differential equations, one describing the fate of the particles' radii, the other describing the concentration along the intestinal tract. The equations can be solved using a mathematical or engineering software package such as MathCad or Mathematica, and total fraction absorbed calculated. The three dimensionless variables are described below and deal respectively with the solubilization process, the absorption process, and the dissolution process.

Solubilization Process

The solubilization process is described by the relative solubility of the dose compared to the capacity of the medium. The calculated variable is termed the dose number, D_o, and is determined by:

$$D_o = \frac{M_o/V_o}{C_s}$$

Where,

M_o is the mass of the dose

V_o is the internal volume into which the dose is dissolved

C_s is the saturation solubility of drug in the intestinal medium

As dose number varies from 0 to 1, the medium is increasingly saturated with solubilized drug until at $D_o = 1$, complete saturation occurs at equilibrium. For $D_o > 1$, there is excess solid present since the dose has exceeded the capacity of the medium to solubilize the drug. The dose M_o is normally dictated by experimental protocol. The volume V_o is dependent on the physiology and the dosing protocol, and can be estimated from the sum of resting volume in the stomach, the liquid ingested at time of dosing, plus a quantity to account for dilution after gastric emptying occurs. Determining the relevant value for saturation solubility C_s can be difficult. In reality, there is most likely a range of solubilities that impact the incremental absorption potential as the compound moves down the intestine, especially for ionizable compounds.

Simple biopharmaceutic models (e.g., MAD and MiMBA) attempt to capture this in one solubility value, whereas more sophisticated models (e.g., IDEA and GastroPlus) account for this by varying solubility as a function of pH in different regions of the intestine. The tradeoff is complexity. For the discussion here, we are interested in a qualitative or at most semi-quantitative evaluation, so we opt for simplicity. As a starting point, aqueous solubility at pH 6.5 is often used. This is probably a reasonable estimate for non-ionizable compounds with log D < ~2 to 3. The intestinal milieu is not, however, pH 6.5 phosphate buffer. It is a complex mix of bile salts, lecithin and other solubilizing components. For compounds with a high log D, solubility is greatly enhanced by partitioning into mixed micelles. Ionization also plays a role. The intestinal pH ranges from about 6.6 ± 0.5 in the jejunum to 7.5 ± 0.5 in the ileum. Ionization of a weakly acidic compound could increase the effective solubility by a factor of about ten over this range.

Of even greater consequence, weakly basic compounds are protonated at stomach pH, yet deprotonated at intestinal pH. Solubilization occurs within the stomach environment, and upon gastric emptying the intestine is presented with a relatively large fraction of the drug in solution. Precipitation competes kinetically with absorption to remove compound from solution. Depending on the rapidity of the precipitation process, the effective solubility could be several orders of magnitude higher than what would be determined by an equilibrium measurement at pH 6.5. To mimic *in vivo* solubilization, several artificial or simulated intestinal fluids (SIFs) have been proposed and these media have been shown to significantly increase the equilibrium solubility of some poorly aqueous-soluble compounds. One or more of these media should be used to evaluate compound solubility, and that value used in the MiMBA calculation. If the compound is ionizable, solubility could be evaluated over a range of pHs, say 4 to 7.5, with the values then used as boundaries for subsequent analysis.

Absorption Process

The absorption process is described by the ratio of the absorption rate to the GI transit rate. The calculated variable is termed the absorption number A_n and is given by:

$$A_n = \frac{P_{eff} \cdot \pi \cdot R \cdot L}{Q}$$

where

P_{eff} is the effective permeability coefficient in the species of interest (cm/s)

R is the radius of the intestine (cm)

L is the intestinal length (cm)

Q is the volume flow rate down the GI tract (cm^3/min)

In a relative sense, as the absorption rate becomes slower than the transit rate ($A_n < 1$), material passes through the intestine before it can be absorbed. The variables R, L, and Q are assigned values based on species physiology and estimated intestinal transit time. The more sophisticated biopharmaceutic models account for regional permeability differences, but again, for the sake of simplicity, we assume permeability, P_{eff}, is constant.

As a first approximation, one assumes that absorption takes place in the small intestine only. If colonic absorption occurs, the effect is one of increasing A_n by a factor proportional to the relative increase in transit time. Effective permeability is the variable into which a compound's behavior gets incorporated. An estimate for P_{eff} can be determined by linking a permeability measurement in a cell-based (e.g., Caco2) or an artificial membrane (e.g., PAMPA) assay to a species-specific value through the means of standards. Metoprolol (P_{eff} in human = 1 .34e-6 cm/s) is often used as one of the standards since its permeability is the boundary for highly permeable compounds in the biopharmaceutics classification scheme (BCS). Permeability may be reported as an absorption rate constant 'k' (min^{-1}) as opposed to a P_{eff} value. The conversion between the two is merely geometric surface to volume ratio of the intestine:

$$P_{eff} = k \cdot \frac{Vol}{SA} = k \cdot \frac{(\pi \cdot R^2 \cdot L)}{(2 \cdot \pi \cdot R \cdot L)} = k \cdot \frac{R}{2}$$

where R is the intestinal radius of the species of interest.

Dissolution Process

The *dissolution process* is described by the ratio of the GI residence time to the time for complete dissolution. The calculated variable is termed the dissolution number D_n and is given by:

$$D_n = \left(\frac{\pi \cdot R^2 \cdot L}{Q}\right) \Big/ \left(\frac{r_0^2 \cdot \rho}{3 \cdot D \cdot C_s}\right)$$

where

r_0 is the initial particle radius (pm)

D is the diffusivity (diffusion coefficient) of the compound (cm^2/sec) ρ is the density of the compound solid (g/cm^3)

C_s, Q, R, L are as previously defined

As the residence time exceeds time for complete dissolution ($D_n > 1$), all material is in solution and is available for absorption. If $D_n > 10$, then all drug particles are dissolved in a tenth of the transit time and we can comfortably assume that particle size has a limited influence on the fraction absorbed. For MiMBA, the approximation is made that all particles are of the same size, i.e., mono-dispersed. While in reality this is not the case since a powder contains a distribution of particle sizes, an upper boundary can be established for the initial radius r_0 from a microscopic image.

The eye is naturally drawn to the larger particles, and these can be used to estimate a 'worst case' r_0. The diffusivity 'D' is a function of molecular size. For a typical small molecule (MW of 300 to 500), D will be about 7 to 8 × 10^{-6} cm^2/sec. The density 'ρ' of an organic solid is typically between 1 to 1.5 g/cm^3.

Fraction Absorbed

The calculation of fraction absorbed involves solving a system of two differential equations.

$$\frac{dr^*}{dz^*} = -\frac{D_n}{3}\frac{1-C^*}{r^*}$$

$$\frac{dC^*}{dz^*} = D_o \cdot D_n \cdot r^* \cdot (1-C^*) - 2 \cdot A_n \cdot C^*$$

The first equation determines the decrease in radius (i.e., dissolution) as the particles transit the intestine. This rate is proportional to the dissolution number (the greater D_n, the faster the dissolution) and the difference between the saturation solubility and concentration in solution (given by $1 - C^*$). The second equation determines the change in concentration down the intestinal tract and is the difference between the amount of drug getting in solution via dissolution and the amount of drug being removed from solution via absorption. Total fraction absorbed is determined by evaluating these equations at the end of the intestine (mathematically when $z^* = 1$) using the expression

$$F = 1 - \{r^*(1)\}^3 - \frac{C^*(1)}{D_o}$$

Conceptually, this can be considered as the initial dose fraction (by definition = 1) minus what has not dissolved and is left as a solid at the end of the intestine ($= r^*(1)^3$) minus what has dissolved, but has not been absorbed ($= C^*(1)/D_o$).

THREE DIMENSIONAL BIOPHARMACEUTICS LANDSCAPE

For a given absorption number A_n, if fraction absorbed is calculated over a range of dose numbers D_o and dissolution numbers D_n, one can generate the three-dimensional biopharmaceutics landscape for the compound. This contour plot demonstrates the effect on fraction absorbed of increasing the dose (D_o therefore increases) shown by the vertical arrow on the right-hand side of the plot, and the effect of increasing particle size (D_n decreases) shown by the horizontal arrow. The cross-sections of the contour plot along the arrows are also shown.

Dose Linearity

If we put some real numbers to the dimensionless variables, we can see how increasing the dose effects oral exposure. Figure 9.3 plots fraction absorbed and the actual amount absorbed for escalating doses of a low permeability, low solubility compound. There is a region of dose linearity, which corresponds to the flat region of the F_{abs} tracing. As the dose increases, linearity is no longer predicted and at high dose, the amount absorbed plateaus such that even though ever more drug is given, no more is absorbed.

The physical explanation for this plateau is that at some dose, the intestinal milieu becomes saturated with drug. As the dose moves down the intestinal tract, the surrounding medium remains saturated throughout the entire absorption region. Above this dose, any more solid material added never dissolves, but just passes through the intestine. This phenomenon is the basis for the MAD (maximum absorbable dose) calculation. During drug development, this behavior is often observed in toxicity studies when high exposure is desired. If the exposure (AUC or C_{max}) vs. dose

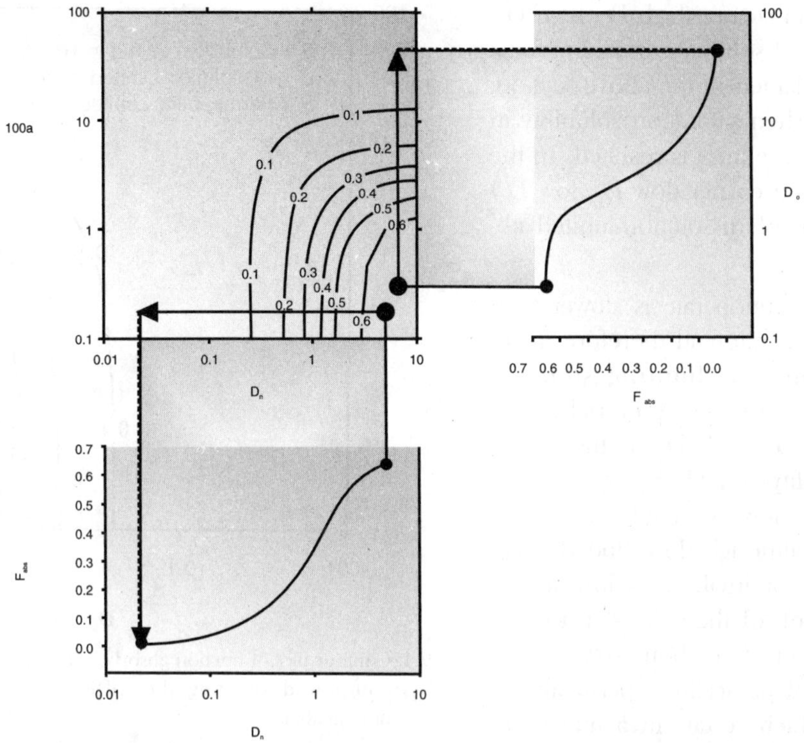

Fig. 9.2. Contour plot of fraction absorbed (F_{abs}) as a function of dose number (D_o) and dissolution number (D_n) for a low permeability compound. As dose increases (D_o increases), F_{abs} decreases. As particle size increases (D_n decreases), F_{abs} decreases.

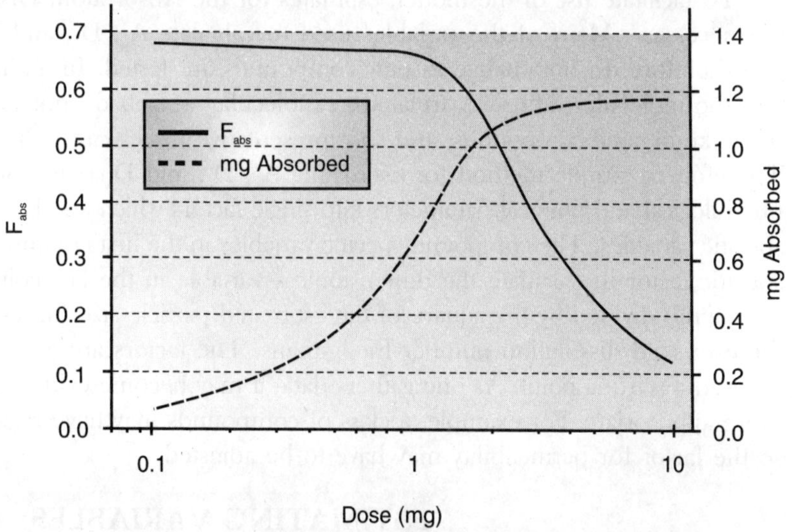

Fig. 9.3. An example of the relationship between fraction absorbed and amount absorbed as dose increases. A plateau in amount absorbed is eventually reached as the intestine becomes saturated with solubilized drug.

curve has flattened, a formulation approach may be employed to help increase the amount absorbed, although there must be a trade-off between the desire for increased exposure and the addition of components that might influence a toxicity outcome.

Regions of Solubility Dissolution and Permeability

From the contour plot in Figure 9.4, we can identify three reasons for limited bioavailability. The upper right-hand corner (high D_o, high D_n) is the region of solubility-limited absorption. As explained in the above section, this occurs when saturation solubility in the intestinal medium is reached. In the lower left-hand corner (low D_o, low D_n) is the region of dissolution-limited absorption.

The dissolution rate is slower than the absorption rate and therefore dissolution is the rate-limiting step for bioavailability. In the lower right-hand corner (low D_o, high D_n) is the region of permeability-limited absorption. Low permeability (low A_n) controls bioavailability and although dose and dissolution rate are optimal, one can only absorb a fraction of the dose – in the example of Figure 9.4, about 70%. At the lower limit of paracellular permeability, $A_n \approx 0.3$ which results in a maximum $F_{abs} \approx 0.45$ for permeability- limited absorption.

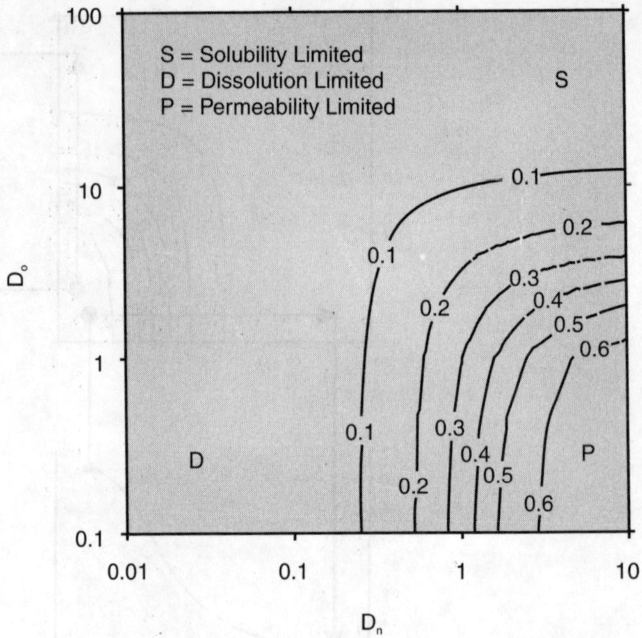

Fig. 9.4. Contour plot of fraction absorbed (F_{abs}) for a low permeability compound showing the different regions of limiting bioavailability.

Calculating the Values of A_n, D_n, and D_o

To facilitate use of the model, estimates for the Absorption, Dissolution and Dose Numbers are necessary. Many of the variables used to calculate A_n, D_n, and D_o are physiologically based and therefore do not change as new compounds are tested. In addition, the density of the solid and the molecular diffusivity (related to molecular weight) do not vary significantly across typical drug-like organic compounds, and so representative values can be used with reasonable accuracy. Therefore, a simple method for estimating A_n, D_n, and D_o is to combine the relatively constant physiological and physical parameters into single factors which are then multiplied by the compound specific variables. The compound specific variables in the first column are multiplied by the species-specific factor to calculate the dimensionless variable in the last column. For example, solubility (in mg/ml) divided by the square of the estimated particle size (in μm) multiplied by 14850 yields the estimated dissolution number for humans. The factors are approximations and are meant to be used a starting point. As one gathers data, it may become evident that a different value may be more appropriate. For example, a class of compounds may have regional permeability issues, and so the factor for permeability may have to be adjusted.

ESTIMATING VARIABLES

To use MiMBA in a practical way, first, estimate the dimensionless variables A_n, D_n, and D_o. Next, choose an absorption contour plot based on the absorption number A_n represent low, moderate, high and very high permeability compounds (A_n = 0.6, 1.3, 8.5 and 20 respectively). As a point of reference, metoprolol, the compound used in the biopharmaceutic classification scheme (BCS) to define the boundary between low and high permeability compounds, has an A_n of 1.33 in humans. Graphically place the calculated D_n and D_o on the contour plot to determine the starting point and the limitation (i.e., dissolution, solubility, or permeability) to bioavailability. Then, choose an appropriate formulation approach as outlined in the following section to enhance bioavailability.

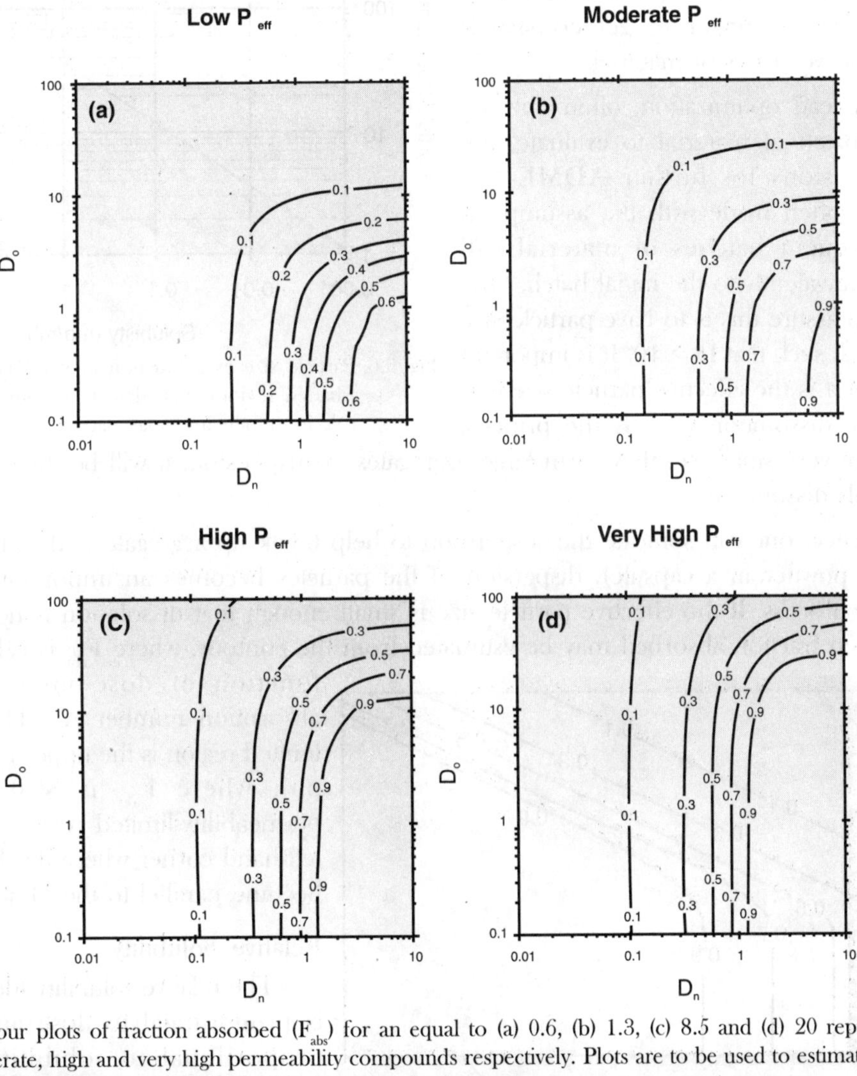

Fig. 9.5. Contour plots of fraction absorbed (F_{abs}) for an equal to (a) 0.6, (b) 1.3, (c) 8.5 and (d) 20 representing low, moderate, high and very high permeability compounds respectively. Plots are to be used to estimate the location of an unknown compound in the biopharmaceutics landscape.

Rate of Dissolution

Dissolution rate is proportional to solubility and inversely proportional to the square of the particle radius. To enhance bioavailability, one must either reduce particle size or increase the solubility (note that increasing solubility effects both D_n and D_o). As D_n becomes greater than 10, the enhancement of $_{Fabs}$ due to particle size reduction decreases. One can deduce this from the contour, as the contour lines begin to run parallel to the D_n-axes at or about $D_n = 10$. Note that this occurs in each of the plots, and so this phenomenon is independent of the permeability. The physical interpretation of $D_n > 10$ is that the time to complete dissolution is at least ten-fold less than the intestinal transit time. From a formulation perspective, it is of benefit for particle size not to be a major determinant of bioavailability in order to get consistent exposure between lots of material.

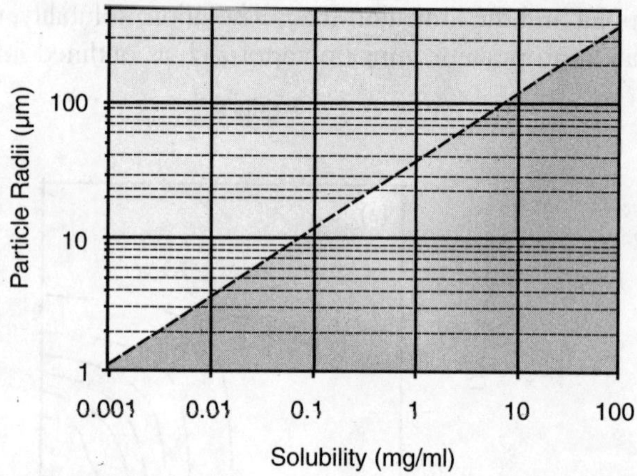

Fig. 9.6. Particle size as a function of solubility for $D_n > 10$ (shaded region) such that dissolution plays a minor role in influencing bioavailability.

During lead optimization, often one has only one batch of material to evaluate, but dosing decisions for further ADME-Tox studies are often made with the assumption that subsequent batches of material will perform equivalently to the initial batch. One way to help assure this is to have particle size small enough such that $D_n > 10$. It is important to note that it is the effective particle size that dictates the dissolution rate. If the primary particles are very small, yet they form large aggregates in suspension, it will be the aggregate size that controls dissolution.

In practice, one can sonicate the suspension to help break up aggregates. Also, if dosed as a solid (e.g., powder in a capsule), dispersion of the particles becomes an important step in the dissolution process. If the effective particle size is small enough that dissolution is not at all rate-limiting, then fraction absorbed may be estimated from the contour, where F_{abs} is calculated as a function of dose number D_o and absorption number A_n. The solubility-limited region is the upper portion of the plot where F_{abs} is < 0.9 and the permeability-limited region is the lower left-hand corner where the F_{abs} contours become parallel to the D_o-axis.

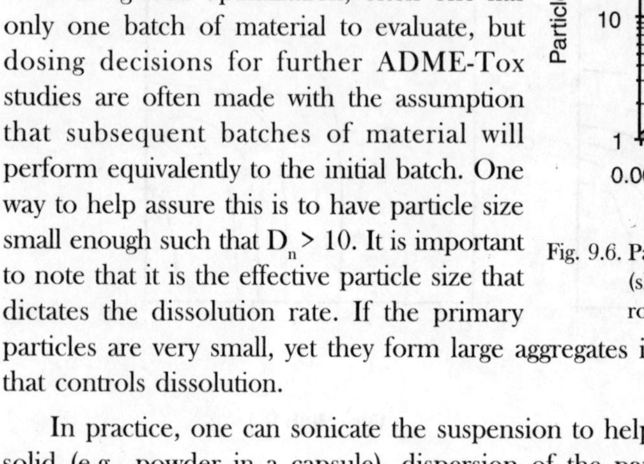

Fig. 9.7. Contour plot of fraction absorbed (F_{abs}) as a function of dose number (D_o) and absorption number (A_n) if dissolution plays no role in influencing bioavailability.

Relative Solubility

The relative solubility (dose number) is proportional to dose and inversely proportional to solubility. Dose is dictated by study design, so lowering it to decrease the dose number is typically not an option. Therefore, increasing solubility becomes the only way to reduce the dose number and increase

F_{abs}. Formulation strategies fall into two general categories. The first is solubilization in a medium, which upon dilution by the intestinal mileau, generates a supersaturated solution. Adding pH-modifying agents to enhance solubility through compound ionization and dosing with co-solvents are common examples of this approach. The second category involves modifying the lattice energy of the solid through disorder (low-crystalline or amorphous materials), or changing the intermolecular bonding network (polymorphs or salts). Other chapters in this section provide more detail on these types of formulations.

Influence of Permeability

Permeability influences the flux across the intestinal membrane (Flux $\propto P_{eff} \cdot$ Concentration), and it is the integrated flux that determines the total amount absorbed. Flux is incorporated into the differential equation describing concentration as the second set of terms (underlined):

$$\frac{dC^*}{dz^*} D_o \cdot D_n \cdot r^* \cdot (1 - C^*) - 2 \cdot A_n \cdot C^*$$

If one is in a permeability-limited region of the biopharmaceutics landscape, then to increase bioavailability, either P_{eff} or solubility needs to be increased. The most effective way to increase permeability is to change the molecule. This is typically not a practical solution, but a pro-drug strategy might be considered if an overall class of compounds has poor permeability. If the molecule is a P-gp substrate, co-administration of a P-gp inhibitor may reduce efflux and increase the effective permeability. Suggested inhibitors include Cremophor EL, Vitamin E-TGPS, and Polysorbate 80.

While this approach has been demonstrated *in vitro*, and there is strong evidence that this mechanism can explain certain drug-drug interactions, verification of its utility as a formulation strategy *in vivo* is limited. The alternative for enhancing flux is to increase solubility via a technology that provides supersaturation compared to the crystalline material at the absorption site, as described in the previous section. The most effective formulation strategies are those that provide the highest degree of supersaturation, yet do not induce precipitation. Materials that inhibit precipitation (typically polymeric) may be added to the formulation. One must keep in mind that while

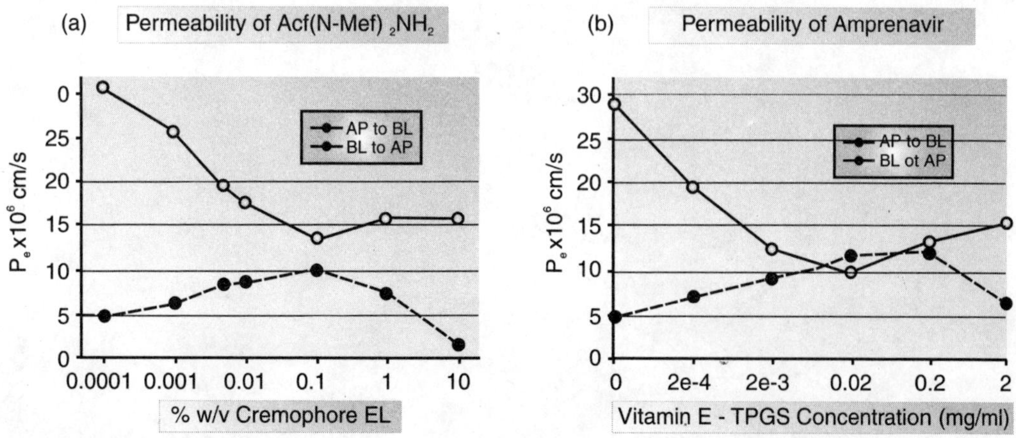

Fig. 9.8. Effect of solubilizing agents on permeability in Caco2 assays demonstrating P-gp efflux inhibition at low surfactant concentrations and decreased chemical potential at higher surfactant concentrations.

concentration is the factor used in the flux equation, it is the chemical potential that dictates the true driving force.

The chemical potential is modified by the presence of solubilizing agents competing with the membrane for drug. The attenuation of the driving force is most pronounced when there is an excess of the solubilizing agent relative to the amount needed to completely solubilize the drug. The effect has been demonstrated *in vitro* in Caco2 assays. In both examples surfactant added acted as a P-gp efflux inhibitor. At low concentrations, the AP (apical) to BL (basolateral) flux increased with a corresponding decrease in the BL to AP flux. As the concentration of solubilizing agent further increased, a decrease in the AP to BL flux was observed and was attributed to the competition of the surfactant micelles for drug molecules and therefore a decrease in the chemical potential across the Caco2 membrane.

CONCLUDING REMARKS

To enhance bioavailability through formulation technologies, determine where you sit in the biopharmaceutics landscape via the contour plots presented in Figures 9.5a-d, then select the appropriate formulation technology to move up the slope to higher $_{\text{Fabs.}}$ This scheme is meant to be qualitative, a starting point to estimate the direction one must take. As *in vivo* data is collected, one should develop a feedback system to adjust the model parameterization. To estimate the compound's effective permeability, one must have an established correlation of an *in vitro* or ex vivo method to P_{eff}.

Reducing effective particle size will never hurt bioavailability. If dosed as a suspension, agglomeration of primary particles increases the effective particle size. If dosed as a solid, rapid dispersion of the formulation to primary particles is important for assuring maximum exposure. If using solubility enhancers, too much can decrease bioavailability via a reduction in chemical potential. This is likely to happen only after complete solubilization of the compound.

10

Pathophysiology of Epilepsy

Epilepsy is a disease of high prevalence, being well known since thousands of years as "*morbus sacer*". In spite of intensive investigations, the pathophysiology of epilepsy is still poorly understood. Studies with various animals models have provided ample evidence for heterogeneity in the mechanisms of epileptogenesis. New evidence derives from investigations of kindling, which involves the delivery of brief, initially subliminal, electrical or chemical stimuli to various areas of the brain. After 10 to 15 days of once-daily stimulation, the duration and intensity of after-discharges reach a stable maximum and a characteristic seizure is produced. Subsequent stimulation then regularly elicits seizures. Surveys of methods being used to test compounds with anticonvulsant properties have been provided by Toman and Everett (1964), Woodbury (1972), Hout et al. (1973), Swinyard (1973), Koella (1985), Meldrum (1986), Rump and Kowalczyk 1987; Loscher and Schmidt (1988), Fisher (1989), Rogawski and Porter (1990), Porter and Rogawski (1992).

Epilepsy becomes drug resistant in 20%-30% of patients. Out of the animal models, the amygdalakindled rat seems to be a suitable approach. Furthermore, the rat cortical dysplasia model is recommended. Several biochemical hypotheses have been advanced, involving the inhibitory GABAergic system and the system of the excitatory amino acids glutamate and aspartate. Excitatory receptors have been divided into subtypes according to the actions of specific agonists or antagonists. Agents which reduce GABAA synaptic function provoke convulsions. A convulsive state is induced by the direct blockade of GABAA receptors (e. g. to the action of bicuculline) or a reduction in the GABA-mediated opening of the chloride ion channel (e. g., by picrotoxin).

One major factor in epileptogenesis seems to be a decreased function of GABAA synapses. More recently, research has focused on the therapeutic potential of blocking excitatory amino acids – in particular glutamate. Of the three receptors of glutamate, the NMDA (N-methyl-D-aspartate)-receptor is considered the one of most interest in epilepsy and competitive NMDA receptor antagonists are proposed as potential anti-epileptic drugs. Excessive excitatory amino acid neurotransmission is thought to be associated with the neuropathologies of epilepsy, stroke and other neurodegenerative disorders. Antagonism of NMDA receptor function appears to be the mechanism of action of some novel anticonvulsant and neuroprotective agents. Excitatory amino acid receptors have been classified into at least three subtypes by electrophysiological criteria: NMDA, *quisqualic acid* (QA) and *kainic acid* (KA). Fabene and Sbarbati (2004) underlined the value of *in vivo* MRI in different models of experimental epilepsy.

INHIBITORY ACTION OF GABA

Roberts (1974) and others have proposed that the inhibitory action of the amino acid γ-aminobutyric acid (GABA) is the fine tuning control for pacemaker neurons. Disruption of this interplay due to inadequacies of the GABA system result in various disorders, in particular convulsive seizures. The nonspecific action of GABA-mimetics makes inhibition of the uptake mechanism, which terminates the neurotransmitters action, the ideal choice for increasing GABA's concentration at specific sites. Demonstration of the high affinity mechanism that best reflects the *in vivo* condition utilizes GABA depleted cerebral cortex synaptosomes. Although the physiological role of GABA transport systems is still unclear, uptake inhibitors such as THPO [4,5,6,7-tetrahydroisoxazolo-(4,5-C)pyrid-3-ol], nipecotic acid, cis-4-hydroxynipecotic acid, and guvacine exhibit anticonvulsant effects. Furthermore, a number of neuroleptics have been shown to inhibit GABA uptake. In particular, fluspirilene was found to be equivalent to the most potent uptake inhibitors known. The assay is used as a biochemical screen for potential anticonvulsants or GABA (γ-aminobutyric acid) mimetic compounds that act by inhibiting GABA uptake.

Method

Requirements

1. 0.5 M Tris buffer, pH7.4
2. Ringer's solution +10 mM Tris buffer, pH 7.4 containing
   - glucose 10.0 mM,
   - NaCl 150.0 mM,
   - KCl 1.0 mM,
   - $MgSO_4$ 1.2 mM,
   - Na_2HPO_4 1.2mM.
3. Depolarizing Ringer's solution, pH7.4 reagent 2 containing
   - KCl 56 mM,
   - $CaCl_2$ 1 mM.
4. 0.32 M sucrose
5. ^{3}H-GABA is diluted to 2.5×10^{-4} M with distilled water. Forty µl of this solution in 1 ml of reaction mixture will yield a final concentration of 10^{-5} M.
6. *Test compounds.* A 10 mM stock solution is made up in distilled water, ethanol, or DMSO and serially diluted, such that the final concentration in the assay ranges from 10^{-3} to 10^{-8} M. Total and nonspecific controls should use solvent of test compound.

Sampling

Male Wistar rats are decapitated and the brains rapidly removed. Cerebral cortex is weighed and homogenized in 9 volumes of ice-cold 0.32 M sucrose using a Potter-Elvejhem homogenizer. The homogenate is centrifuged at 1000 *g* for 10 min. The supernatant (S_1) is decanted and recentrifuged at 1000 *g* for 10 min. The pellet (P_2) is resuspended in 9 volumes of 0.32M sucrose and centrifuged at 24,000 *g* for 10 min. The washed pellet is resuspended in 15 volumes of depolarizing Ringer's solution, incubated at 25°C for 10 min and centrifuged at 3000 *g* for 10 min. The resulting pellet is resuspended in 15 volumes of Ringer's solution and is ready for use.

Analyzing process

60 μl Ringer's solution,

100 μl vehicle or appropriate drug concentration,

800 μl tissue suspension.

Microcentrifuge tubes are set up in triplicate. Non-specific controls are incubated at 0°C and totals at 25°C for 10 min. Forty μl of ^{3}H-GABA are added and the tubes are reincubated for 10 min. All tubes are centrifuged at 13,000 *g* for 1 min. The supernatant is aspirated and 1 ml of solubilizer (Triton X-100 + 50% EtOH, 1:4, v/v) is added and mixed to dissolve pellets. Tubes are incubated at 90°C for 3 min, then centrifuged at 13,000 *g* for 15 min. 40 μl of supernatant is counted in 10 ml Liquiscint scintillation cocktail.

Assessment

Active uptake is the difference between cpm at 25°C and 0°C. The percent inhibition at each drug concentration is the mean of three determinations. IC_{50} values are derived from log-probit analysis.

GABA Transporter

The *GABA transporter*, the subsynaptic $GABA_A$-receptor, and the $GABA_B$-autoreceptor are therapeutically the most relevant targets for drug actions influencing GABAergic synaptic transmission. Uptake inhibitors are potential anticonvulsants.

Method

For measurement of GABA uptake, rat hippocampal slices are cut with a McIlwain tissue slicer (100 μm-thick prisms) and dispersed in ice-cold Krebs-Ringer solution with HEPES buffer (pH 7.4). Following two washes, slices (15 mg) are incubated at 37°C for 15 min in the presence or absence of test compound. [^{3}H]-GABA is added and samples are incubated for an additional 5 min before filtration through Whatman GF/F filters. Samples are then washed twice with 5 ml ice-chilled 0.9% saline. Distilled water is added and samples are allowed to sit at least 60 min before measured for radioactivity by liquid scintillation spectroscopy. Blanks are treated in an identical manner but are left on ice throughout the incubation. *For measurement of GABA release,* rat hippocampal slices are prepared and dispersed in ice-cold HEPES-buffered (pH 7.2) Krebs Ringer solution and incubated with 0.05 μM [^{3}H]-GABA for 15 min at 37°C. Following two washes, the slices are incubated for an additional 15 min and finally resuspended in medium. Tissue (10 mg) is incubated at 37°C for a 15 min release period in the presence or absence of test compound. At the end of the release period, the medium is separated from tissue by centrifugation at 500 *g* for approximately 1 min and poured into 0.5 ml of perchloric acid (0.4 N). The tissue is homogenized in 0.13 N perchloric acid. Radioactivity in the samples is measured by using liquid scintillation spectroscopy.

Assessment

For GABA-uptake, IC_{50}-values (μM) are determined. In GABA-release experiments, results are expressed as the amount of radioactivity released as a percent of the total radioactivity.

Latest technique

Roskoski (1978) studied the net uptake of GABA by high affinity synaptosomal transport systems. Nilsson et al. (1990, 1992) tested GABA uptake in astroglial primary cultures. The *isolated nerve-bouton preparation* was used to study GABA release. The technique was developed by Drewe et al. (1988), Vorobjev (1991), Haage et al. (1998), Rhee et al. (1999), and Koyama et al. (1999). The method is based on the local application of mechanical vibration directly to the chosen site of a brain slice and does not require the enzymatic pretreatment of the tissue. The mechanical vibration is applied via a glass rod (0.5 mm in diameter) mounted on a *piezoelectric* bimorph crystal at the site of the chosen brain tissue. The dissociated cells are allowed to settle at the bottom of a Petri dish for 20 min.

The cell bodies are usually 10-15 μm at their longest axis, rounded or elongated in shape. Some cells had remaining neurites up to 100 μm long. The majority of cells had neurites less than 15 μm long. In other studies, a custom-built vibrating stylus was placed in the appropriate region for mechanical dissociation. The glass capillary (1.5 mm o.d.) was pulled to a fine tip and fire-polished. The tip was placed within the appropriate region by a manipulator. The vibrating stylus was driven by an electronic relay and the tip was horizontally moved (excursions of 2-3 mm at 0.5-2 Hz) for 2 min. Neurons with adherent functional synaptic terminals were investigated by tight-seal whole-cell recordings from the postsynaptic cells.

Ionotropic Glutamate Receptors

The *ionotropic glutamate receptors* are ligand-gated ion channels that mediate the vast majority of excitatory neurotransmission in the brain. The family comprises three pharmacologically defined classes that were originally named after reasonably selective ligands: *N*-methyl-D-aspartate (NMDA), α-amino-3-hydroxy-5-methylisoxazole-4-propionic acid (AMPA), and kainate. It turned out that NMDA, AMPA and kainate receptor subunits are encoded by at least six gene families as defined by sequence homology: a single family of AMPA receptors, two for kainate, and three for NMDA. The NMDA subtype is a hetero-oligomer consisting of an NR1 subunit combined with one or more NR2 subunits and a third subunit, NR3.

The receptor has two amino acid recognition sites, one for glutamate and one for glycine, both of which must be occupied to promote channel opening. A variety of drugs have been identified which block the channel selectively. The AMPA subtype is a hetero-oligomer formed from combinations of iGluR 1-4. Selective agonists and competitive antagonists acting at the glutamate recognition site have been useful for defining the physiological and pathophysiological roles played by the receptor. AMPA receptor modulators have been discussed as cognitive enhancers. The kainate subtype consists of hetero-oligomers, comprising five subunits. Excessive excitatory amino acid neurotransmission has been associated with the neuropathologies of epilepsy, stroke and other neurodegenerative disorders.

Antagonism of NMDA receptor function appears to be the mechanism of action of some anticonvulsant and neuroprotective agents. The binding site for [^{3}H]2-amino-4-phosphonobutyric acid (AP4) may represent a fourth site which is less well characterized. NMDA receptors are believed to be coupled to a cation channel which converts to an open state with NMDA receptor activation. The opening and closing of this cation channel are also modulated by glycine, Mg^{2+} and Zn^{2+}. Dissociative anesthetics, such as phencyclidine (PCP) and ketamine, and novel anticonvulsants, such as MK-801, block the ion channel and are noncompetitive NMDA receptor antagonists. Competitive NMDA receptor antagonists, such as CPP and the phosphono analogs of L-glutamate, AP7 and AP5 (2-amino-5- phosphonopentanoic acid), are inhibitors at the excitatory amino acid binding site. The following assay is used to assess the affinity of compounds for the excitatory amino acid binding site of the NMDA receptor complex. [^{3}H]CPP 3-[(±)-2- carboxypiperazin-4-yl]-1-phosphonic acid is a structurally rigid analog of the selective NMDA receptor antagonist 2-AP7 (2-amino-7-phosphonoheptanoic acid).

Method

Requirements

1. Buffer A: 0.5 M Tris HCl,pH7.6

   60.0 g Tris HCl

   13.9 g Tris Base

   q.s. to 1 liter with distilled water

2. Buffer B: 50 mM Tris HCl, pH 7.6

   Dilute buffer A 1:10 with distilled water

3. L-Glutamic acid, 5×10^{-3} M

   Dissolve 7.36 mg of l-glutamic acid with 10.0 ml distilled water. Aliquots of 20 µl to the assay tube will give a final concentration of 10^{4} M.

4. [^{3}H]CPP is obtained from New England Nuclear, specific activity 25-30 Ci/mmol. For IC_{50} determinations, a 200 nM stock solution is made with distilled water. Aliquots of 50 µl are added to each tube to yield a final concentration of 10 mM.

5. Test compounds. A stock solution of mM is made with a suitable solvent and serially diluted, such that the final concentration in the assay ranges from 10^{-5} to 10^{-8} M. Higher or lower concentrations may be used, depending on the potency of the drug.

6. Triton-X 100, 10% (v/v) A stock solution of Triton-X100, 10%, can be prepared and stored in the refrigerator. Dilute 1.0 ml of Triton-X100 to 10.0 ml with distilled water. On the day of the assay, the tissue homogenate (1:15 dilution) is preincubated with an aliquot of Triton-X 100, 10%, to give a final concentration of 0.05% (v/v).

Sampling

Cortices of male Wistar rats are dissected over ice and homogenized in ice-cold 0.32 M sucrose, 15 volumes of original wet weight of tissue, for 30s with a Tissumizer setting at 70. The homogenate is centrifuged at 1000 *g* for 10 min. (SS34, 3000 rpm, 4°C). The supernatant is centrifuged at 20,000 *g* (SS34, 12,000 rpm, 4°C) for 20 min. Resuspend the pellet in 15 volumes of ice-cold distilled water (Tissumizer setting 60, 15 s) and spin at 7600*g* (SS34, 8000 rpm, 4°C) for 20 min. Save the supernatant, swirl off the upper buffy layer of the pellet and add to the supernatant. Centrifuge the supernatant at 48,000 *g* (SS34, 20,000 rpm, 4°C) for 20 min.

Resuspend the pellet with 15 volumes of cold distilled water and centrifuge. Discard the supernatant and store the pellet at -70°C. On the day of the assay, resuspend the pellet in 15 volumes ice-cold 50 mM Tris buffer, pH7.6. Preincubate the homogenate with Triton-X in a final concentration 0.05% (v/v) for 15 min at 37°C with agitation. Centrifuge the homogenate at 48,000 *g* (SS 34, 20,000 rpm, 4°C for 20min. Wash the pellet an additional 3 times by resuspension with cold buffer and centrifugation. The final pellet is resuspended in a volume 20 times the original wet weight.

Analyzing process

1. Prepare assay tubes in triplicate.

   380 µl distilled water

   50 µl buffer A, 0.5 M Tris HCI, pH7.6

   20 µl L-glutamic acid, 10^{-4} M, or destilled water, or appropriate concentration of inhibitor

   50 µl [^{3}H]CPP

   500 µl tissue homogenate.

2. Following the addition of the tissue, the tubes are incubated for 20 min at 25°C with agitation. Place the tubes in an ice bath at the end of the incubation. Terminate the binding by centrifugation (HS4, 7000 rpm, 4°C) for 15 min. Return the tubes to ice. Aspirate the discard the supernatant. Carefully rinse the pellet three times with 1 ml ice-cold buffer, avoiding disruption of the pellet. Transfer the pellet to scintillation vials by vortexing the pellet with 2 ml scintillation fluid, rinse the tubes twice with 2 ml and add an additional 4 ml scintillation fluid.

Assessment

Specific binding is determined from the difference of binding in the absence of presence of 10^{-4} M L-glutamic acid and is typically 60-70% of total binding. IC_{50} values for the competing drug are calculated by log-probit analysis of the data.

Latest methodology

Binding of glutamate

The assay measures the binding of glutamate, which binds non selectively to ionotropic glutamate receptors including the NMDA, AMPA, and kainate subtypes. In addition, glutamate binds to a family of metabotropic glutamate receptors. Whole brains (except cerebellum) are obtained from male Wistar rats. A membrane fraction is prepared by standard techniques. Ten mg of membrane preparation is incubated with 1.6 nM [^{3}H]L-glutamate for 10 min at 37°C. Non-specific binding is estimated in the presence of 50 μM L-glutamate. Membranes are filtered and washed 3 times to separate bound from free ligand and filters are counted to determine [^{3}H]L-glutamate bound. Convulsions induced in mice by intravenous injections of 2.0 mmol/kg L-glutamic acid can be inhibited by glutamate antagonists.

Binding of glutamate AMPA

The assay measures the binding of [^{3}H]AMPA (α-amino-3-hydroxy-5-methyl-4-isoxazole propionic acid), a selective agonist which binds to the AMPA receptor subtype of glutamate-gated ion channels. Membranes are prepared from male rat brain cortices by standard techniques. Fifteen mg of membrane preparation is incubated with 5 nM [^{3}H]AMPA for 90 min at 4°C. Non-specific binding is estimated in the presence of 1 mM L-glutamate. Membranes are filtered and washed 3 times and the filters are counted to determine [^{3}H]AMP bound. Mutel et al. (1998) recommended [^{3}H]Ro 48-8587 as specific for the AMPA receptor. Fleck et al. (1996) described AMPA receptor heterogeneity in rat hippocampal neurons. AMPA receptor antagonists were described by Kohara et al. (1998), Wahl et al. (1998), Kodama et al. (1999), Nielsen et al. (1999) and reviewed by Chimirri et al. (1999).

Binding of glutamate kainate

The assay measures the binding of [^{3}H]kainate, a selective agonist that binds to the kainate subtype of the ionotropic glutamate receptors in rat brain. Whole brains (except cerebellum) are obtained from male Wistar rats. Fifteen mg of a membrane fraction prepared by standard techniques is incubated with 5.0 nM [^{3}H]kainate for 1 h at 4°C. Non-specific binding is estimated in the presence of 1 mM L-glutamate. Membranes are filtered and washed 3 times to separate free from bound ligand and filters are counted to determine [^{3}H]kainate bound. Toms et al. (1997), Zhou et al. (1997) recommended [^{3}H]-(2S,4R)-4-methylglutamate as kainate-receptor selective ligand. Irreversible inhibition of high affinity [^{3}H]kainate binding by a photoactivatable analogue was reported by Willis et al. (1997). Worms et al. (1981) described the behavioral effects of systemically administered kainic acid. Hu et al. (1998) described neuronal stress and seizure-induced injury in C57/BL mice after systemic kainate administration.

Binding of glutamate NMDA-agonist site

The assay measures the binding of CGS 19755, a selective antagonist, to the agonist site of the NMDA receptor.

Binding Affinity of NMDA

The purpose of this assay is to determine the binding affinity of potential noncompetitive NMDA antagonists at the phencyclidine (PCP) binding site which is believed to be within or near the NMDA-regulated ion channel. TCP, 1-[1-(2-thienyl) cyclohexyl]-piperidine, is a thienyl derivative of PCP. Excessive activity of excitatory amino acid neurotransmitters has been associated with the neuropathologies of epilepsy, stroke and other neurodegenerative disorders. Antagonism of NMDA receptor function appears to be the mechanism of action of some novel anticonvulsant and neuroprotective agents. Excitatory amino acid receptors have been classified into at least three subtypes by electrophysiological criteria: NMDA, *quisqualic acid* (QA) and *kainic acid* (KA). The binding site for [^{3}H]2-amino-4-phosphonobutyric acid (AP4) may represent a fourth site which is less well characterized. NMDA receptors are believed to be coupled to a cation channel which converts to an open state following activation.

The opening and closing of this cation channel are also modulated by glycine, Mg^{2+}, Zn^{2+} and polyamines. Dissociative anesthetics, such as phencyclidine (PCP) and ketamine, and the neuroprotective agent MK-801 block the ion channel and are noncompetitive NMDA receptor antagonists. Competitive NMDA receptor antagonists, such as 3-[(±)-2-carboxypiperazin-4-yl]-1-phosphonic acid (CPP), and the phosphono analogs of L-glutamate, 2-amino-7-phosphonoheptanoic acid (2-AP7) and 2-amino-5-phosphonopentanoic acid (2-AP5), are inhibitors at the excitatory amino acid recognition site. Molecular cloning and functional expression of rat and mouse NMDA receptors, a family of AMPA-selective glutamate receptors and the metabotropic glutamate receptors mGluR1-mGluR6 have been reported.

Method I

Requirements

1. Buffer A: 0.1MHEPES,pH7.5

   Weigh 23.83 g HEPES

   Add approximately 900 ml distilled water.

   Adjust pH to 7.5 with 10 N NaOH.

   q.s. to 1 liter with distilled water.

2. Buffer B: 10 mM HEPES, pH 7.5

   Dilute buffer A 1:10 with distilled water and adjust pH to 7.5.

3. L-Glutamic acid, 5×10^{-3} M

   Dissolve 7.36 mg with 10.0 ml distilled water.

   Aliquots of 20 μl to the assay tube will give a final concentration of 10^{-4} M.

4. Glycine, 5×10^{-4} M

   Dissolve 3.75 mg with 10.0 ml distilled water. Dilute 1:10 with distilled water.

   Aliquots of 20 μl to the assay tube will give a final concentration of 10^{-5} M.

5. Phencyclidine HCl (PCP) is used for nonspecific binding.

   Dissolve 0.7mg in 0.5 ml distilled water.

   Aliquots of 20 μl to the assay tube will give a final concentration of 10^{-4} M.

6. [^{3}H]TCP is obtained from New England Nuclear, specific activity 42-60 Ci/mmole. For IC_{50} determinations, a 50 nM stock solution is made with distilled water. Aliquots of 50 μl are added to each tube to yield a final concentration of 2.5 nM.

7. Test compounds. A stock solution of 5 mM is made up with a suitable solvent and serially diluted, such that the final concentration in the assay ranges from 10^{-5} to 10^{-8} M. Higher or lower concentrations may be used, depending on the potency of the drug.

Sampling

Cerebral cortex of male Wistar rats, 7-10 weeks of age, is dissected over ice and homogenized in ice-cold 0.32 M sucrose, 30 volumes of original tissue weight, for 60s with a Tissumizer setting at 70. The homogenate is centrifuged at 1000 *g* for 10 min. (SS34, 3000 rpm, 4°C). The supernatant is centrifuged at 20,000 *g* for 20 min. (SS34, 12,000 rpm, 4°C). The pellet is resuspended with cold distilled water, to 50 volumes of original tissue weight, using the Tissumizer, 60s at setting of 70. The homogenate is incubated at 37°C for 30 min, transferred to centrifuge tubes, and centrifuged at 36,000 *g* for 20 min. (SS34, 16,500 rpm, 4°C). The pellet is again resuspended in 50 volumes distilled water, incubated and centrifuged.

All resuspensions with the Tissumizer are for 60s at a setting of 70. The resulting pellet is resuspended in 30 volumes of ice-cold 10 mM HEPES buffer, pH 7.5, centrifuged, and washed once again (resuspension and centrifugation) with buffer. Following resuspension in 30 volumes of buffer, the homogenate is frozen in the centrifuge tube and stored at –70°C until the day of the assay. On the day of the assay, the homogenate is thawed and centrifuged at 36,000 *g* for 20 min. (SS34, 16,500 rpm, 4°C). The pellet is washed three times by resuspension with ice-cold 10 mM HEPES buffer, pH 7.5, centrifuged and finally resuspended in 30 volumes of buffer. Aliquots of 500 µl are used for each assay tube, final volume 1000 µl, and correspond to approximately 0.2 mg protein.

Analyzing process

1. Prepare assay tubes in triplicate. For each test compound, inhibition of [^{3}H]TCP binding is measured both in the absence (basal) and presence (stimulated) of 100 µM L-glutamic acid and 10 mM glycine.

Basal	*Stimulated*	
380 µl	340 µl	Distilled water
50 µl	50 µl	Buffer A, 0.1 M HEPES, pH 7.5
20 µl	20 µl	PCP (reagent A5) or distilled water, or appropriate concentration of inhibitor
0 µl	20 µl	L-glutamic acid (reagent A3)
0 µl	20 µl	Glycine (reagent A4)
50 µ	50 µl	[^{3}H]TCP (reagent A6)
500 µl	500 µ	Tissue homogenate

2. Following the addition of the tissue, the tubes are incubated for 120 min at 25°C with agitation. The assay is terminated by separating the bound from non- bound radioligand by rapid filtration with reduced pressure over Whatman GF/B filters, pre-soaked in 0.05% polyethylene-imine, using the Brandell cell harvesters. The filters are rinsed once with buffer before filtering the tubes, and rinsed two times after filtration. The filters are counted with 10 ml Liquiscint.

Assessment

Specific binding is determined from the difference of binding in the absence or presence of 10^{-4} M PCP. Specific binding is typically 50% of total binding in basal conditions, and 90% of total binding when stimulated by L-glutamic acid and glycine. L-Glutamic acid and glycine typically increase specific binding to 300% and 200% of basal binding, respectively. The combination of L-glutamic acid and glycine typically produce a greater than additive effect, increasing specific binding to 700% of basal binding. IC_{50} values for the competing drug are calculated by log-probit analysis of the data.

Latest approach

This modified procedure for the preparation of membrane homogenates does not use extensive lysing and washing of the tissue to remove endogenous L-glutamate, glycine and other endogenous compounds which enhance [^{3}H]TCP binding. This procedure may be used for rapid screening of compounds for inhibition of [^{3}H]TCP binding site without specifically defining an interaction at the ion channel or modulatory sites of the NMDA receptor complex.

Method II

Requirements

1. Buffers A and B are prepared as described above.
2. Phencyclidine HCl is used for nonspecific binding and is prepared as described above.
3. [^{3}H]TCP is prepared as described above.
4. Test compounds are prepared as described above.

Sampling

Cortical tissue is dissected, homogenized in 30 volumes of 0.32 M sucrose and a crude P_2 pellet is prepared as described above. The pellet is resuspended in 30 volumes of 10 mM HEPES, pH7.5, centrifuged at 36,000*g* (SS34, 16,500 rpm, 4°C) for 20 min, and again resuspended in 100 volumes of buffer. This homogenate is used directly in the assay in aliquots of 500 µl.

Analyzing process

1. Prepare assay tubes in triplicate.

Volume	*Solution*
380 µl	Distilled water
50 µl	Buffer A, 0.1 M HEPES, pH 7.5
20 µl	PCP (reagent IA5) or distilled water, appropriate concentration of inhibitor
50 µl	[^{3}H]TCP (reagent IA6)
500 µl	Tissue homogenate

2. Following the addition of the tissue, the tubes are incubated for 120 min at 25°C with agitation. The assay is terminated by rapid filtration as described above. The filters are rinsed and counted for bound radioactivity as above.

Assessment

Specific binding is determined from the difference of binding in the presence or absence of 10^{-4} M PCP. Specific binding is typically 90% of total binding. IC_{50} values for the competing drug are calculated by log-probit analysis.

Latest technology

Instead of [^{3}H]TCP, radiolabeled [^{3}H]MK-801 has been used as ligand. Sills et al. (1991) described [^{3}H]CGP 39653 as a N-methyl-D-aspartate antagonist radioligand with low nanomolar affinity in rat brain. Nowak et al. (1995) reported that swim stress increases the potency of glycine to displace 5,7-[^{3}H[dichlorokynurenic acid from the strychnine-insensitive glycine recognition site of the *N*-methyl-D-aspartate receptor complex. NMDA receptor cloning studies have shown that NMDA receptors contain at least one of seven different NMDAR1 subunits (NR1A-NR1G) and at least one of four NMDAR2 subunits (NR2A-NR2D). While the NR1 subunits are generated by alternative splicing of a single gene, the NR2 subunits are products of four highly homologous genes. Thus, there are thousands of potential subunit combinations yielding complexes of four or five subunits.

Grimwood et al. (1996) reported generation and expression of stable cell lines expressing recombinant human NMDA receptor subtypes, two cell lines expressing NR1a/NR2A receptors and one cell line expressing NR1a/NR2B receptors. NR2B selective NMDA antagonists were described by Fischer et al. (1997), Kew et al. (1998), Reyes et al. (1998), Chenard and Menniti (1999). For discovery of novel NMDA receptor antagonists, Bednar et al. (2004) developed a high-throughput functional assay based on fluorescence detection of intracellular calcium concentrations. Mouse fibroblasts L(tk-) cells expressing human NR1a/NR2B NMDA receptors were plated in 96-well plates and loaded with fluorescence calcium indicator fluo-3 AM. NR2B antagonists were added after stimulation of NMDA receptors with 10 μM glutamate and 10 μM glycine. Changes in fluorescence after addition of the antagonists were fitted with a single exponential equation providing k_{obs}.

Native Receptors

In addition to ionotropic (AMPA, kainate and NMDA) receptors, glutamate interacts with a second family of receptors, metabotropic or mGlu receptors. Three groups of native receptors are distinguishable on the basis of similarities of agonist pharmacology, primary sequence, and G protein-effector coupling: Group I (mglu$_1$ and mglu$_5$ and splice variants) are coupled via $G_{q/11}$ to phosphoinositide hydrolysis. Group II (mglu$_2$ and mglu$_3$) are negatively coupled via G_i/G_o to adenylyl cyclase and inhibit the formation of cAMP following exposure of cells to forskolin or activation of an intrinsic G_s-coupled receptor (e. g. adenosine A_2 receptor).

The group III receptors (mglu$_4$, mglu$_6$, mglu$_7$, and mglu$_8$) also inhibit forskolin-stimulated adenylyl cyclase. Various agonists and antagonists for metabotropic glutamate receptors were described. Schoepp et al. (1999) reviewed pharmacological agents acting at subtypes of metabotropic glutamate receptors. Gssparini et al. (2002) described allosteric modulators of group I metabotropic glutamate receptors as novel subtype- selective ligands and their therapeutic perspectives. Several radioligands for metabotropic glutamate receptors were described:

1. for subtype mGluR4a receptor by Eriksen and Thomsen (1995),
2. for group II mGlu receptors by Cartmell et al. (1998), by Ornstein et al. (1998), and by Schaffhauser et al. (1998).

Riedel and Reymann (1996) discussed the role of metabotropic glutamate receptors in hippocampal long-term potentiation and long-term depression and their importance for learning and memory. Furthermore, possible roles in the treatment of neurodegenerative disorders, and of Parkinson's disease were discussed. Anticonvulsive properties as well as anxiolytic properties of metabotropic glutamate receptor ligands were reported. Christoffersen et al. (1999) found a positive effect on short-term memory and a negative effect on long-term memory of the class I metabotropic glutamate receptor antagonist, AIDA, in rats.

Method

Cultured cells are prepared from cerebral cortex of 17-day embryos of Wistar rats. Prior to the experiments, the culture is maintained for 8-12 days with minimum essential medium (MEM) containing 5% fetal calf serum and 5% horse serum. For *cyclic AMP assays,* the cultured cells are preincubated with HEPES-buffered Krebs-Ringer solution containing 5.5 mM glucose (HKR) for 1-1.5 h, then exposed to various agonists for 15 min in the absence or presence of 10 μM forskolin. The content of cyclic AMP is measured using a radioimmuno-assay kit after homogenization with 0.1 M HCl. For phosphoinositide turnover assays, the cultured cells are prelabeled with myo-1,2-[^{3}H]inositol in MEM for 8-10 h. The cells are washed twice with HKR containing 10 mM LiCl, and then exposed to various agonists in HKR containing 10 mM LiCl for 30 min. The reaction is terminated with 2% trichloroacetic acid, and the homogenized samples are analyzed for inositol constituents by anion exchange chromatography. The extracts are applied to columns containing 1 ml of Dowex 1 in the formate form.

The phosphate esters are then eluted by the step-wise addition of solutions containing increasing concentrations of formate. Glycerophosphoinositol and inositol 1:2-cyclic phosphate are eluted with 5 mM-sodium tetraborate plus 150 mM sodium formate. The penultimate solution contains 0.1 M-formic acid plus 0.3 M-ammonium formate, followed by 0.1 M-formic acid plus 0.75 M-ammonium formate, each of which removes more polar inositol phosphates. The 1 ml fractions eluted from the columns are counted for radioactivity after addition of 10 ml of Biofluor. The percentage of radioactivity of inositol phosphates to the total applied to the column is calculated.

Assessment

Dose-response curves for inhibition of forskolin-stimulated cAMP formation and for percentage of phosphoinositide hydrolysis are established for each test compound.

Latest technology

Thomsen et al. (1993, 1994) used baby hamster kidney (BHK) cells stably expressing $mGluR_{1\alpha}$, $mGluR_2$ or $mGluR_4$ for measurements of phosphoinositol hydrolysis or cAMP formation. Varney and Suto (2000) recommended functional high throughput screening assay for the discovery of subtype-selective metabotropic glutamate receptor ligands.

Glutamate as Excitatory Transporter

Glutamate is not only the predominant excitatory neurotransmitter in the brain but also a potent neurotoxin. Following release of glutamate from presynaptic vesicles into the synapse and activation of a variety of ionotropic and metabotropic glutamate receptors, glutamate is removed from the synapse. This is achieved through active uptake of glutamate by transporters located pre- but also post-synaptically, or glutamate can diffuse out of the synapse and be taken up by transporters located on the cell surface of glial cells. The excitatory amino acid transporters form a gene family out of which at least 5 subtypes were identified. A role for glutamate transporters has been postulated for acute conditions such as stroke, CNS ischemia, and seizure, as well as in chronic neurodegenerative diseases, such as Alzheimer's disease and amyotrophic lateral sclerosis. Glutamate transport is coupled to sodium, potassium and pH gradients across the cell membrane creating an electrogenic process. This allows transport to be measured using electrophysiological techniques.

Method

Complementary DNAs encoding the human glutamate transporters, EAAT1 and EAAT2, are subcloned into pOTV for expression in X. laevis oocytes. The plasmids are linearized with BAMHI and cRNA is transcribed from each of the cDNA constructs with T7 RNA polymerase and capped with 5′,7-methyl guanosine using the mMES SAGE mMACHINE (Ambion, Austin TX). cRNA (50 ng) encoding either EAAT1 or EAAT2 is injected into defolliculated Stage V X. laevis oocytes. Two to 7 days later, transport is measured by two-electrode voltage-clamp recording using a Geneclamp 500 amplifier and a MacLab 2e recorder and controlled using a pCLAMP 6.01 interfaced to a Digidata 1200. Oocytes are voltage-clamped at –60 mV and continuously superfused with ND96 buffer (96 mM NaCl, 2 mM KCl, 1.8 mM $CaCl_2$, 1 mM $MgCl_2$ and 5 mM HEPES, pH 7.5). For transport measurement, this buffer is changed to one containing the indicated concentration of substrate and/or blocker.

The voltage dependence of block of glutamate transport is measured by clamping the membrane potential at –30 mV and then applying a series of 100 ms voltage pulses from –100 mV to 0 mV and measuring the steady state current at each membrane potential. This protocol is applied both before and during the application of the compound in question and then the base-line current at each membrane potential is subtracted from the current in the presence of the compounds to get a measure for the transport specific current at the various membrane potentials.

Assessment

Current (I) as a function of substrate concentration ([S]) is fitted by least squares to

$$I = I_{max}[S]/(K_m + [S])$$

where I_{max} is the maximal current and K_m is the Michaelis transport constant. The I_{max} values for the various substrates are expressed relative to the current generated by a maximal dose of L-glutamate in the same cell. I_{max} and K_m values are expressed as mean ± standard error and are determined by fitting data from individual oocytes. The potent competitive blockers are characterized by Schild analysis and the K_b estimated from the regression plot. The less potent blockers are assumed to be competitive and K_i values calculated from IC_{50} values using the equation

$$K_i = IC_{50}/(1 + [\text{glutamate}]/K_m)$$

where K_i is the inhibition constant, IC_{50} is the concentration giving half maximum inhibition, K_m is the transport constant and [glutamate] is 30 μM. The fraction of the membrane electric field sensed by transport blockers when bound to the transporters is estimated using the Woodhull equation,

$$K_i = K_i^0 \exp(-\zeta\delta\, FE/RT)$$

where K_i is the inhibition constant, K_i^0 is the inhibition constant at 0 mV, ζ is the charge on the blocker, δ is the fraction of the membrane field, F is Faraday's constant, E is the membrane potential, R is the gas constant and T is temperature in K.

Anticonvulsant Agents

To screen potential anticonvulsant agents which interact at the convulsant binding site of the benzodi-azepine/GABA/chloride ionophore complex by measuring the inhibition of binding of [^{35}S]TBPS to rat cortical membranes. TBPS, t-butylbicyclophosphorothionate, is a potent convulsant which blocks GABAergic neurotransmission by interacting with the convulsant (or picrotoxin) site of the GABA/benzodiazepine/chloride ionophore receptor complex. Picrotoxin, pentylenetetrazol and the so-called "*cage convulsants*" are believed to change the state of the chloride channel to a closed conformation, and thereby block GABA-induced increases in chloride permeability. Anticonvulsants such as the barbiturates and the pyrazolopyridines, cartazolate, etazolate and tracazolate, appear to interact at depressant sites allosterically coupled to the convulsant sites, and facilitate the effects of GABA on chloride permeability, by converting the ionophore to the open conformation.

Benzodiazepines interact at a separate recognition site to modulate the actions of GABA. Convulsant compounds and some anticonvulsants can inhibit [^{35}S]TBPS binding. These two classes can be differentiated by their effects on dissociation kinetics. [^{35}S]TBPS dissociates slowly, half-life approximately 70 min, in a monophasic manner in the presence of convulsant compounds; anticonvulsants produce a biphasic dissociation, with rapid and slow phase components. It has been postulated that the rapid and slow phases of [^{35}S]TBPS dissociation may correspond to the open and closed conformation of the chloride ionophore.

Method

Requirements

1. Buffer A 0.05 M Tris with 2 M KCl, pH 7.4

   6.61g TrisHCl

   0.97 g Tris Base

   149.1 g KCl

   q.s. to 1 liter with distilled water

2. Buffer B

   a 1:10 dilution of buffer A in distilled water (5 mM Tris, 200 mM KCl, pH 7.4)

3. [^{35}S]TBPS is obtained from New England Nuclear with a high initial specific activity, 90-110 Ci/mmol. For an inhibition assay with a 2 nM final concentration of TBPS, a specific activity of 20-25 Ci/mmol will provide sufficient counts due to a high counting efficiency (87%) for ^{35}S. The specific activity of [^{35}S]TBPS can be reduced with the addition of 3-5 volumes (accurate measurement with a Hamilton syringe) of an equimolar ethanolic solution of non-radiolabeled TBPS (7.9 × 10^{-6} M). The new specific activity (Ci/mmol) is calculated by dividing the number of Curies by the number of mmoles TBPS. Since [^{35}S]TBPS has a relatively short half-life, 87.1 days, the specific activity is calculated for each assay, based on the exponential rate of decay:

A_0 = initial specific activity

A = specific activity at time t

t = days from date of initial calibration of specific activity

$t_{1/2}$ = half-life of [^{35}S] in days (87.1)

For IC_{50} determinations, a 40 nM stock solution is made with distilled water and 25 μl is added to each tube to yield a final concentration of 2 nM in the assay.

4. Unlabeled TBPS is available from New England Nuclear. A stock dilution of 7.923×10^{-6} M is prepared in ethanol.
5. Picrotoxin, is obtained from Aldrich Chemical Company. A solution of 5×10^{-4} M is prepared with distilled water, with sonication if necessary. Aliquots of 10 μl are added to assay tubes to give a final concentration of 10^{-5} M.
6. Test compounds. A stock solution of 1 mM is made up with a suitable solvent and serially diluted, such that the final concentration in the assay ranges from 10^{-5} to 10^{-8} M. Higher or lower concentrations may be used, depending on the potency of the drug.

Sampling

Whole cerebral cortex of male Wistar rats is dissected over ice and homogenized with a Tekmar Tissumizer, 20s at setting 40, in 20 volumes of 0.32 M sucrose, ice-cold. The homogenate is centrifuged at 1000 *g* for 10 min (SS34, 3000 rpm, 4°C). The supernatantis then centrifuged at 40,000 *g* for 30 min (SS34, 20,000 rpm, 4°C). The resulting pellet is resuspended in 20 volumes of ice-cold distilled water with two 6-s bursts of the Tissumizer, setting 40. The homogenate is centrifuged at 40,000 *g* for 30 min. The pellet is washed (resuspended and centrifuged) once with 20 volumes ice-cold buffer (Tris HCl 5 mM, KCl 200 mM, pH 7.4).

The resulting pellet is resuspended with 20 volumes buffer and frozen at –70°C overnight. The following day, the tissue homogenate is thawed in a beaker of warm water, approximately 15 min, and then centrifuged at 40,000 *g* for 30 min (SS34, 20,000 rpm, 4°C). The pellet is washed twice with 20 volumes of ice-cold buffer, and then resuspended and frozen at –70°C for future use. On the day of the assay, the homogenate is thawed and centrifuged at 40,000 *g* for 30 min. The resulting pellet is washed once with 20 volumes ice-cold buffer and finally resuspended in 30 volumes buffer. Aliquots of 250 μl are used for each assay tube, final volume 500 μl, and correspond to 8.35 mg original wet weight tissue per tube, approximately 0.2 mg protein.

Analyzing process

1. Prepare assay tubes in triplicate:

   190 μl distilled water

   25 μl Tris 0.05M, KCl 2M, pH7.4

   10 μl picrotoxin, 10^{-5} M final concentration or distilled water or inhibitor 25 μl [^{35}S]- TBPS, final concentration 2 nM

   250 μl tissue preparation, 1:30 homogenate

2. Following the addition of the tissue, the tubes are incubated at 25°C for 150 min with agitation. The assay is terminated by rapid filtration over Whatman GF/B filter circles, presoaked in buffer, with 5 × 4 ml rinses of ice-cold buffer. Vacuum filtration is performed with the 45-well filtration units to avoid contamination of the Brandell harvesters with [^{35}S] . The filters are counted with 10 ml Liquiscint.

Assessment

Specific binding is determined from the difference between binding in the absence or presence of 10 mM picrotoxin, and is typically 85-90% of total binding. The percent inhibition at each drug concentration is the mean of triplicate determinations. IC_{50} values for the competing drug are calculated by log-probit analysis of the data.

Latest methodology

1. Prepare assay tubes as follows:
   185 µl distilled water

   25 µl Tris 50 mM, KCl_2 M, pH7.4

   10 µl test compound or vehicle
2. Add 250 ml tissue homogenate to tubes. Vortex. Pre-incubate 30 min at 25°C.
3. Add 25 ml [^{35}S]-TBPS. Vortex. Incubate 180 min at 25°C.
4. Add 5ml picrotoxin (10^{-3} M) to give a final concentration of 10^{-5} M. Vortex.
5. At various times after the addition of picrotoxin (0-120 min) tubes are filtered and rinsed as described above.

Advanced autoradiography

1. Sections of rat brain, 20 mm thickness, are collected onto gel-chrome alum subbed slides, freeze-dried for approximately 1 h and stored at –70°C until used.
2. After thawing and drying at room temperature, the sections are preincubated for 30 min in buffer B.
3. Preparation of slide mailers for incubation:
   (a) for scintillation counting:

   2.47 ml distilled water

   0.325 ml buffer A

   3.25 ml buffer B

   0.13 ml picrotoxin, 10^{-5} M final concentration or distilled water or inhibitor

   0.325 ml [^{35}S]TBPS, final concentration 2 nM

   6.50 ml final volume

   (b) for autoradiography:

   4.56 ml distilled water

0.60 ml buffer A

6.00 ml buffer B

0.24 ml picrotoxin, 10^{-5} M final concentration or distilled water or inhibitor

0.60 ml [^{35}S]TBPS, final concentration 2 nM]

12.0 ml final volume

4. Sections are incubated in slide mailers at room temperature with [^{35}S]TBPS in the absence or presence of appropriate inhibitors for 90 min.
5. Slides are transferred to vertical slide holders and rinsed in ice-cold solutions as follows: dip in buffer B, two 5 minute rinses in buffer A and a dip in distilled water.
6. Slides are dried under a stream of cool air and desiccated overnight at room temperature.
7. Slides are mounted onto boards with appropriate [^{35}S] brain mash standards.
8. In the dark room under safelight illumination (GBX filter), slides are opposed to Kodak X-OMAT AR film and stored in cassettes for 7-10 days.
9. Develop films as described in "X-OMAT AR Film Processing".

Glycine as Inhibitory Transmitter

The amino acid glycine is a major inhibitory transmitter in the vertebrate system. Glycinergic synapses are particularly abundant in spinal cord and brain stem, but are also found in higher regions, including the hippocampus. The inhibitory actions of glycine are potently blocked by strychnine. Glycine modulates and may activate the excitatory amino acid receptors of the NMDA-subtype. The strychnine-sensitive, postsynaptic glycine receptor is a ligand-gated chloride channel protein that belongs to the nicotinic acetylcholine receptor family. It is a pentameric transmembrane protein composed of α and β subunits.

Glycine has been shown *in vitro* to potentiate the effects of L-glutamate or NMDA on the stimulation of [^{3}H]TCP binding and [^{3}H]norepinephrine release, and *in vivo* to act as a positive modulator of the glutamate-activated cGMP response in the cerebellum. The activation of NMDA receptors requiring the presence of glycine is necessary for the induction of long-term potentiation (LTP), a type of synaptic plasticity which may be fundamental to learning processes. A [^{3}H]glycine binding site in the brain has been identified and characterized as a strychnine-insensitive site associated with the NMDA receptor complex. Autoradiographic studies have shown a similar distribution of [^{3}H]glycine and [^{3}H]TCP (NMDA ion channel radioligand) binding sites. Compounds which interact with the glycine site offer a novel mechanism of action for intervention with NMDA receptor function. Schmieden and Betz (1995) reviewed the pharmacology of the inhibitory glycine receptor, the agonist and antagonist actions of amino acids and piperidine carboxylic compounds.

Hyperekplexia is a hereditary neurological disorder in humans characterized by an excessive startle response which can be caused by mutations in the α_1-subunit of the heteropentameric inhibitory glycine receptor. Becker et al. (2002) generated transgenic mice resembling this disease. The following assay is used to assess the affinity of compounds for the glycine binding site associated with the *N*-methyl-D-aspartate (NMDA) receptor complex using [^{3}H] glycine as the radioligand.

Method

Requirements

1. Buffer A: 0.5 M Tris maleate, pH 7.4

   59.3 g Tris maleate

   q.s. to 0.51

   Adjust pH to 7.4 with 0.5 M Tris base.
2. Buffer B: 50 mM Tris maleate, pH 7.4

   Dilute buffer A 1:10 with distilled water; adjust pH with 50 mM Tris maleate (acid) or 50 mM Tris base.
3. Glycine, 5×10^{-2} M

   Dissolve 3.755 mg of glycine with 1.0 ml distilled water. Aliquots of 20 µl to the assay tube will give a final concentration of 10^{-3} M
4. [^{3}H]Glycine is obtained from New England Nuclear, specific activity 45-50 Ci mmol. For IC_{50} determinations, a 200 nM stock solution is made with distilled water. Aliquots of 50 µl are added to yield a final assay concentration of 10 nM.
5. Test compounds. A stock solution of 5 mM is prepared with a suitable solvent and serially diluted, such that the final concentrations in the assay ranges from 10^{-4} to 10^{-7} M. Higher or lower concentrations may be used, depending on the potency of the compound.
6. Triton-X 100, 10% (v/v). A stock solution of Triton-X 100, 10% can be prepared and stored in the refrigerator. Dilute 1.0 ml of Triton-X 100 to 10.0 ml with distilled water. On the day of the assay, the tissue homogenate (1:15 dilution) is preincubated with an aliquot of the 10% solution to give a final concentration of 0.04% (v/v).

Sampling

Cortices of male Wistar rats are dissected over ice and homogenized in ice-cold 0.32 M sucrose, 15 volumes of original wet weight of tissue, for 30 s with a Tissumizer setting at 70. Three cortices are pooled for one preparation. The homogenate is centrifuged at 1000 *g* for 10 min (SS34, 3000 rpm, 4°C). The supernatant is centrifuged at 20,000 *g* (SS34, 12,000 rpm, 4°C) for 20 min. Resuspend the pellet in 15 volumes of ice-cold distilled water (Tissumizer setting 70, 15 s) and spin at 7600 *g* (SS34, 8000 rpm 4°C) for 20 min. The pellet is resuspended with 15 volumes of cold distilled water and centrifuged. Discard the supernatant and store the pellet at –70°C. On the day of the assay, the pellet is resuspended in 15 volumes ice-cold 50 mM Tris maleate, pH7.4. Preincubate the homogenate with Triton-X in a final concentration of 0.04% (v/v) for 30 min at 37°C with agitation. Centrifuge the suspension at 48,000 *g* (SS34, 20,000 rpm, 4°C) for 20 min. Wash the pellet an additional 3 times by resuspension with cold buffer and centrifugation. The final pellet is resuspended in a volume 25 times the original wet weight.

Analyzing process

1. Prepare assay tubes in quadruplicate.

380 μl distilled water

50 μl buffer A, 0.5 M Tris maleate, pH 7.4

20 μl glycine, 10^{-3} M final concentration, or distilled water or appropriate concentration of inhibitor

50 μl [^{3}H] glycine, final concentration 10 nM

500 μl tissue homogenate.

1000 μl final volume

2. Following the addition of the tissue, the tubes are incubated for 20 min in an ice-bath at 0-4°C. The binding is terminated by centrifugation (HS4, 7000 rpm, 4°C) for 20 min. Aspirate and discard the supernatant. Carefully rinse the pellet twice with 1 ml ice-cold buffer, avoiding disruption of the pellet. Transfer the pellet to scintillation vials by vortexing the pellet with 2 ml scintillation fluid, rinse the tubes twice with 2 ml and add an additional 4 ml scintillation fluid.

Assessment

Specific binding is determined from the difference of binding in the absence or in the presence of 10^{-4} M glycine and is typically 60-70% of total binding. IC_{50} values for the competing compound are calculated by log-probit analysis of the data.

Latest technique

Baron et al. (1996), Hofner and Wanner (1997), Chazot et al. (1998) described [^{3}H]MDL 105,519 as a high affinity ligand for the NMDA associated glycine recognition site.

Strychnine-Sensitive Glycine Receptor

The strychnine-sensitive glycine receptor is a member of the family of ligand-gated ion channel receptors. Within this family, the glycine receptor is most closely related to the GABA-receptor. Like the $GABA_A$-receptor, the glycine receptor has an inhibitory role, mediating an increase in chloride conductance. However, in contrast to the $GABA_A$-receptor, the glycine receptor is located mainly in the spinal cord and lower brainstem, where glycine appears to be the major inhibitory neurotransmitter. Purification and molecular cloning has shown that the glycine receptor is an oligomeric transmembrane protein complex composed of three α and two β subunits. The inhibitory actions of glycine are potently blocked by strychnine. In addition to strychnine, the steroid derivative RU5 135, phenylbenzene-ω-phosphono-α-amino acid, and 5,7-dichloro-4-hydroxyquinoline-3-carboxylic acid antagonize glycine responses in cultured neurons or cells expressing recombinant glycine receptors. A glycine receptor agonist may be a potential anti-spastic agent.

Method

Male Wistar rats weighing about 200 g are sacrificed. About 220 mg of frozen pons and medulla are homogenized in 2 × 10 ml ice-cold 50 mM potassium phosphate buffer, pH 7.1, by an Ultra-Turrax homogenizer. The homogenate is centrifuged for 10 min at 30,000 *g* at 0-4°C in a refrigerated centrifuge. The pellet is rehomogenized in another 2 × 10 ml portion of the same buffer and recentrifuged as before. This washing procedure is repeated a total of four times. The final pellet is resuspended in 200 vol/g original tissue in ice-cold 50 mM potassium phosphate buffer, pH 7.1, with or without 1000 mM NaCl, and used directly for binding assays. Binding assays consist of 1 ml tissue homogenate, 50 μl test solution (water or 5% v/v ethanol/water is used for serial dilutions), 50 μl water, 5% ethanol/ water or glycine solution (40 mM final concentration), and 25 μl [^{3}H]strychnine working solution, final concentration 2 nM.

The samples are mixed well and incubated for 20 min in an ice bath. Free and bound radioactivity are separated by filtration through Whatman GF/C glass fiber filters followed by washing with 2 × 10 ml ice-cold 50 mM potassium phosphate buffer, pH 7.1. Tritium on the filters is monitored by convential scintillation counting in 3 ml Hydroluma. Nonspecific binding is binding in the presence of 40 mM glycine and is always subtracted from total binding to give specific binding.

Assessment

K_i values are calculated as

$$K_i = (IC_{50}/1 + [K_D]/[L])$$

whereby: IC_{50} are the concentrations that inhibit by 50% the specific binding of [^{3}H]strychnine determined in two independent experiments using at least 3 concentrations of the agent in duplicate assays, [L] is the concentration of the radioligand, K_D is the affinity constant in the absence or the presence of 1000 mM NaCl. NaCl shift used for differentiating glycine agonists from glycine antagonists is the ratio K_i 1000 mM NaCl versus K_i^0 mM NaCl.

Transverse Hippocampal Slices in Vitro

The *transverse hippocampal slice* has been described as a well-defined cortical structure maintained *in vitro*. The hippocampus slice has the advantage that each slice may contain all hippocampal structures: The chain of neurons goes from the perforant path to granule cells of the dentate gyrus, through mossy fibres to CA-3 pyramidal cells and then through Schaffer collaterals to CA-1 cells with their axons leaving the hippocampus through the alveus. The pyramidal cells lie close together and can be easily seen and penetrated with fine microelectrodes.

Method

Male guinea pigs weighing 300-400g are anesthetized with ether, the brains removed and the hippocampi dissected. Transverse slices of the hippocampus (300-400 μm thick) are cut in parallel to the alvear fibres. After preparation, the slices are submerged in 28°C warm saline which is equilibrated with 95% O_2 and 5% CO_2. After a preincubation period of 2 h, slices are transferred in a Perspex chamber (1.5 × 4cm) and attached to the bottom consisting of optically plain glass. The chamber is mounted on an inverted microscope allowing detailed inspection of the excised tissue.

The slices are superfused by an approximately 3 mm thick layer of 32°C warm saline. Intracellular recordings are achieved by means of micropipettes with tip diameters of less than 0.5 μm which are filled with 3 mol/l potassium chloride. Under microscopic control the tips of the micropipettes are placed within the stratum pyramidale and moved by means of a step motor driven hydraulic microdrive. For intracellular injections of drugs, e. g., pentylenetetrazol, via the recording microelectrode, a passive bridge is used. Alternatively, drugs are added to the incubation bath.

Assessment

The resting membrane potential and paroxysmal depolarizations are recorded before and after application of drugs. The hippocampal slice has been one of the most useful models for the study of basic mechanisms underlying the epilepsies. The model has also been recommended for screening of putative anticonvulsant drugs.

Latest technique

Harrison and Simmonds (1985) performed quantitative studies on some antagonists of N-methyl-D-aspartate in slices of rat cerebral cortex consisting of cerebral cortex and corpus callosum. Tissue culture models of epileptiform activity were described by Crain (1972). Oh and Dichter (1994) studied the effect of a GABA uptake inhibitor on spontaneous postsynaptic currents in cultured rat hippocampal neurons by the whole cell patch clamp method. Blanton et al. (1989) described whole cell recordings from neurons in slices of reptilian and mammalian cerebral cortex. Synaptic currents and membrane properties could be studied in voltage and current clamps in cells maintained within their endogenous synaptic currents.

Gahwiler (1988), Stoppini et al. (1991) described methods for organotypic cultures of nervous tissue. Hippocampal slices from 2-23-day old rats were maintained in culture at the interface between air and the culture medium. They were placed on a sterile, transparent and porous membrane and kept in Petri dishes in an incubator. This yielded thin slices which remained 1-4 layers thick and were characterized by a well preserved organotypic organization. Excitatory and inhibitory synaptic potentials could be analyzed using extra- or intracellular recording techniques. After a few days in culture, long-term potentiation of synaptic responses could reproducibly be induced.

Using this method, Liu et al. (1995) studied dopaminergic regulation of transcription factor expression in organotypic cultures of developing striatum of newborn rats. Stuart et al. (1993) reported the implementation of infrared differential interference contrast video microscopy to an upright compound microscope and a procedure for making patch-pipette recordings from visually identified neuronal somata and dendrites in brain slices. Bernard and Wheal (1995) described an *ex vivo* model of chronic epilepsy using slices of rat hippocampus previously lesioned by stereotactic injections of kainic acid.

Extracellular population spikes were recorded from the stratum pyramidale of CA1 after stimulation by bipolar twisted wire electrodes placed in the stratum radiatum of CA1 area proximally to stratum pyramidale near the recording electrode. Using hippocampal slices prepared from brain tissue of patients undergoing neurosurgery for epilepsy, Schlicker et al. (1996) showed that the serotoninergic neurons of the human hippocampus are endowed with presynaptic inhibitory autoreceptors.

Clamp Configuration

The use of the cell-attached patch clamp configuration to record action potential currents has shown to have utility in the testing for drug actions on ion channels in excitable cell membranes.

Method

Sampling

The cultured cells are obtained from the hippocampus or the hypothalamus of rat brain. The isolation of the hippocampal CA1 neurones is performed according the procedure of Banker and Cowan (1977). The dissociated hypothalamic neurons are prepared according to Jirikowski et al. (1981). The hippocampal and hypothalamic neurons that are selected for electrophysiological recording are bipolar in shape with the long axis dimension between 10-15 p.m. The neurons are studied over a period of 5-10 days after isolation.

Electrophysiology

The cell-attached patch clamp configuration is used to record spontaneous action potentials in the cultured neurons. The bath solution contains 140 mM NaCl, 5 mM KCl, 0.5 mM $CaCl_2$, 1 mM $MgCl_2$, 5 mM HEPES, pH 7.3. The composition of the patch pipette solution is the same as the bath solution. The drugs used in the experiments are added to the bath solution. The patch pipettes are fabricated using a specific patch pipette puller are fire-polished and filled immediately prior to use. The resistance of the pipettes is in the range 4-8 MΩ and the tip diameters are between 1 and 2 μm. An axopatch amplifier, with low-pass filter set at 5 kHz, is used to record the capacitative currents. After recording, at a sampling frequency of 5 kHz, the data are stored on hard disc or video tape for subsequent analysis. All data are obtained at room temperature (21-24°C).

Assessment

The capacitative component of current recorded by the patch pipettes is proportional to the rate of change of membrane potential and can be expressed as $I_C = C dV/dt$, where C is the specific membrane capacitance. Assuming a value of C of 1 μF/cm² and a tip diameter of the patch pipette of 2 μm, the membrane area isolated by the patch pipette is about 3×10^{-8} cm². Using a value of dV/dt of 100 mV/ms gives an approximate expected magnitude of I_C near 3 pA. When a class III antiarrhythmic drug that blocks a delayed rectifier K^+ channel is added to the bath, the portion of I_C corresponding to the after-hyperpolarization component of the action potential is completely abolished. The Na^+ spike is not altered by the drug. The cell-attached recordings of I_C can also be used to determine effects on the Na^+ spike when tetrodotoxin is included in the bath solution. Thus, the spontaneous action potential can be used for evaluation of drug effects on both K^+ and Na^+ channels in excitable membrane.

Latest technique

Chen et al. (1990) measured current responses mediated by $GABA_A$ receptors in pyramidal cells acutely dissociated from the hippocampus of mature guinea pigs according to the procedure of Kay and Wong (1986) using whole-cell voltage-clamp recordings. Caulfield and Brown (1992) studied inhibition of calcium current in NG108-15 neuroblastoma cells by cannabinoid receptor agonists using whole-cell voltage-clamp recordings. Gola et al. (1992, 1993) performed voltage recordings on non-dissociated sympathetic neurones from rabbit coeliac ganglia using the whole-cell configuration of the patch clamp technique.

Stolc (1994) used the voltage-clamp technique in internally dialyzed single neurones isolated from young rat sensory ganglia to study the effects of pyridoindole stobadine on inward sodium and calcium currents and on slow non-inactivating components of potassium outward current. McGivern et al. (1995) examined the actions of a neuroprotective agent on voltage dependent Na^+ currents in the neuroblastoma cell line, NIE-115, using the whole cell variant of the patch clamp technique. Smith (1995) reviewed the use of patch- and voltage clamp procedures to study neurotransmitter transduction mechanisms. Using whole-cell and perforated-patch recordings, Delmas et al. (1998) examined the part played by endogenous G-protein $\beta\gamma$ subunits in neurotransmitter- mediated inhibition of N-type Ca^{2+} channel current in dissociated rat superior cervical sympathetic neurones. Gonzales et al. (1985) registered membrane potentials with intracellular electrodes in cultured olfactory chemoreceptor cells.

Neonatal Rat Spinal Cord

The spinal cord of the neonatal rat is a useful *in vitro* preparation, originally proposed by Otsuka and Konishi (1974). In this preparation, ventral root potentials of ten seconds of duration can be recorded after supramaximal electrical stimulation of the lumbar dorsal root. Variously implicated in the generation of these slow ventral root potentials are tachykinins, such as substance P and neurokinin B and agonists at the glutamate receptor sites. These long lasting reflexes are thought to reflect a nociceptive reflex for several reasons; the threshold of activation corresponds to that of C fibre primary afferents; they can be depressed by opioids and α_2-adrenoceptor agonists and a similar response can be evoked by peripheral noxious stimulation.

Method

Sampling

Male Wistar rats aged 6-9 days are used. Under ether anesthesia the spinal column is quickly removed from the animal and placed in a Petri dish, filled with oxygenated physiological solution. A laminectomy is performed on the dorsal surface of the spinal column at room temperature. The spinal cord of the mid-thoracic to mid-sacral level is then carefully removed from the column and hemisected in the longitudinal plane under a dissecting microscope. After removal of the dura mater, the hemisected cord is completely submerged in the recording chamber (total volume: approximately 0.5 ml), which is perfused with physiological solution (124 mM NaCl, 5 mM KCl, 1.3 mM $MgSO_4$, 2.5 mM $CaCl_2$, 1.2 mM KH_2PO_4, 15 mM $NaHCO_3$, 11 mM glucose) at a flow rate of 1.5 to 2.5 ml/min.

The perfusion medium is continuously bubbled with a gas mixture of 95% O_2 and 5% CO_2 and the temperature is kept at 25 ± 0.5°C. The cut ends of the corresponding dorsal and ventral roots in an L_{3-5} segment are fixed to a pair of suction electrodes for stimulating and recording. The preparation is stabilized in the recording chamber for at least 90 min to allow recovery from the dissection and the sealing of the roots to suction electrodes.

Testing of monosynaptic reflexes

Test stimulations, composed of square wave pulses of 0.05-0.2 ms duration and 5-30 V, are applied to the dorsal root every 10s. The discharges of the corresponding ventral root are recorded with a suction electrode, amplified and monitored on an oscilloscope and stored on an analog data recorder or computer discs for later analysis. The mean values for the waveform of the monosynaptic reflex (amplitude, area and latency) are obtained from 6-18 successive responses in each experiment before and during application of drugs.

Testing of single motoneuron activity

Test pulses (0.01-0.1 ms duration and 5-15 V) are applied to the dorsal or ventral root every 2s. The activity of single motoneurons is recorded extracellularly using glass microelectrodes (electrical resistance approximately 10-30 MΩ) filled with 3 M sodium chloride or 2 M sodium acetate. The microelectrode is inserted into the ventral part of the cord through the hemisected surface while monitoring the field potential. The motoneurons in the ventral horn are identified by the short and consistent latency of antidromic spikes (1.66 ± 0.46 ms, n = 5), following the stimulation of the ventral root.

The motoneurons also produce transsynaptic spikes with orthodromic stimulation of the dorsal root, of which the latency is 10.26 ± 1.05 ms upon supramaximal stimulation. The spike generation of motoneurons is displayed on an oscilloscope and stored on magnetic tapes. The spontaneous firing of the motoneuron is also monitored on an oscilloscope and recorded through a window discriminator and spike counter. The mean number and latency of spikes and latency of the dorsal root-elicited spikes are obtained from 20-40 successive responses in each experiment. Comparisons are made before and 3-5 min after application of drugs.

Assessment

All data are expressed as the mean ± SEM. Statistical significance of the data is determined by repeated measures analysis of variance (ANOVA) and, when appropriate, Student's t-test. A P value of less than 0.05 is considered statistically significant.

Latest technique

Smith and Feldman (1987), Wong et al. (1996) described an *in vitro* neonatal rat brainstem/ spinal cord preparation. The brainstem and cervical spinal cord were isolated from 0-4 days old ether anesthetized Sprague Dawley rats. The en bloc neuraxis was pinned down with ventral surface upward in a recording chamber and superfused continuously with artificial cerebrospinal fluid. Respiratory activity was recorded with suction electrodes from the C4 ventral root.

Neuroculture

Cell culture of neurons, especially of hippocampal neurons, has become a widely used tool in pharmacological studies. The basic information on methodology of cell culture of rat hippocampal neurones was given by Banker and Cowan (1977). One modification used by Skaper et al. (1990, 2001) studying the role of mast cells on potentiation by histamine of synaptically mediated excitotoxicity in cultured hippocampal neurones is described below.

Method

Sampling

Timed pregnancies are obtained in female Sprague Dawley rats by daily checking vaginal washings for sperm, the day on which sperm is found being regarded as day 0. At the appropriate stage of gestation the pregnant rats are anesthetized and the uterus removed to a sterile dish. The remainder of the cell preparations is performed in a sterile hood. The brains are removed from the fetuses with a pair of fine scissors, and the cerebral hemispheres separated from the brain stem. When the hemisphere of an 18- to 19-day-old fetus in viewed in a dissecting microscope the hippocampus can be clearly seen on its medial surface.

The hippocampal fissure, usually marked by a conspicuous group of blood vessels, indicates the approximate junction between the hippocampus and the adjoining subicular and entorhinal cortex. The developing fimbria is seen as a white translucent band along the free margin of the hippocampus. Before separating the hippocampus from the hemisphere, the meninges and adherent chorioid plexus are carefully pulled off with fine forceps. At this stage the full depth of the hippocampal fissure can be seen. Then with iridectomy scissors the hippocampus is separated from the adjoining cortex by a cut parallel to the hippocampal fissure, and by transverse cuts at its rostral and caudal ends.

Cytoculture

Hippocampi isolated from embryonic rats (gestational age 17.5 days) are incubated with 0.08% trypsin, and dissociated in Neurobasal medium containing 10% heat-inactivated calf serum. Cells are pelleted by centrifugation (200 *g*, 5 min) and resuspended in Neurobasal medium containing B27 supplements (with antioxidants), 25 μM glutamate, 1 mM sodium pyruvate, 2 mM L-glutamine, 50 U/ml penicillin, and 50 μg/ml streptomycin. The cell suspension is plated onto poly-D-lysine (10 μg/ml) coated 48-well culture plates at a density of 4.5×10^4 cells per cm^2. Cultures are maintained at 37°C in a humidified atmosphere of 5% CO_2-95% air. After 5 days, one-half of the medium is replaced with an equal volume of maintenance medium (plating medium but containing B27 supplements without antioxidants, and lacking glutamate). Additional medium exchanges (0.5 volume) are performed every 3-4 days thereafter.

Cells are used between 14 and 16 days in culture. During this period, neurones develop extensive neuritic networks, and form functional synapses. Mast cells are collected from the peritoneal lavage of male Sprague Dawley rats and isolated over a bovine serum albumin gradient to >90% purity, as judged by toluidine blue and safranin staining.

Neurotoxicity assays

Cultures are washed once with Locke's solution (pH 7.0-7.4) with or without 1 mM $MgCl_2$. Drug treatments are carried out for 15-30 min (25°C) in a final volume of 0.5 ml. In the case of mast-cell-neurone co-cultures, transwell inserts (3-μm pore size, 9 mm diameter) are seeded with 5×10^4 mast cells in RPMI-1640 medium and placed in 24-well plates overnight. Inserts with mast cells are then placed into wells with hippocampal cells. Mast cells activation is achieved using an antigenic stimulus (0.3 μg/ml anti-DNP IgE/0.1 μg/ml DNP albumin). The mast cell containing inserts are removed at the end of the Mg^{2+}-free incubation. After this time all cell monolayers are washed with complete Locke's solution and returned to their original culture medium for 24 h.

Cytotoxicity is evident during 24 h after the insult. Viable neurons have phase-bright somata of round-to-oval shape, with smooth, intact neurites. Neurones are considered nonviable when they exhibit neurite fragmentation and somatic swelling and vacuolation. Cell survival is quantified

24 h after the insult by a colorimetric reaction with 3-(4,5-dimethylthiazol-2-yl)-2,5- diphenyltetrazolium bromide (MTT).

Assessment

Data are analyzed by one-way ANOVA with Student-Newman-Keuls *post hoc* test for differences between groups.

Latest technique

Brewer (1997) reported the isolation and culture of adult rat hippocampal neurons. Using different proteases and special separation techniques, about 90000 viable neurons could be isolated from each hypothalamus at any age rat from birth to 24-36 months. Neurons were cultured for more than 3 weeks. Flavin and Ho (1999) found that propentofylline protects hippocampal neurons in culture from death triggered by macrophage or microglia secretory products. To study neurite outgrowth in cultured hippocampal cells from Wistar rat embryos, 5000 cells/well were seeded in 8-well LabTec tissue culture slides with a grown surface of permanox plastic and grown in Neurobasal medium supplemented with B27, 20 mM HEPES, 0.4% bovine serum albumin, penicillin (100 IU/ml), and streptomycin (100 μg/ml). For image analysis, cells were fixed in 4% paraformaldehyde and stained for 20 min with Coomassie Blue R250.

Cover slides were observed in an inverted microscope using phase contrast optics. To measure neurite out-growth from hippocampal neurons, an unbiased counting frame containing a grid with a number of test lines was superimposed on the images of cells. The number of intersections of cellular processes with the test lines was counted and related to the number of cell bodies, thereby allowing quantification of neurite length per cell. Cell culture experiments were also performed with neuronal cells from other areas of the brain besides the hippocampus. Brain tissue samples of rat embryos containing either septum plus preoptic area or retrochiasmatic hypothalamus were dissociated and cultured for 14 and 21 days by Jirikowski et al. (1981). By means of immunofluorescence, LHRH, α-MSH, vasopressin and neurophysin-containing hormones could be identified.

Sinor et al. (2000) studied NMDA and glutamate evoked excitotoxicity at distinct cellular locations in rat cortical neurones *in vitro.* Canals et al. (2001) examined neurotrophic and neurotoxic effects of nitric oxide on neuronal enriched fetal midbrain cultures from embryonic Sprague Dawley rats. Lopez et al. (2001) investigated the release of amino acid neurotransmitters in cultured cortical neurons obtained from gestation day 19 rats by nicotine stimulation. Ehret et al. (2001) studied the modulation of electrically evoked acetylcholine release in cultured septal neurones from embryonic Wistar rats. Tang et al. (2001) found a lack of replicative senescence in cultured rat oligodendrocyte precursor cells. Yamagishi et al. (2001) used cultured rat cerebellar granule neurons as a model system for studying neuronal apoptosis.

Noh and Koh (2000) prepared mixed mouse cortical cultures containing both neurons and astrocytes, and pure astrocyte cultures, from fetal (15 d of gestation) and neonatal (1-3 postnatal days) mice. Saluja et al. (2001) found that PPAR δ agonists stimulate oligodendrocyte differentiation in glial cell culture of mouse cerebra. Ushida et al. (2000) succeeded to directly isolate clonogenic *human* central nervous system stem cells from fresh human brain tissue, using antibodies to cell surface markers and fluorescence-activated cell sorting.

ELECTROSHOCK IN MICE

The electroshock assay in mice is used primarily as an indication for compounds which are effective in grand mal epilepsy. Tonic hindlimb extensions are evoked by electric stimuli which are suppressed by anti-epileptics but also by other centrally active drugs.

Method

Groups of 6-10 male NMRI mice (18-30g) are used. The test is started 30 min after i.p. injection or 60 min after oral treatment with the test compound or the vehicle. An apparatus with corneal or ear electrodes is used to deliver the stimuli. The intensity of stimulus is dependent on the apparatus, e. g., 12 mA, 50Hz for 0.2 s have been used. Under these conditions all vehicle treated mice show the characteristic extensor tonus.

Assessment

The animals are observed closely for 2 min. Disappearance of the hindleg extensor tonic convulsion is used as positive criterion. Percent of inhibition of seizures relative to controls is calculated. Using various doses, ED_{50}-values and 95% confidence interval are calculated by probit analysis. ED_{50}-values after oral administration are:

1. Diazepam 3.0 mg/kg
2. Diphenylhydantoin 20.0 mg/kg

The electroshock test in mice has been proven to be a useful tool to detect compounds with anticonvulsant activity.

Latest methodology

Cashin and Jackson (1962) described a simple apparatus for assessing anticonvulsant drugs by the electroshock seizure test in mice. Kitano et al. (1996) developed the increasing-current electroshock seizure test, a new method for assessment of anti- and pro-convulsant activities of drugs in mice. A single train of pulses (square wave, 5 ms, 20Hz) of linearly increasing intensity from 5 to 30 mA was applied via ear electrodes. The current at which tonic hindlimb extension occurred was recorded as the seizure threshold. The method allows the determination of seizure threshold current for individual animals.

Antagonized Bicuculline

Seizures can be induced by the $GAGA_A$-antagonist bicuculline and are antagonized by known anti-epileptics.

Method

Female Sprague-Dawley rats are injected i.v. with 1 mg/kg bicuculline. At this dose, a tonic convulsion appears in all treated rats within 30s after injection. Test compounds are administered orally 1 or 2 h before bicuculline injection. Dose-response curves can be obtained.

Assessment

Percentage of protected animals is evaluated. ED_{50} values and 95% confidence limits are calculated by probit analysis. Like the electroshock test, the bicuculline test is considered to be relatively specific for anti-epileptic activity.

Latest methodology

Czuczwar et al. (1985) studied the antagonism of N-methyl-D,L-aspartic acid-induced convulsions by anti-epileptic drugs and other agents.

K^+ Channel Antagonist in Mice

The K^+ channel antagonist 4-aminopyridine is a powerful convulsant in animals and in man. The drug readily penetrates the blood-brain barrier and is believed to induce seizure activity by enhancing spontaneous and evoked neurotransmitter release. Although both excitatory and inhibitory synaptic transmission is facilitated by 4-aminopyridine, the epileptiform activity induced by the drug is predominantly mediated by non-NMDA type excitatory amino acid receptors. In mice, parenterally administered 4-aminopyridine induces clonic-tonic convulsions and lethality.

Method

Male NIH Swiss mice weighing 25-30 g are allowed to acclimatize with free access to food and water for a 24- h period before testing. Test drugs are administered in various doses intraperitoneally 15 min prior to s.c. injection of 4-aminopyridine at a dose of 13.3 mg/kg which was found to be the LD_{97} in this strain of mice. Controls treated with 4-aminopyridine only exhibit characteristic behavioral signs, such as hyperreactivity, trembling, intermitted forelimb/hindlimb clonus followed by hindlimb extension, tonic seizures, opisthotonus and death. The mean latency to death at the LD_{97} is about 10 min. Groups of 8 mice are used for each dose.

Assessment

The percentage of protected animals at each dose is used to calculate ED_{50} values. Phenytoin-like anticonvulsants such as carbamazepine and broad spectrum anticonvulsants such as phenobarbital and valproate are effective whereas GABA-enhancers such as diazepam, several NMDA antagonists, and CA^{2+} channel antagonists such as nimodipine are not. The profile of drugs effective in this seizure model is distinct from other chemoconvulsant models and more similar to those that prevent tonic hindlimb extension in the maximal electroshock seizure test. The test is useful to differentiate the mode of action of anticonvulsant drugs.

Latest technique

Morales-Villagran et al. (1996) described protection against seizures induced by intracerebral or intra-cerebroventricular administration of 4-aminopyridine by NMDA receptor antagonists.

Effect of 3-Nitropropionic Acid-Induced Seizures in Mice

3-Nitropropionic acid is a naturally occurring toxin demonstrated to impair energy metabolism via irreversible inhibition of a mitochondrial complex II component, succinate dehydrogenase. 3-Nitropropionic acid evokes seizures in mice after i.p. injection of 100-200 mg/k. Urbanska et al. (1998) and Zuchora et al. (2005) evaluated anticonvulsants for their protective effect against 3-nitropropionic-acid-induced seizures.

Method

Male albino Swiss mice weighing 20-25 g were injected i.p. with 210 mg/kg 3-nitropropionic acid, which is equal to the ED_{97} dose (i. e., the dose required to evoke seizures in 97% of the animals). Groups of eight mice received in addition various doses of the anticonvulsant drugs. Percentage of animals with seizures and latency until occurrence of seizures were determined. Mortality rate was determined 2 h after injection of 3-nitropropionic acid.

Assessment

ED_{50} and LD_{50} values together with their confidence limits were estimated by computerized fitting of the data by linear regression analysis according to Litchfield and Wilcoxon. Statistical comparisons of latency data were performed by means of one-way analysis of variance (ANOVA) followed by adjustment of *P* value by the Bonferroni method.

Focal Lesions

Intrahippocampal injections of noxious agents or certain cerebral lesions can induce seizures in animals. Cavalheiro et al. (1982) studied the long-term effects of intrahippocampal kainic acid injections in rats.

Method

Adult male Wistar rats are anesthetized with a chloral hydrate/Nembutal mixture and placed in a stereotactic apparatus. For injections, a 0.3 mm cannula is inserted through a burr hole in the calvarium. The coordinates for hippocampal injections are based on a stereotactic atlas, e.g., Albe-Fessard et al. (1971). Kainic acid is dissolved in artificial serum and infused in various doses (0.1 to 3.0 µg) in a volume of 0.2 µl over a period of 3 min. For recording, bipolar twisted electrodes (100 µm) are positioned stereotaxically and fixed on the skull with dental acrylic cement. Depth recording sites include the dorsal hippocampus and amygdala ipsilateral to the injected side. Surface electrodes are guided from jeweler's screws over the occipital cortex. An additional screw in the frontal sinus serves as indifferent electrode for grounding. Signals are recorded by an EEG polygraph.

Assessment

EEG recordings and observations of convulsive seizures are performed during the acute phase and during the chronic phase (up to 2 months) with and without drug treatment.

Latest technology

Several agents have been used as convulsants after topical administration, e. g., application of alumina cream, implantation of cobalt powder, injection of a colloidal gel of tungstic acid, topical application of penicillin, subpial injection of saturated $FeCl_3$ solution, intracerebral injections of zinc sulfate, intracerebral injection of antibodies to brain gangliosides, microinjections of cholinergic agonists, topical application of atropine, injection of tetanus toxin into the hippocampus, injection of strychnine in the visual or somato-sensory cortex, electrophoretic application of bicuculline from a fluid filled microelectrode. Bernhard et al. (1955, 1956) evaluated the anticonvulsive effect of local anaesthetics in cats and monkeys. The head was fixed in light Nembutal anaesthesia, the parietal areas exposed and covered with paraffin oil. Stimulating electrodes were placed at the surface of the parietal region.

The cortex was stimulated with repetitive square wave shocks (duration 1-3 ms) with a frequency of 25 per s for 5s. In order to avoid muscular movements, d-tubocurarine was given. Cortical afterdischarge was registered before and after injection of local anaesthetics. Cortical epileptic lesions were produced by local freezing.

Repetitive electrical stimulation of discrete regions of the central nervous system has been used as a convenient method for reproduction of the ictal phenomena of epilepsy. Remler and Marcussen (1986), Remler et al. (1986) studied the pharmacological response of systemically derived focal epileptic lesions. A defined area of left hemisphere of rats was radiated by α-particles from a cyclotron destroying the blood brain barrier. After a period of 150 days following irradiation, bicuculline was injected intraperitoneally resulting in focal lesions with EEG spikes and convulsions. Anticonvulsant drugs decreased these effects. Walton and Treiman (1989) and Walton et al. (1994) described a model of cobalt-lesioned rats in which status epilepticus was induced by injection of homocysteine thiolactone. Anderer et al. (1993) pointed out that restriction to a limited set of EEG-target variables may lead to misinterpretation of pharmaco-EEG results. Krupp and Löscher (1998) developed a cortical ramp-stimulation model allowing repeated determinations of seizure threshold at short time intervals in individual rats without inducing postictal threshold increases.

Model of Goddard et. al.

Kindling, first described by Goddard et al. (1969), results from repetitive subconvulsive electrical stimulation of certain areas of the brain. Initially, local after-discharge is associated with mild behavioral signs; however, with continued stimulation electrical activity presumably spreads, and generalized convulsions occur. Although the pathogenesis of kindled seizures is not fully understood, it serves as a useful tool for investigating the efficacy of experimental anticonvulsant agents.

Method

Adult female Sprague-Dawley rats (270-400 g) are used. The rats are implanted with an electrode in the right amygdala according to the coordinates of Pellegrino et al. (1979): frontal, 7.0; lateral, -4.7; horizontal, 2.5. At least 1 week has to elapse before electrical stimulation of the brain is started. After discharge threshold is determined for each rat. Duration and amplitude, behavioral seizure duration and seizure stage are recorded with increased stimuli afterdischarges. Seizure severity is classified into 5 stages. Rats are considered to be kindled on the first stimulation causing a stage 5 seizure which is followed by at least 2 consecutive stage 5 seizures. The animals are tested on the day before and after treatment with the test compound (i.p. or orally). Amygdala stimulation is applied at various time intervals.

Assessment

The occurrence and the degree of seizures are compared between control results and the those after administration of the test compound. The kindled seizure model offers an approach to study anticonvulsive drugs on the basis of a pathophysiological model. This method may give more relevant results than the simpler methods using maximal electroshock or chemically induced convulsions.

Latest technique

Generalized convulsive seizures have been induced by daily amygdaloid stimulation in baboons, and in rhesus monkeys. The kindling effect can be produced by intermittent administration of small doses of pentylenetetrazol. Durmuller et al. (1994) tested a competitive (NBQX) and a non-competitive (GYKI 52446) AMPA antagonist, and a competitive NMDA antagonist (DCPPene) against the development of kindling and against fully kindled seizures in amygdala-kindled rats. Croucher et al. (1996) described a chemical kindling procedure in rats by daily focal microinjection of NMDA into the right basolateral amygdala and the inhibition of seizures by a NMDA receptor antagonist.

Suzuki et al. (1996) studied the anticonvulsant action of metabotropic glutamate receptor agonists in kindled amygdala of rats. Loscher et al. (1993), Ebert et al. (1997), and Ebert and Loscher (1999) studied the effect of phenytoin on the spread of seizures in the amygdala kindling model in rats. Sprague Dawley rats implanted with a stimulation and recording electrode in the basolateral amygdala showed an increase in current intensity necessary for eliciting after-discharges of about 200% after administration of phenytoin, while seizure severity at threshold was increased compared to controls. Phenytoin-resistant kindled rats are considered as a model of drug-resistant epilepsy. Loscher (1998) discussed the pharmacology of glutamate receptor antagonists in the kindling model of epilepsy. The kindling procedure can also be used to evaluate antidepressant drugs.

Investigating Posthypoxic Myoclonus in Rats

The syndrome of posthypoxic myoclonus in man was described by Lance and Adams (1963). Lance (1968), Fahn (1986). Troung et al. (1994), Jaw et al. (1994, 1995, 1996) reported on a model in rats resembling this human disorder.

Method

Male Sprague Dawley rats fasted 12-24 h prior surgery are anesthetized with 100 mg/kg ketamine i.p., supplemented by 0.4 mg/kg atropine. The animal is placed on a circulating water pad and kept at a constant body temperature by a heating lamp. The rat is intubated and ventilated with 30% O_2 in N_2O. A femoral artery and vein are cannulated for monitoring blood pressure and delivery of drugs, respectively. Electrocardiogram and blood pressure are recorded with a polygraph. The rat is then paralyzed with 2 mg/kg succinly-choline i.v. and ventilator settings are adjusted to a rate of 60 strokes/min and a volume of 7.5 ml/kg, which yields blood gases of >150 mm Hg pO_2, 35-40 mm Hg pCO_2, and a pH of 7.35-7.40. N_2O is replaced with N_2 and an equilibrium period of 5 min is allowed.

Cardiac arrest is accomplished with a trans-thoracic intracardiac injection of KCl and cessation of the respiration. Resuscitation is begun 10 min after the arrest by turning on the ventilator (100% O_2), manual thoracic compressions, and i.v. injections of 20 µg/kg epinephrine hydrochloride and sodium bicarbonate (4 mEq/kg). The rat is then weaned from the ventilator over 2-4 h and extubated.

Auditory-induced myoclonus

Rats are presented with a series of 45 clicks from a metronome (1 Hz, 95 dB, 40 ms), and the response to each click is scored as follows: 0 = no response, 1 = ear twitch, 2 = ear and head jerk, 3 = ear, head, and shoulder jerk, 4 = whole body jerk, 6 = whole body jerk of such severity that it causes a jump. The total myoclonus score of each rat is determined by summing up the scores yielded over 45 clicks. Since rats ranging from 3 to 14 days post cardiac arrest show similar susceptibility to audiogenic stimulation, animals within this period are used for pharmacological tests. Myoclonus scores are assessed 30 min before and 60 min after intraperitoneal drug application.

Assessment

Changes in myoclonus scores are analyzed by paired two-tailed Student's *t*-test. Some anticonvulsant drugs, such as clonazepam and valproic acid were reported to be active in this test; however, phenytoin is not. Posthypoxic myoclonus may present a special pathological condition.

Investigating Rat Kainate Epilepsy

Temporal lobe epilepsy is characterized by complex partial seizures that involve and apparently originate in the mesial temporal structures of the limbic system. These complex partial seizures can evolve into secondarily generalized, tonic-clonic seizures. Patients become resistant to the treatment with the usual anti-epileptic drugs. The kainate-treated rat is one of several models used to study temporal lobe epilepsy. Examination of the hippocampus and dentate gyrus from kainate-treated rats has revealed a similar pattern of neurodegeneration in the hippocampus and the presence of mossy fiber sprouting in the inner molecular level of the dentate gyrus. Several authors used this model to find drugs for treatment-resistant epilepsy. Maj et al. (1998) tested the activity of several drugs against kainate-induced status epilepticus and hippocampal lesions in the rat.

Method

Male Wistar rates weighing 225-250 g are anesthetized with sodium pentobarbital (50 mg/kg i.p.). The are implanted extradurally with electrodes over the frontal and parietal cortex and with a reference electrode on the cerebellum. Caution is taken not to break the inner table of the diploe. All the electrodes are connected to plugs and held to the skull with dental acrylic cement. At least 7 days after surgery, rats are treated with either saline or test drugs intraperitoneally. Then, 15 min later, the rats receive a single i.p. dose of kainic acid (10 mg/kg). EEG recordings and behavioral observations are performed up to 240 min after kainic acid administration. Status epilepticus is defined as a sustained ictal EEG pattern lasting 20 min or longer without any interruption longer than 1 min. Seven days later, the rats are sacrificed, the brains removed and immersed for 48h in 10% formalin. Coronal sections (4 μm) are stained with hematoxylineosin. Hippocampal injury is assessed by counting the number of histologically normal CA4 pyramidal neurons.

Assessment

The percentage of animals protected from status epilepticus is analyzed using Fisher's exact test. For calculation of the latency to status epilepticus (min) and duration of status epilepticus (min), all animals are included regardless of whether they showed status epilepticus or not. The data are evaluated by analysis of variance (ANOVA) followed by Dunnett's test. Neuronal counts are analyzed using the Mann-Whitney non-parametric test.

Latest technique

Cilio et al. (2001) used immature rats to test the anticonvulsant action and long-term effects of gabapentin. Hellier et al. (1998) used repeated low-dose systemic treatment in order to reduce the mortality associated with single injections with kainate. Since intracerebroventricular administration of kainic acid decreases hippocampal neuronal number and increases dopamine receptor binding in the nucleus accumbens, kainic lesions have been discussed as an animal model of schizophrenia. Humphrey et al. (2001) described methods for inducing neuronal loss in preweanling rats using an intracerebroventricular infusion of kainic acid. Hu et al. (1998) investigated neuronal stress and injury in C57/BL *mice* after systemic kainic acid administration. Bouilleret et al. (1999) tested recurrent seizures and hippocampal sclerosis following intrahippocampal kainate injection in adult *mice*.

Investing Pilocarpine Epilepsy

Several post-status models are described in which epilepsy develops after a chemically induced status epilepticus, such as the kainate, the pilocarpine and the pilocarpine-lithium model. Several modifications of the pilocarpine and the lithium-pilocarpine model are reported in the literature. When rats are pretreated with lithium chloride, status epilepticus can be produced with a substantially lower dose of pilocarpine, and rats display the same clinical and EEG features of status epilepticus as with pilocarpine alone and Rigoulot et al. (2004) tested antiepileptic drugs in the lithium-pilocarpine model of epilepsy.

Method

Male Wistar rats weighing 225-250 g were anesthetized for electrode implantation by an i.p. injection of 2.5 mg/kg diazepam and 1 mg/kg ketamine chlorhydrate. Two single-contact recording electrodes were placed on the skull, one on each side of the parietal cortex, and one bipolar deep recording electrode was placed in the right hippocampus. One week after surgery, rats received 3 mEq/kg lithium chloride i.p. On the following day, 1 mg/kg methylscopolamine bromide was administered s.c. to limit the peripheral effects of the convulsant. Status epilepticus was induced by injecting pilocarpine (25 mg/kg s.c.) 30 min after methylscopolamine. Various doses of test drug (i.p.) or 2.5 mg/kg diazepam (i.m.) were injected at 1 h after the onset of status epilepticus. The onset of status epilepticus corresponds to the moment at which rats experience successive seizures without recovery.

Continuous spiking of the EEG occurs 30-60 min after pilocarpine administration. The bilateral EEG cortical activity and the unilateral EEG hippocampal activity were recorded during the whole duration of status epilepticus and concurrent behavioral changes were noted. Quantification of neuronal damage was performed 14 days after status epilepticus. Brains of rats sacrificed in pentobarbital anesthesia were removed and coronal sections containing the hippocampus from the anterior to the posterior level were prepared. Quantification of cell density was performed with a microscopic grid. The numbers of cells obtained in 12 counted fields were averaged.

Assessment

Statistical analysis of neuronal damage and epilepsy between the different groups was performed by means of analysis of variance followed by a posthoc Dunnett's test for multiple comparisons.

Latest methodology

Hort et al. (1999) studied the relation between spontaneous recurrent seizures and the derangement of cognitive function in pilocarpine-induced status epilepticus, Tang et al. (2004)

recorded EEG in freely moving mice after pilocarpine-induced status epilepticus. A transmitter was fixed on the electrode socket by plug connection with wires attached to the skull by two screws 3 days before pilocarpine induction. The EEG signals were telemetrically received via an HF receiver which passed the signals to the computer.

Investigating Self-Sustained Status Epilepticus

Status epilepticus causes neuronal damage that is associated with cognitive impairment. *Self-sustained status epilepticus* (SSSE) can be induced in rats by electrical stimulation of the perforant pathway. This model is used to find antiepileptic drugs for patients with therapy-resistant epilepsy. Pitkänen et al. (1996), Halonen et al. (1996, 1999, 2001), and Mazarati et al. (2004) studied the effect of drugs on status epilepticus in rats.

Method

Under ketamine (60 mg/kg) and xylazine (15 mg/kg) anesthesia, male Wistar rats weighing 260-280 g were implanted with a bipolar stimulation electrode into the angular bundle of the perforant path (0.5 mm anterior and 4.5 mm left to lambda) and a bipolar recording electrode into the ipsilateral dentate gyrus (3 mm posterior and 2.5 mm left to bregma). The depth of the electrode was 3.5-4 mm from the brain surface and was optimized by finding the maximal population spike evoked from the dentate gyrus by stimuli applied to the perforant path. For induction of self-sustained status epilepticus perforant path stimulation was delivered using a Grass stimulator model 8800, for 30 min with the following parameters: 10-s, 20-Hz trains for 1 ms, 30-V pulses delivered every minute, together with continuous 2Hz stimulation using the same parameters.

Test drugs were injected i.v. into the tail vein either 20 min before perforant path stimulation, or 10 or 40 min after the end of perforant path stimulation. Control animals were treated with saline. Electrographic activity was acquired and analyzed off-line using Harmonie software, configured for automatic detection and saving spikes and seizures. Analysis of EEG was performed by a "blinded" unbiased investigator.

Assessment

The following indices were used to quantify seizure activity: duration of self-sustained status epilepticus (= time between the end of perforant path stimulation and the end of the last electrographic seizure); cumulative seizure time (the sum of the duration of all individual seizures); number of seizure episodes; average duration of individual seizures (cumulative seizure time divided by number of seizures); number of spikes per hour. Statistical analysis was carried out with one-way ANOVA followed by Newman-Keuls post hoc test, or, if the normality test failed, ANOVA on ranks followed by Mann-Whitney post hoc test.

Latest technique

Brown et al. (1953) and Barton et al. (2001) characterized the 6Hz psychomotor seizure model of partial epilepsy in rats. Nissinen et al. (2000) described a model of chronic temporal lobe epilepsy induced by electrical stimulation of the lateral nucleus of the amygdala in rats. Walton et al. (1996) induced status epilepticus in rats with actively epileptogenic cortical cobalt lesions by administration of homocysteine thiolactone. Lauren et al. (2003) described selective changes in gamma-aminobutyric acid type A receptor subunits in the hippocampus in spontaneously seizing rats with chronic temporal lobe epilepsy. Brandt et al. (2003) studied epileptogenesis and neuropathology after different types of status epilepticus induced by prolonged electrical stimulation of the basolateral amygdala in rats.

Investigating Cortical Dysplasia in Rats

Epilepsy becomes drug resistant in 20%-30% of patients. Cortical dysplasia is implicated as a major contributing factor of many types of epileptic disorders that are resistant to pharmacological intervention. Several animal models of cortical dysplasia with specific clinical pathologies have been described. Baraban and Schartzkroin (1995, 1996), Baraban et al. (2000), and Smyth et al. (2002) exposed rats *in utero* to methylazoxymethanol (MAM).

Method

Dysplastic and control rats were generated by injecting pregnant Sprague Dawley rats on day 15 of gestation with 25 mg/kg i.p. MAM or vehicle (10% DMSO in 0.3 ml 0.9% saline). For *in vitro* studies, recordings were performed using acute hippocampal slices from adult vehicle or MAM-treated rats. Hippocampi were not dissected out, and all slices included entorhinal cortex and other overlying cortical structures. After cutting, slices remained submerged in a holding chamber containing oxygenated recording medium (nACF) consisting of (in mM): 124 NaCl, 3KCl, 1.25 NaH_2PO_4, 2 $MgSO_4$, 26 $NaHCO_3$, 2 $CaCl_2$ and 10 dextrose. A slice was then transferred to a gas-interface recording chamber and perfused with oxygenated nACF at a flow rate of 2.5 ml/min at 33.5°C. Borosilicate glass electrodes were pulled, filled with 2 M NaCl (2-8 MΩ) and placed in the CA1 region of stratum pyramidale and/or within neuronal heterotopias under visual microscopic control. A mono-polar stimulating electrode was placed in stratum radiatum.

Voltage was recorded with a Neurodata IR-283 amplifier and monitored on a PC running pClamp software. Spontaneous field activity and evoked population spikes were stored on hard disk for later blinded analysis. Interictal epileptiform burst activity was initiated with perfusion of nACF containing 4-aminopyridine (100 μM), a potassium channel blocker known to cause seizures in humans and spontaneous epileptiform activity in hippocampal slice preparations. The 4-aminopyridine *in vitro* seizure model is based on blockade of A-type potassium channels leading to the appearance of giant excitatory post-synaptic potentials generated by the prolonged firing of pyramidal neurons in CA3 burst-generating regions of hippocampus. Burst frequency was determined by counting the number of interictal epileptiform events per second during a 3-min epoch before and after 60 min of antiepileptic drug co-perfusion, and was expressed as Hz. Burst amplitude (1.5-6 mV) was determined by measuring the average peak-to-peak interval for ten consecutive representative bursts during the same epoch.

Evoked synaptic responses were analyzed by averaging the number of population spikes obtained on ten consecutive sweeps recorded after stratum radiatum stimulation (0.3- to 3-mA pulses 100 μs pulse-width). A downward voltage deflection ≥ 0.5 mV superimposed on the population *excitatory post-synaptic potential* (EPSP) was defined as a "*population spike*"; the number of population spikes was compared for each slice during perfusion with normal ACSF (baseline), ACSF plus 4-aminopyridine and ACSF plus 4-aminopyridine and antiepileptic drug. For each slice experiment, the population spike was monitored every 15 min. For *in vivo* studies, control and MAM-exposed rats were administered 15 mg/kg kainic acid, a concentration that reliably produces acute seizure activity. Behavioral activity was scored on a 6-stage scale. Animals were treated with 400 mg/kg i.p. valproate 30 min before kainate injection. Latencies to the first sign of hyperexcitability and to the first tonic-clonic seizure were recorded.

Assessment

Date were plotted graphically as "survival" curves, and differences in mean latencies were ranked and analyzed using a non-parametric Kruskal-Wallis one-way ANOVA. Since the MAM-exposed rats exhibit a dramatically reduced sensitivity to commonly prescribed antiepileptic drugs, this model is considered to be relevant for drug- resistant epilepsy.

Latest technique

Lere et al. (2002) described a model of "epileptic tolerance" for investigating neuroprotection, epileptic susceptibility and gene expression-related plastic changes. Expression of status epilepticus was triggered by infusion of the excitotoxic agent kainate in the right hippocampus of adult rats. An appropriate dose of kainate was used to cause brain damage to the homolateral, but not contralateral, hippocampus. At various times following the preconditioning insult, kainite was then re-administered into the lateral ventricle and neuroprotection was observed in the contralateral side between 1 and 15 days later.

Heredity of Epilepsy

Several animal species exhibit epilepsy with spontaneous recurrent seizures such as dogs, rats, and mice. Serikawa and Yamada (1986) described spontaneous epileptic rats which are double mutants and exhibit both tonic and absence-like seizures.

Method

Spontaneous epileptic rats are obtained by mating the tremor heterozygous rat (*tm/+*) with the zitter homozygous rat (*zi/zi*) found in a Sprague-Dawley colony. The behavior of the spontaneous epileptic rats is recorded weekly for 2 h on videotapes. The frequency of tonic convulsions and wild jumping occurring in the absence of external stimuli are recorded. Under anesthesia silver ball-tipped and monopolar stainless- steel electrodes are chronically implanted in the left frontal cortex and hippocampus. An indifferent electrode is placed on the frontal cranium. The frequency of absence-like seizures and tonic convulsions, as well as the duration of each seizure, are measured on the EEG. A mild tactile stimulus is given on the back of the animal every 2.5 min to induce consistent tonic convulsions. Compounds are given i.p. or orally.

Assessment

The number of seizures and the duration of each seizure are obtained and the total duration of the seizures (number × duration) is calculated every 5 min before and after injection of the drug. Percent changes between values before and after drug administration are calculated. Studies in spontaneous epileptic rats and other genetic models are of value for an in depth investigation of a potential anti-epileptic drug.

Latest technique

The tremor rat (*tm/tm*) was described as a model of petit mal epilepsy. Seki et al. (2002) attempted to determine whether gene transfer of aspartoacyclase inhibited absence-like seizures in tremor rats using recombinant adenovirus. Noda et al. (1998) and Iida et al. (1998) described the *NER rat strain* (Noda epileptic rat) as a genetic model in epilepsy research, which was developed by inbreeding rats with spontaneous tonic-clonic seizures in a stock of Crj:Wistar. The genetic epileptic WAG/RiJ rat has been recommended as a useful model for general absence epilepsy in humans. Danober et al. (1995, 1998), Deransart et al. (2000), Lakaye et al. (2002), and Nehling and Boehrer (2003) studied the GAERS rat, the genetic absence epilepsy rat from Strasbourg, which shows generalized non-convulsive absence seizures characterized by the occurrence of synchronous and bilateral spike and wave discharges.

Amano et al. (1996) developed an epileptic rat mutant with spontaneous limbic-like seizures by successive mating and selection from an inherited cataract rat. Racine et al. (1999) reported selective breeding of kindling-prone and kindling-resistant rats. The selection of these strains was based on their rates of amygdala kindling. From a parent population of Long Evans hooded and Wistar rats, the males and females that showed the fastest and slowest amygdala kindling rates were selected and bred. Sarkisian et al. (1999) described seizures in the Flathead (FH) rat as a genetic model in early postnatal development. Tsubota et al. (2003) identified the Wakayama epileptic rat (WER) in a colony of Wistar rats, a mutant exhibiting both tonic-clonic seizures and absence-like seizures.

Investigating Epilepsy in Transgenic Animals

The availability of transgenic animals has stimulated research on pathogenesis of epilepsy. Several surveys on this topic are available. Several studies contribute to the understanding of pathology of epilepsy. Luthi et al. (1997) found that mutant mice over-expressing protease nexin-1 (PN-1) in brain under the control of the Thy-1 promoter (Thy 1/PN-1) or lacking PN-1 (PN-1 -/-) develop epileptic activity *in vivo*. An endogenous serine protease inhibitor modulated epileptic activity and hippocampal long-term potentiation. Kunieda et al. (2002) recommended mice with systemic overexpression of the alpha 1B-adrenergic receptor as an animal model of epilepsy. Some of the studies gave hints for further development of antiepileptic drugs, such as the neuropeptide galanin or the neuropeptide-Y. Several studies were devoted on the role of *brain-derived neurotrophic factor* (BDNF).

11

New Drugs Against Mycobacterium

Approximately 30 years ago, the prevailing opinion was that the fight against infectious diseases was all but over. Optimistic predictions were made in the belief that a combination of improved public health, vaccination, and the current variety of antimicrobial agents would be sufficient to combat most, if not all, infectious diseases. Synthetic analogs of known antibiotics that were effective against a broad range of organisms had been developed, and thus there was little perceived need for the development of new drugs. In fact, until very recently, no new classes of antibiotics had been licensed for clinical use against bacterial infections. The search for new drugs was limited to well-known compound classes with insufficient variability to prevent the escalation of clinical resistance. With the increasing incidence of infection with multiple-drug-resistant tuberculosis strains, the World Health Organization (WHO) declared a worldwide emergency and responded with the *directly observed therapy*, short-course (DOTS) program. Streptomycin, isoniazid, and pyrazinamide, which were developed in the 1940s, are still integral parts of the chemotherapeutic regimens used today.

The most recent introduction of an antituberculosis agent into the modern treatment scheme of tuberculosis was rifampin in the 1960s. The current treatment recommendations for tuberculosis from the Centers for Disease Control and the WHO for pan-sensitive strains involves a 6-month regimen, beginning with a 2-month (intensive phase) treatment cycle of isoniazid, rifampin, and pyrazinamide, followed by a 4-month cycle of isoniazid and rifampin. Although protracted, this is considered short-course chemotherapy. Alternative agents such as ethambutol and streptomycin are included in the intensive phase if there is a possibility of drug resistance. The post-genomic era of expression profiling affords insight into the global biological response induced under diseaselike conditions in addition to drug pressure. Through the identification of gene and operon responses, specific pathways can be identified that are unique to a particular disease. To study bacteria in a diseaselike state has always been a difficult undertaking to achieve.

Recently studies have been performed looking at the global gene response to altered growth states using microarray. These studies have been performed under artificial growth conditions to control the number of variables. In each case, specific changes were made, such as oxygen and nutrient limitation. Although much information has been obtained at the level of the gene and related metabolic pathway, there is still not a clear picture of the impact this information has on the biology of the bacterium and thus the pathology of the disease. However, when these genes and pathways are analyzed, a picture develops of how the cell alters related regulons in order to compensate for the change in growth environment.

The identification of these regulon networks results in information on the metabolome of the bacillus. This knowledge can be used to establish gene essentiality, conditional expression requirements, and determine genes associated with virulence. Although this information is extensive for the simple design of high-throughput screens, it affords predictive value in terms of the targets affected and the circumventive measures the bacillus uses to avert the pressure of drugs under different growth conditions such as the intracellular environment. To some, high-throughput screening (HTS) has come to mean screening thousands of compounds with a pass–fail criterion. Ironically, none of the current antitubercular drugs could have passed the stringent conditions that are in place today. Therefore, we now realize that HTS is not intended to screen-out candidates but, rather, define rapidly and systematically the properties of many compounds so that they can be prioritized and further characterized through relevant secondary screens. It is important to clearly define the criteria that will be used to judge the candidates at each step of the drug discovery process. The overall goal is to make HTS a rational process such that assumptions made in the design do not compromise the quality of the data and therefore adversely impact the conclusions drawn from that data.

PHARMACOGENOMIC STUDIES

The sequencing of complete genomes has had a profound influence on drug discovery, with its main promise being the potential to understand disease processes at a genetic level and determine optimal targets for drug intervention. Thus far over 1100 complete genomes have been sequenced, including 98 from bacteria and 16 from archaea. This list includes many important human pathogens, although the bias is toward organisms that affect individuals in the developed countries. The complete *M. tuberculosis* genome was added to the list in 1998. It is obvious that this sequence contains information on all possible drug targets encoded therein. Genomic information is now an integral part of the modern drug discovery process and can be used interpretatively to understand protein and pathway function, allowing rapid target identification. This understanding is then used to establish a potential pathway to support the target of interest and virtually validate it. In addition to interpretive analysis, the greatest benefit of genome sequence is that it affords predictions toward establishing the metabolism of the bacterium.

Usually researchers study *M. tuberculosis* on an artificial medium. Although the transcriptional regulators and associated networks can be used to predictively develop assays, they may not be reflective of the metabolism occurring in the organism during an infection and disease. The availability of the entire sequence information provides insight into the global metabolic capabilities and tendencies, which can be used to guide inhibitor design in a manner that may bias development toward relevant aspects of disease. This approach has been deemed "*structural genomics*," and the aim is to structurally characterize proteins and thus pathways from sequence information. There is considerable effort in the postgenomic era to validate protein predictions with experimental determination *via* crystallography and NMR spectroscopy and synthetic pathways *via* experimental enzymology. However, currently the functional annotation of proteins and thus metabolic roles is based primarily on similarities to homologs in the databases.

The primary assumption is that proteins with homologous sequences catalyze the same reactions. Although this is generally true, gene sequence cannot always be a reliable way to determine protein activity. Frequently, homologous enzymes do not catalyze the same chemical reaction and the same chemical reaction is not catalyzed by homologous enzymes. Although homologs can maintain function, there is mounting evidence that divergent evolution is an active process; a progenitor gene is duplicated and its copy assumes a new function in response to selective pressure. Because the outcome of divergent evolution can be quite different, functional assignment on the basis of homology could be grossly misassigned. Although gene and protein homology can provide information about the nature of the reaction that is catalyzed, homology alone will provide very little information about the specific identity of the reaction.

Therefore, genomic enzymology requires a correlation between structure (sequence) and chemical reaction (enzymatic function); neither structure-only-based nor function-only-based analysis will be totally accurate. This could be an underlying factor contributing to the high percentage (~40% in the case of *M. tuberculosis*) of genes that have no known function. However, regardless of the difficulties associated with predicting all of the diverse chemical reactions present in the life of bacteria, including *M. tuberculosis*, genomic information has already proven invaluable in addressing a host of important research questions. A primary advantage of the availability of sequence information is that it has effectively made classical "*reverse genetics*" highly abbreviated. With genomic information, genes can be rapidly identified and cloned to evaluate function, and this approach is proving particularly useful when coupled with classical biochemical knowledge. Below we provide examples of this principle in the context of target definition and drug discovery.

Pathways that can be exploited as novel and potent antibacterial targets are being mined *in silico* and this information is then applied to actual validation experiments. Identifying genes, predicting protein function and pathway elucidation from preexisting biochemical data, is proving useful in determining the role of hypothetical and unknown genes through linkage. This linkage may be established based solely on location in the genome and proximity to known genes or by known function established through homology or experimentation. Ultimately, the information that is acquired contributes to the elucidation of specific and associated pathways, which may provide complete metabolome information. This then can provide clues about essential genes involved with disease, virulence factors, and conditional responses. Experiments designed to elucidate specific metabolic responses are rapidly developing into potential options for drug screening. In an attempt to understand the disease process at a genomic level, studies have been performed to explore the altered metabolism of *M. tuberculosis* grown under defined conditions thought to mimic their intracellular niche.

Classical biochemical studies have demonstrated that *M. tuberculosis* can survive extended periods of nonreplicating, stationary phase states, induced by limited growth conditions such as reduced oxygen and nutrient deprivation. However, until recently, these studies have not addressed the bacterial response at the gene-protein level. Using microarray and proteome analysis, under nutrient starvation growth conditions, differential gene and protein expression were investigated. Proteome analysis derived from bacteria under these conditions revealed alterations in several proteins. Among repressed proteins were the 45-kDa and MPT64 antigens. It is interesting that these very proteins are the ones primarily recognized by the immune response of the majority of tuberculosis patients and their contacts.

This phenomenon of the repression of the expression of two important immunogens under nutrient starvation may be explained by the fact that these are secreted proteins, and thus the artificial nutrient starvation conditions may affect protein secretion. It is not at all clear whether such artificially induced nutrient starvation conditions reflect long-term persistence or reactivation of disease. Several protein spots were shown to increase in intensity under starvation conditions. Among these are two conserved hypothetical proteins that have no identifiable functional motifs or domains when an *in silico* sequence-based analysis was performed. In a more sensitive, perhaps more relevant and global approach to analyze the adaptive response of *M. tuberculosis* to nutrient starvation, the transcriptional profile was monitored at multiple time points using DNA microarrays.

Differential gene expression was detected for all of the major functional gene classes. Genes associated with energy metabolism such as *aceAb* were upregulated, which may reflect altered carbon sources, although other tvilarboxylic acid cycle (TCA) related genes did not show consistent alteration. In accordance with reduced respiration rates and metabolic shutdown, genes thought to encode proteins involved in a shift from aerobic to anaerobic metabolism had altered expression profiles. Genes that are categorized into other functional groups, such as lipid and polyketide synthesis, translation, and transcription, responded accordingly to the slowdown in bacterial growth rate and metabolism.

Of interest was the altered expression response of several unknown regulatory genes, and this may have the greatest potential in understanding the way *M. tuberculosis* responds and postures itself to surviving long-term starvation. Although a number of these genes have homologs and have been studied outside mycobacteria, the full understanding of their role in metabolic adaptation is as-yet unknown. Sherman and colleagues have also used DNA microarray technology to explore the adaptive response of *M. tuberculosis.* In these studies, gene response to hypoxic growth conditions were investigated. The gene expression changes that were noted were indicative of broad adaptations to reduced metabolic activity. Among the repressed genes, 40-60% (range is due to genes of unknown function) encode proteins with roles in biosynthesis, cell division, and aerobic metabolism. Genes induced by hypoxia tell a subtler story.

Genes classically assigned to play a roles in *M. tuberculosis* latency, such as genes encoding glyoxylate shunt enzymes and alternative sigma-factors, were essentially unaltered. However, of notable interest was the hypoxic induction of a two-component regulatory operon encoding Rv3134c/Rv3133c/Rv3132c. Providing more evidence that this regulatory operon is involved in hypoxic adaptation is the fact that targeted disruption of the upstream gene Rv3134c, eliminated hypoxic regulation of *hspx.* This is consistant with previous studies which clearly indicate the role of *hspx* in stationary-phase growth and growth in macrophages. It is clear that the availability of genomic information has fostered the ability to delve into the intricacies of the bacterial response to a number of environmental changes.

However, the current challenge is how to delineate expression profiles to provide an understanding of the disease state. Thus far, microarray and protein profiling have provided evidence for induction of the stringent response and downregulation of aerobic respiration, translation, cell division, and lipid biosynthesis, in response to hypoxic and nutrient-limited growth conditions. Induced genes such as *acr*, alanine dehydrogenase, and fumerate reductase may facilitate survival under these conditions and therefore may represent relevant drug targets for persistent organisms. In addition, the numerous genes of unknown function induced under these conditions may perform specific mycobacterial functions associated with invasion, pathology, and latent disease and warrant further investigation in terms of their relevance as potential drug targets.

TARGETED SCREENING

Targeted screening allows the researcher to take advantage of the power of bioinformatics, which can be used to select potential targets from known or previously unexplored biochemical pathways. Once the target is selected, molecular biology techniques are used to clone and express recombinant forms of the protein. The resulting highly purified protein (usually an enzyme or a receptor) is then the foundation for the development of a high-throughput screen. This method facilitates rational drug design, since it is often possible to utilize the purified recombinant protein for three-dimensional structural studies, which, in turn, allow *structure-activity relationship* (SAR) analysis. With a computational representation of the preferred molecular interaction, drug candidates can be screened, *in silieo*, for their ability to interact within this protein, which streamlines library screening.

The power of target-based drug discovery is that very quickly the structure and the putative activity of a protein can be revealed. Then the specific characteristics can be applied to either understanding the pathway of interest to pursue related steps as targets or to developing an HTS-style drug screen. Since target-based assays are completely synthetic situations in which all of the mechanism kinetics can be predetermined, the assay can inform the researcher very easily about the inhibitory properties of a compound or class of compounds. This information can then be applied to combinatorial chemistry approaches to optimize inhibitory activity, as is discussed later. An additional advantage of target-based methods is that they allow detection of inhibitors that are not able to cross permeability barriers. Although these compounds are obviously not useful as therapeutics because they are not bioavailable nor able to penetrate the bacterium, they do provide useful information that can be applied to discover other novel targets that are to be used to optimize new classes of inhibitors to improve their druglike properties. Target-based screening does have some drawbacks. The method presupposes that a robust assay amenable to HTS can be developed.

Development of assays of this nature can be time consuming and requires a considerable level of expertise. For this reason, target-based screening cannot address potential targets that have no known function. Thus, ~40% of the *M. tuberculosis* genome, some of which may be potential targets, is unavailable for target-based drug discovery. In addition, the essentiality of the potential target must be established *in vivo*, and this is not a trivial consideration when dealing with *M.*

tuberculosis. Finally, compounds that inhibit an assay *in vitro* may not have druglike properties or may not be able to cross the permeability barrier imposed by the complex mycobacterial cell wall; turning an enzyme inhibitor into a drug is a very expensive endeavor in terms of both resources and time. This approach in the context of tuberculosis is exemplified by targeting the L-rhamnose biosynthetic pathway in *M. tuberculosis*.

L-Rhamnose plays an essential structural role in the cell wall of *M. tuberculosis* as a key sugar component of the disaccharide "*linker unit*" (P-GlcNAc(α1-3)Rha), which connects the peptidoglycan heteropolysaccharide to the remainder of the cell wall "*core*," the galactan-arabinan-mycolic acid complex. In addition to its structural role in the cell wall, rhamnose biosynthesis was

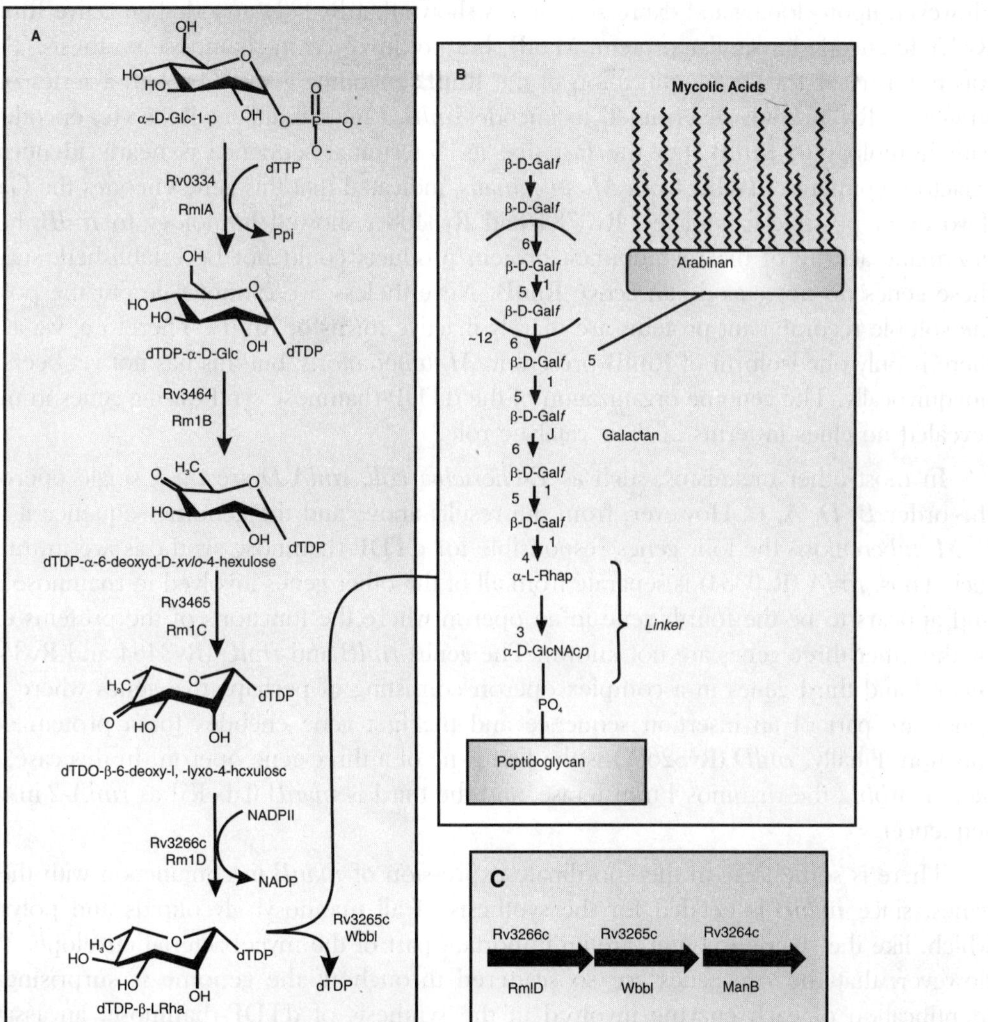

Fig. 11.1. Anabolic pathway and proteins associated with rhamnose biosynthesis in M. tuberculosis. A-The metabolic conversion of glucose-1-P to the rhamnose nucleotide. B-The macromolecular structure of the rhamnose containing "inker region". C-Operon of genes.

also chosen because it does not exist in mammalian cells, and, hence, inhibitors of its biosynthesis should have a lesser chance of being toxic to humans. dTDP-Rhamnose is the precursor of the rhamnose found in the linker unit, and it is known that four enzymatic steps are required for its biosynthesis. Accordingly, four potential enzymes (RmlA-D) were sought through homology via *in silico* analysis of the *M. tuberculosis* genome. This analysis revealed eight ORFs, which could encode for proteins that catalyze these four enzymatic steps. Only one copy of *rmlC* (Rv3465) and *rmlD* (Rv3266c) were identified, and, upon cloning, the hypothesized enzymatic activity for each protein was demonstrated.

The situation for *rmlA* was ambiguous because two candidate genes showed possible homology. However, upon cloning and expression, it was shown that Rv0334 encoded an active RmlA, whereas Rv3264c encoded a similar protein, ManB, but not involved in rhamnose synthesis. The situation was not as clear for the identification of the RmlB encoding gene. Through a series of molecular strategies, Rv3464 was determined to encode *rmlB*. One candidate, Rv3634c, encodes a protein with homology to RmlB, but the fact that its N-terminal sequence is nearly identical to UDP-galactose epimerase (GalE) from *M. smegmatis*, indicated that this gene encodes the GalE protein. Two other possible candidates, Rv3784 and Rv3468c, showed homology to *rmlB*; however, the enzymatic activity of the recombinant protein products could not be established, suggesting that these genes do not encode an active RmlB. Nevertheless, we cannot rule out the possibility that the soluble recombinant proteins are merely inactive forms of RmlB. Therefore, we assumed that there is only one isoform of RmlB present in *M. tuberculosis*, but this has not yet been established unequivocally. The genome organization of the dTDP-rhamnose synthesizing genes in mycobacteria revealed no clues in terms of their catalytic role.

In most other organisms, such as *Escherichia coli*, *rmlA-D* are on a single operon, often in the order *B, D, A, C*. However, from the results above and the genome sequence it is clear that in *M. tuberculosis* the four genes responsible for dTDP-rhamnose synthesis are in three different loci. Thus, *rmlA* (Rv0334) is separate from all of the other genes involved in rhamnose metabolism and appears to be the fourth gene in an operon where the functions of the proteins encoded for by the other three genes are not known. The genes *rmlB* and *rmlC* (Rv3464 and Rv3465) are the second and third genes in a complex operon consisting of perhaps five genes where the last two genes are part of an insertion sequence and the first gene encodes for a protein of unknown function. Finally, *rmlD* (Rv3266c) is the first gene of a three-gene operon. In this case, the second gene is *wbbL*, the rhamnosyl transferase, and the third is *manB* (labeled as *rmlA-2* in the genome sequence).

There is some logic in the coordinate expression of *manB* in conjunction with the rhamnose genes, since *manB* is needed for the synthesis of all mannosyl glycolipids and polysaccharides, which, like the rhamnosyl unit, are an important part of the mycobacterial envelope. The finding, however, that the *rml* genes are so scattered throughout the genome is surprising. With the identification of each enzyme involved in the synthesis of dTDP-rhamnose, an assay was then developed. This assay took advantage of the fact that RmlD converts dTDP-6-deoxy-L-*lyxo*-4-hexulose to dTDP-rhamnose with the concomitant oxidation of NADPH to NADP, which makes it possible to couple the enzyme activity of each step in a facile microtiter plate assay amenable to screening inhibitors of the rhamnose biosynthetic pathway. Increased resolution specific to each enzymatic step was achieved by titrating the enzyme of interest within the assay system, thus altering required inhibitor concentrations. Upon analysis, it became apparent which enzymatic step was being inhibited. This target-based assay, developed from genomic information, was used to screen 8000 compounds in an HTS format.

Prior to testing, *in silico* analysis of these candidates was performed based on Lipinski's "rule of 5," which categorizes candidate's in terms of molecular weight, logP, hydrogen bond donors, and hydrogen bond acceptors. These characteristics have been experimentally determined and have been shown to correlate to the effective bioavailability of a drug from current orally available medicines. With respect to these four criteria, 89% of the compounds of the library had "druglike" properties and 9% of the compounds were within the acceptable range, satisfying three of the four criteria. The compounds were also selected for the presence of "druglike" functional groups and the lack of reactive groups such as aldehydes.

The candidates that survived were screened at a concentration of 10 μM. Eighteen compounds, which showed activity initially, were rescreened in triplicate. To confirm the inhibitory activity of these 18 candidates, new samples were reexamined. However, upon rescreening, only 11 of the 18 compounds were inhibitory in the target-based assay. These 11 compounds were then assayed for their ability to specifically inhibit RmlB, RmlC, or RmlD, and all were shown to inhibit RmlC (dTDP-6-deoxy-D-xylo-4-hexalose epimerase). With specific inhibitory activity determined, these drug candidates will now proceed into whole cell-based assays to determine essentiality and inhibitory activity.

DISCOVERY OF ANTITUBERCULOSIS DRUGS

Classical antituberculosis drug discovery relied on whole-cell-based assays in which the primary readout was cell death. Although this approach has been responsible for the few anti-TB drugs in use today, the extended culturing times of *M. tuberculosis* (~4 weeks) made high-throughput screening impossible. However, since the genome sequence has become available, whole-cell-based screening has entered the era of HTS. A number of advances in our understanding of the biochemistry of the tubercle bacillus *via* genomic information have provided a scientific platform to design quick (~24 hr) bioluminescent-based HTS assays. A primary breakthrough was the demonstration of the ease of expression profiling under stress conditions. The basic rationale in terms of modernized whole-cell-based screen development is that the selective inhibition of a metabolic pathways causes an accumulation of precursors and a concomitant depletion in products. These gross changes can be expected to selectively induce changes in the transcription of genes, thus altering the expression of the proteins that comprise the affected pathway, particularly before a more generalized stress response ensues.

The resulting expression profile then serves as a "*biological fingerprint*" for that specific metabolic perturbation. While regulated genes themselves may or may not be targets for drug intervention, they, with the help of software algorithms linking genes with proteins and known functions, can point researchers toward known pathways. In addition, this information can be used to incriminate the affected pathway and classify the target for unknown inhibitors. In terms of drug discovery and screening development, the expression of each gene in the genome or each protein in the proteome needs to be interrogated simultaneously in the presence and absence of the inhibitor. If an inhibitor of the pathway of interest is not available, then a manipulatable alteration needs to be introduced. One approach that we are currently employing is genetic engineering to construct temperature-sensitive strains of *M. tuberculosis* which can then be studied. This approach allows us to directly perturb the intended target protein, thereby introducing a "drug-like" effect. Another strategy that has had limited success is the introduction of an inducible promoter to control the gene of the target protein.

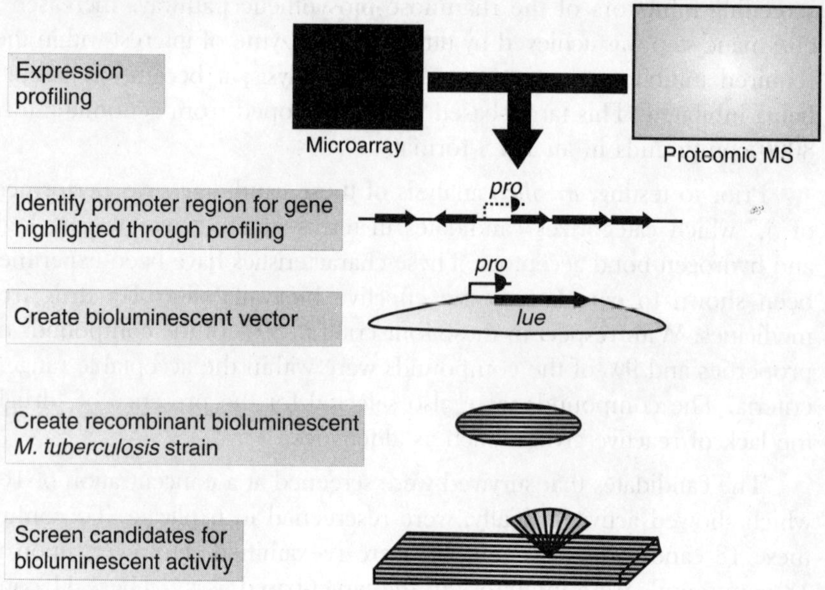

Fig. 11.2. Design of high-throughput bioluminescence reporter strains used in *M. tuberculosis* drug discovery.

The limitation in the case of *M. tuberculosis* is attributable to the lack of an available good promoter that can be turned off when needed. Currently, microarray technology is at the forefront of expression profiling in tuberculosis research because it is able to report on differential expression on a global scale that has not been realized by proteomic techniques. A typical *M. tuberculosis* array is fabricated with either amplicons or oligonucleotides and covers only the predicted ORFs. In terms of expression profiling of *M. tuberculosis* and drug discovery, the utility of transcriptional response profiling has been validated with the use of the *M. tuberculosis*-specific inhibitor, isoniazid. Although isoniazid has been thoroughly studied with arrays, it illustrates the important aspect of interpreting microarray data for screen development. The mode of action of isoniazid, arguably, has been studied more than any other antimicrobial agent. Although the actual proteomic target has been debated, it is agreed that the affected macromolecules are the mycolic acids.

When the transcriptional response of *M. tuberculosis* to isoniazid treatment was studied using DNA microarray technology, a subset of genes involved in fatty acid synthesis were differentially regulated. Many of these genes were assigned through biochemical studies and homology to fatty acid metabolism. The list includes genes of the FAS II operon Rv2243-2247, and associated genes *fbpC* (antigen 85c), *fadE23*, and *fadE24*, in addition to isoniazid associated genes such as *efpA* and *ahpC*. To confirm and validate the microarray results, real-time polymerase chain reaction (PCR) was performed. These experiments confirmed the expression profiles of all the genes tested. Our understanding of the molecular consequences of isoniazid on mycolate biosynthesis provides confidence that many of the responsive genes encode proteins that are relevant to the mode of action of the drug. Accordingly, microarray hybridization provides evidence useful in predicting the mode of action of a novel compound based on a physiologically derived interpretation of its expression response to that compound.

Specific drug-responsive promoters identified by expression profiling serve as sensors of the intracellular conditions that are characteristic of the drug's activity which can then be exploited for the design of screens to identify novel compounds that exert similar effects on the bacteria. Building on this approach, we have experimentally characterized the bacterial transcriptional response to over 20 inhibitors that affect several aspects of cell wall and macromolecular biosynthesis and cell division. Although many inhibitors clearly show similar transcriptional profiles, it is not always easy to rationalize the results from other inhibitors. Oftentimes, the highly induced genes are conserved hypothetical genes with no functional assignment.

In such cases, it is hard to link a specific gene with the pathway affected by the inhibitor. Nonetheless, information is still generated that can be used as the basis for a mode of action specific high-throughput screen. We have been able to categorize compounds on the basis of common gene inductions. This not only indicates which drugs may affect similar biosynthetic pathways but also allows for the design of reporter strains of *M. tuberculosis* that can be used in an HTS format to rapidly evaluate the MIC of a molecule and characterize its possible mode of action. One has to remember, as previously discussed, that the premise of whole-cell, genome-based screens is that one is monitoring a unique phenotypic response, a "*biological fingerprint*" and not necessarily the targeted pathway.

MODERN ANTITUBERCULOSIS DRUG DISCOVERY EFFORT

Modern combinatorial chemistry has already contributed to the modern drug discovery revolution. It is now possible to rapidly produce drug candidates with high levels of diversity and complexity, in marked contrast to rational drug design of just a decade ago in which chemists selected specific molecules for synthesis. This new chemical approach leads to compound libraries which have the ability, in theory, to sample all possible three-dimensional space. Barry and colleagues have been instrumental in pioneering the use of combinatorial chemistry and expression profiling to discover new classes of inhibitors for *M. tuberculosis*. The beauty of this approach is that it capitalizes primarily on genomic sequence information which is readily available. The approach is centered on the idea that from genetic information, screens can be designed to report on the activity of drugs against a specific target, strictly from the reactive response of the bacterium.

The advantage of this approach is that reporter strains (based on transcriptional response) can be designed very easily, and these are amenable to HTS automation. Progress is limited by classical structural methods such as crystallography. To validate the whole-cell-based screening approach, Barry and colleagues used the known antimycobacterial drug ethambutol and interrogated the bacterial response to its inhibitory action. Although the target of ethambutol has been identified as arabinose polymerization in the cell wall, its specific protein target or targets are still unknown. In addition, ethambutol's potency could be improved; unlike other mycobacterial inhibitors, the toxicity profile is relatively low. These two factors allow for vast improvement of the drug's activity with a lesser chance of toxicity. With information of the transcriptional response to ethambutol treatment, a bioluminescent reporter strain was created and validated using ethambutol as a control. In addition to developing an ethambutol specific whole-cell-based HTS, a directed compound library was synthesized using combinatorial chemistry strategies.

Ethambutol is a symmetrical diamine, and the directed library focused on exploring the activity of asymmetrical molecules around the diamine scaffold. The overall goal was to identify better inhibitors with the same mode of action as ethambutol using only genomic information. Numerous candidates were identified using HTS formatted MIC and bioluminescent assays that shared the same mode of action of ethambutol, but had better minimal inhibitory concentrations. The inhibitory activities of several of the candidates were confirmed using classical biochemical techniques and tested in the animal model of infection to evaluate efficacy. From analyzing the structure of all of the molecules in the library (compounds with and without inhibitory activity) in terms of their activity, it was possible to develop SAR information. Another round of molecular optimization can then be imitated biased by the SAR information to further improve the potency and the pharmacokinetic characteristics of the compounds. These first-generation molecules demonstrated the power of this approach in the development of new antimycobacterial drugs.

AUTOMATION TECHNOLOGIES

The convergence of chemistry and genomic information in combination with *automation technologies* has fostered a new approach to drug screening which utilizes both whole-cell-based and target-based approaches. In the scheme, a desired phenotypic change is sought by screening initially through high-throughput low-resolution whole-cell assays. These assays are designed based on unique patterns associated with specific pathway inhibition as determined by expression profiling. It is important to reiterate that the expression pattern may not implicate the actual proteomic target; it is simply readout of the bacterial response to the metabolic pressure. These screens have the ability to accommodate large libraries of diverse compounds and drastically reduce the number of candidates progressing through the discovery process.

The "hits" identified from the whole-cell-based screens are then interrogated with higher resolution assays designed around specific enzymatic targets. These high-resolution assays provide high-quality data and establish inhibitory kinetics. Once the library has been narrowed to a small fraction of compounds with the desired effect, they are prioritized by surrogate animal models and finally evaluated for efficacy in an actual animal model. This drug screening process is designed so that the high-throughput low-resolution assays which are robust enough to handle large compound libraries are used before low-throughput high-resolution assays are used to confirm and bias the procession of candidates into model systems of infection. As the number of potential drug candidates increases, so does the need for quick, efficient, and high-volume screens that generate data of high

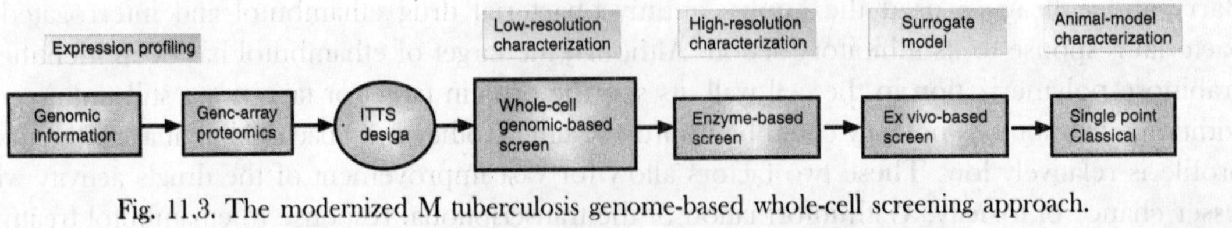

Fig. 11.3. The modernized M tuberculosis genome-based whole-cell screening approach.

quality. The current testing management problem is that screening is not conducted on the same candidate a million times but on a million different candidates at the same time.

The purpose of the drug discovery process is to find and explore significant differences between compounds. Therefore, the screening process has to be designed to categorize quickly the inhibitory characteristics of the drug candidates. The largest challenge in screening is not just the large member libraries, but the low hit rates associated with them. With the latest advances in high-throughput screening using multiwell plates, the initial testing volume can be much greater than 10,000 candidates a day. Modern primary screens are designed to accomplish two things: first, to establish a minimal inhibitory concentration for each candidate and, second, to provide clues as to the potential target. This can be accomplished by multiplexing the assays. The advantage of multiplexed whole-cell-based screens is that a couple of requirements are built into the assay.

The drug candidate has the strict requirement to be able to subterfuge the mycobacterial cell wall and be metabolically stable within the cell. The primary benefit of whole-cell-based screens is to streamline the process and reduce the screening burden on target-based screens. Potential drug candidates coming through the whole-cell-based assays can be prioritized and biased into the target-based assays based on their specific inhibitory profile. With the information on MIC, the ability to kill the bacterium is established and a reasonable idea of target pathway is known. However, the absolute target is not known and therefore the specific kinetic details have not been established. Although, in simplistic terms, this detailed information may not be necessary, it is required for predictive studies in terms of resistance mechanisms and potential microbial spectrum.

There are a number of advantages in target-based screens, particularly if they are not responsible for the primary categorization of potential drug molecules. They provide better resolution in terms of target identification and they are an additional method of target validation. Whole-cell-based assays and target-based assays are very compatible strategies in the quest to rapidly characterize potential molecules as drug candidates. They exploit both genomic information and expression profiling. A screening strategy that uses these two approaches in tandem has the ability to establish essentiality through MIC, elucidate targets, and establish kinetic information of the inhibited reaction. Despite all of these advantages, each approach has its disadvantages. Whole-cell-based screens can be insensitive; many active compounds have associated toxicity issues in general and possible mixed mechanisms of action. Target-based assays have problems associated with turning *in vitro* inhibitors into antibacterial drugs and genetic validation of essentiality. However, in combination, these two approaches compliment each another, and the disadvantages associated with each are essentially eliminated.

SURROGATE SCREENING SYSTEM

The greatest need in the search for new antituberculosis drugs is a *surrogate screening system* for animal testing. Animal models of infection are used in the final stages of the drug development cycle to assess the antimicrobial efficacy in a model of human disease. Recent experience with library testing confirms historical evaluations of the extremely poor correlation between *in vitro* drug potency and *in vivo* drug efficacy. The reasons for this poor correlation may be due to (i) altered physiology of the bulk of the bacterial cells growing in macrophages; (ii) the simultaneous existence of several discrete populations of bacteria each with varying susceptibilities to the drugs being tested; and (iii) the extremely complex pharmacokinetics involved in delivering the drugs to the site of infection, namely the alveolar macrophages. A large number of compounds are being discovered and synthesized that show promise in *in vitro* assays against *M. tuberculosis*.

In fact, many more *in vitro* lead molecules are being generated than can reasonably be assessed through traditional animal models of efficacy. Therefore, the development of reliable surrogate models for chemotherapy of human disease that has a moderately high-throughput format is an urgent matter. Bioluminescent screening systems to address these questions have been attempted using the same basic approaches as the whole-cell assays. Currently, murine-derived macrophage cell culture is used as a possible testing surrogate. More recently, screens have been developed with bacteria derived from infectious animal tissue, such as the lung or spleen of a mouse. Since the bacteria have established an actual infection, they have also adopted the proper physiological state to cause disease and persist.

The purpose is to establish this bacterial metabolic state so that drug candidates can be screened rapidly and easily in a system that mimics the bacterial state of disease. Although current approaches have utilized macrophage lines, it is best if this metabolic state could be induced in artificial medium. This does not seem to be possible at the moment, but, with genomic information and ongoing studies on *in vivo* metabolism of *M. tuberculosis*, a manipulatable, quorum-sensing-based system may be possible in the future. It is important to remember that the primary quality from a surrogate animal system is that it establishes the bacterium in a metabolic state similar to that of true infection. This should not be confused with developing a system that tries to mimic the site of infection. After the efficacy of a compound has been addressed in the surrogate model of infection, it is then tested in an animal model of infection.

Due to the number of candidates that can proceed through a whole-cell-based discovery strategy to this step, low-cost, low-resolution, single-time-point animal experiments are useful. These animal tests are based on a single-time-point analysis, such as 40 days of infection. The animals begin therapy at 2 weeks postinfection, and the gross effects of drug efficacy are determined at the time of harvest. It is important to begin therapy before visible pathology is apparent. The bacterial burden, which reflects the efficacy of the drug candidate, can be qualitatively determined *via* gross pathology, and the quantities of bacteria can be determined through approaches such as bioluminescence and molecular tagging, if desired. This approach is intended, like the other screens, to bias the procession of candidates, sifting the most promising compound to the top of the pile. Candidates that show promising activity can be further analyzed *via* classical animal model experiments designed to determine the efficacy profiles of potential drugs over several weeks, consisting of multiple time points that provide high-quality statistical data.

INHIBITORY MECHANISM

In terms of drug discovery, it is paramount to have functional knowledge derived from genomic sequence or experimental data to fully understand the *inhibitory mechanism*. With the genomes of so many organisms completed, it is becoming possible to globally compare organisms with similar disease niches and pathology. This genomic–metabolic linkage among pathogens is proving to be the aspect that is most promising to understanding the weakness of organisms in general. With this metabolic information, combinatorial libraries can be designed which are biased for unique activity specifically against reactions, not necessarily unique enzymes. The next step to optimizing drug activity is to marry mechanistic and chemical information with the predicted function of the target protein based on genomic information and homology so that chemical libraries can be tailored to specific enzymatic chemistry.

At the moment, whole-cell-based bioluminescent screening strategies have the potential to characterize large sets of candidates against a single enzyme within a pathway. This is slowly producing hits and resulting in the identification of novel inhibitors. However, they usually lead to a single target, and, with the treatment pressure by current drugs, they are rapidly losing their potency due to resistance. Thus, years of work and resources can be lost because a drug that has been designed against a single enzyme which undergoes a mutation event renders the drug useless. Perhaps the next step is to use genomic sequence to inform drug design against specific enzymatic reactions that are represented numerously within the metabolism of an organism or multiple organisms. This would effectively make it impossible for an organism to acquire a single polymorphism that would provide resistance. Microorganisms develop resistance at a frequency as low as 10^5; if a number of targets were inhibited simultaneously, this mutation frequency would be multiplied based on the number of targets.

Organisms have the ability to reach high numbers in an infection, which makes this frequency a very real number. The basic rationale is not to design inhibitors against a reaction catalyzed by a specific enzyme, but rather against a specific reaction catalyzed by protein analog super families. Analysis of the mycobacterial genome sequence reveals that the organism has undergone massive gene duplication. This is particularly evident by the number of fatty acid genes, related polyketide syntheses (PKS), and glycosyl transferases that are responsible for cell wall construction and accessory molecules. Although all of these related enzymes catalyze similar reactions utilizing a variety of related substrates, they clearly produce very different products. This has essentially led to the characteristic heteropolymeric cell wall of the tubercle bacillus, which is made up of diverse polysaccharides and unique lipids. Once a gene is duplicated, it is thought that it undergoes evolution to fulfill a new metabolic role. It has been proposed that metabolic pathways evolve backward.

When the substrate of an enzyme in a pathway is limited, a new enzyme evolves to supply that substrate from an available precursor. Accordingly, evolution is constrained to retain binding specificity, because the original enzyme and the newly evolved enzyme must retain the ability to bind the same substrate/product. The specific mechanism of the reactions need not be conserved. Ultimately, the progenitor and new enzyme may catalyze the same reaction from slightly different substrates or catalyze a slightly different reaction from similar substrates, each case utilizing the same type of intermediate. In *M. tuberculosis*, an example of the first idea may be the ketoacyl syntheses KasA and KasB and possibly the related domain of fatty acid synthase, Type I (FASI). Each one catalyzes the same reaction under the same conditions, the primary difference being substrate specificity; KasA specifically catalyzes the condensation of short-chain lipids and KasB condenses long-chain lipids. If molecules can be tailored to mimic the mechanistic state of the condensation in this case, then all three enzymes in addition to PKSs could be potential targets.

The family of enzymes responsible for the introduction of heterogeneity into mycolic acids may represent the second scenario. It has been suggested that these enzymes introduce different functionality into the hydrocarbon chain under a common transition state. In this case, the substrate and intermediate are the same; the primary difference in the reaction is how the reaction proceeds, therefore resulting in a different functional group. A synergistic strategy utilizing genomic information and evolutionary principles can be expected to provide a more efficient solution to drug development than the more traditional approach of studying a single enzyme at a time.

CONCLUDING REMARKS

It was once thought that the "Holy Grail" in understanding the molecular basis of infectious disease was to know the total genetic sequence of the infectious organism with the promise of an understanding of the disease and pathology of each infectious agent at a genetic level. However, it is becoming obvious that deciphering and understanding all of the information to the point of assigning function is the largest hurdle. Having the genomic sequence of a bacterial organism allows one to mine the genome for information beyond predicting ORFs. The big push now is in correlating expression profiles with metabolic outcome. At the moment, there is a tremendous amount of raw sequence data and related information. Considering all of the preexisting biochemical information and differential expression data from DNA based arrays and proteomic studies, each piece of DNA sequence can now be coupled with this information. Our aim with this chapter is not to rehash the themes of the past, but rather, with history, to spur all of us toward new directions and goals that push the limits of science, to nudge us to step out of the scientific comfort "pale" and explore new dimensions within the drug discovery arena in the postgenomic era. Therefore, the primary goal of genomics-based research in the *M. tuberculsois* drug discovery process is to identify new drug targets.

Drugs emanating from these novel targets will provide great benefits in terms of treatment of MDR strains. This is obtainable because the efficiency of identifying targets from the genome has increased markedly, replacing the old process of identifying proteins, cloning them, and establishing a level of validation of their potential utility. This barrier has essentially been removed as a consequence of the completion of the *M. tuberculosis* genome and the development of *in silico* analysis tools. All proteins now become potential targets and can be prioritized on the basis of information indicating essentiality to cell survival, conserved range in pathogens, and presence in humans.

12

Immunology of Drug

Contrary to what some pharmaceutical "*visionaries*" have predicted, pathogenic microbes have not gone away, but persist as a significant problem in modern medicine. Thanks to years of inappropriate antibiotic therapies, they are back with a vengeance. Antibiotic-resistant pathogens have emerged and continue to wreak havoc on public health worldwide. The common denominator in this developing scenario is the ever-evolving genomes of microorganisms. There are many approaches suitable to monitoring fluctuations in gene expression and microbial adaptation, which have been presented in numerous review articles. However, of these various genomic approaches, the proteomic approach is unique in that a single genome can give rise to qualitatively and quantitatively different proteomes. The term *proteome*, overused as it is, represents the expressed protein complement of the genome.

By examining pathogens at the protein level, one can monitor phenomena such as protein expression levels, posttranslational modifications and processing, as well as study multisubunit protein complexes. As we work with proteins for a living, we would, like to believe that proteomics can do it all. However, the reality is that proteomics is really nothing new but simply a newly coined moniker that represents analytical biochemistry with newer, neater equipment. Many of the technologies that make up what is now the current buzzword, "*proteomics*," have been in existence for some time. The key to obtaining useful information from the "new" field of proteomics lies in an older, stodgier field–that of cell biology.

BIOINFORMATIC ANALYSIS

While the characterization of protein structure and function has long been a traditional part of biochemistry, the advent of modern mass spectrometry techniques, in conjunction with advances in DNA sequence databases and bioinformatic analysis, has certainly given a boost to the field of proteomics. However, every analytical approach has both its advantages and pitfalls. Granted, through proteomics, previous impediments to protein target identification such as degree of purity and the necessity for relatively high quantities of expressed protein have in essence been eliminated.

Using older technologies such as 2D electrophoresis, enhanced technologies such as new-generation mass spectrometers, and novel technologies such as genomic sequence databases, proteomics or what some would call "*real protein biochemistry*" has come a long way in a brief period of time. The beginnings of microbial proteomics can be traced back to the introduction of two-dimensional electrophoresis by O'Farrell, in which proteins were separated initially by isoelectric focusing in low-percentage acrylamide poured in tubes followed by molecular weight separation using sodium dodecyl sulfate-polyacrylamide gel electrophoresis (SDS-PAGE) in the second dimension.

Nearly every graduate student, medical fellow, and post-doc of this era should have fond memories of the many times that an overnight focusing was initiated only to find that the gel had separated from the tube in the wee hours of the morning. Thus the buffers poured into each other and the power source simply gave up. For those of us studying phosphorylation, these frustrations were usually compounded by the creation of a sizable pool of phosphorus-32 on our benchtops. Further, it was often difficult to remove tube gels from the tube if the run was actually successful. Invariably, unless one had the dexterity of a senior neurosurgeon, the flimsy tube gels would get chopped up into a million pieces or they would fly out of the tubes and never be found (unless it was a kinase study and then a good-quality Geiger counter usually picked up the trail). However, when tube gels worked properly, they were very good and the power of 2D electrophoresis as a biochemical separation technique was firmly established as O'Farrell was able to resolve 1100 different proteins from *Escherichia coli.* Improvements to the O'Farrell approach came later in the form of immobilized pH gradients, which significantly enhanced the reproducibility of 2D electrophoresis.

Comparison of the spotting patterns on 2D electrophoretic profiles can allow for the quantification and identification of modified and up- or down-regulated proteins in a biological sample. Then, of course, there was the identification of these 1100 "spots", which at the time would have been a pretty heroic accomplishment (and still would be no small feat!). Now, however, the spot may be extracted, digested with an endoprotease such as tryspin or Lys-C, and subjected to mass spectrometry in a rather direct and unencumbered manner. The experimental peptide masses may then be compared to the theoretical masses (calculated from the genomic information) through a variety of programs such as MASCOT and MS-Fit and a match is obtained. Yates and coworkers, utilizing a program called SEQUEST, developed further refinement of this database-searching procedure. SEQUEST correlates tandem mass spectra of peptides with individual amino acid sequences from protein and nucleotide databases.

While the instrumentation and analytical programs have become significantly refined with reported sensitivities down to the femtomole level, most analysts agree that the 2D gels appear to be the rate-limiting factor in analysis. Another area of concern is that the dynamic ranges of electrophoresis-analytical software have often lead to publications reporting extraordinary numbers of spots isolated from a gel that in reality were either instrumentation limitations or other artifacts. Reproducibility can also be an issue when running replicates of the same sample. An alternative to the much improved, but still laborious, 2D gel approach is that of 2D liquid chromatography – mass spectrometry (LC-MS) in which a protease digested sample is simply injected directly onto a column and has been described by Yates and termed multidimensional protein identification technology (MudPIT). In this procedure, proteins are digested in solution and the peptide mixture is loaded onto a strong cation-exchange column; peptides are sequentially eluted with increasing electrolyte concentration according to their charge and hydrophobicity.

Each ion-exchange fraction is then subsequently separated by conventional reverse-phase liquid chromatography combined into a biphasic column with reverse-phase and strong cation-exchange material packed sequentially in the same column. The eluent is subjected to direct mass spectrometric analysis. This approach is limited by a slow sample-loading rate. Also, it is not possible to quantitate information much less get a "global view" of what is going on in the proteome by this approach. An attempt to resolve the issues of quantitation in LC-MS has been described by Aebersold and colleagues. Described as a method that has the potential to identify and quantify most, if not all, the proteins expressed by a cell or tissue, it is based on a new class of reagents termed isotope-coded affinity tags (ICAT). The ICAT reagent consists of a biotin affinity tag, a polyether linker that can incorporate stable isotopes (i.e., deuterium), and an iodoacetamide-reactive group that specifically reacts with cysteinyl thiols. Stable isotopes are incorporated into proteins by selective alkylation of cysteines with either an isotopically heavy or light reagent.

The proteins in one sample are labeled with the isotopically heavy light reagent and the proteins in the second sample are labeled with the isotopically heavy reagent. The two protein mixtures are then combined. At this point, any optional fractionation technique can be performed to enrich for low-abundance proteins or to reduce the complexity of the mixture, while the stable isotope tag imprinted in the proteins strictly maintains the relative quantities. Prior to analysis, the protein mixture is digested with trypsin and passed over a avidin–agarose column. Because the ICAT label contains the stable isotope information as well as a biotin tag, ICAT-labeled peptides are selectively isolated for analysis by microcapillary liquid chromatography electrospray ionization tandem mass spectrometry (LC-ESI-MS/MS). If required by the complexity of the sample, the peptide mixture can optionally be further fractionated prior to mass spectrometric analysis.

The ratio of ion intensities from coeluting ICAT-labeled peptide pairs of identical sequence precedes accurate quantitation, while a subsequent tandem MS analysis identifies the proteins in the sample by their amino acid sequence. The method is reported to be capable of detecting proteins of very low abundance and is an automatable process. This technology has its own limitations (such as being cysteine specific). In addition, high-quality quantitation (CVs less than 20%) from LC/MS/MS data, where the peptide pairs do not entirely coelute, has its challenges as well as limited sensitivity. Moral of the story: Damn gels—can't live with them, can't live without them.

PROTEOMICS ON A CHIP

Last, but not least, in the technology toolbox is the so-called proteomics on a chip. Obviously, a critical need exists for the development of novel technologies that will enable high-throughput studies of the proteins encoded by the approximately 30,000-100,000 genes encoded by the human genome. "*Traditional proteomics*" methodologies, while adequate for basic laboratory approaches, are inadequate for the detection and characterization of large sets of protein-protein interactions. Protein microarray technology recently developed by several labs utilizes protein microchannel microarrays to characterize antibody profiles in a variety of infectious and other disease states. Using microfabrication technologies borrowed from the microelectronics industry, these systems offer the ability to realize on-chip microchannel networks that can be computer controlled to transport, mix, separate, and detect protein fragments.

The advantage is that an automated series of reactions can be performed on 1 to 1,000-pl samples with all reactants and products of the assay remaining suspended in a buffered solution at all times. This is in contrast to the solid-phase detection systems of microarrays typically used in combinatorial proteomics research, where the small spot sizes dictated by the needs for high sample density result in rapid evaporation of solvents.

Other purported advantages of the microchannel approach include the ability to combine assays with simple binding event detections as well as the ability purify candidate species of known reactivity for subsequent testing off-chip. Enough, however, with the "*wonders of technology.*" What is the application of proteomics to disease and, in particular, infectious disease? There are three areas of antimicrobial research that come to mind where proteomics may be able to provide a technological advantage to the discovery of new therapeutics: vaccines, thereapeutics for resistant organisms, and cell-based screening. In the following three "real-world" applications, the opportunities for proteomic approaches are discussed.

GRAM NEGATIVE BACTERIUM (*HELICOBACTER PYLORI*)

Helicobacter pylori is a gram-negative bacterium that chronically infects the gastric mucosa of more than half of all humans worldwide and is a major cause of gastritis and peptic ulcer disease and an early risk factor for gastric cancer. Only some 10 to 20% of infections, however, result in overt disease. DNA typing has established that *H. pylori* is extremely diverse as a species, and it is likely that the varied outcomes of infection reflect differences in bacterial genotype, human host genotype, and physiologic, immunologic, and environmental factors. These considerations make it valuable to thoroughly characterize the proteins and other antigens that *H. pylori* produces and the human responses to them. There is a great need for an effective anti-*H. pylori* vaccine, especially in Third World, high-risk populations where *H. pylori* eradication by standard antimicrobial therapies is often followed by reinfection.

Much attention has been focused on urease-based vaccines because of the essentiality of urease and some encouraging results with mouse *Helicobacter felis* models. VacA has also been considered a candidate based on results with a mouse *H. pylori* model. These mouse models, however, may not adequately mimic the human condition and clinical trials of urease vaccines have been only marginally encouraging. This reinforces the sense that other or additional antigens may be needed for a truly effective vaccine. Other factors important for *H. pylori* colonization or virulence have been identified. Some of the more prominent factors include

1. Flagellae, which allow the organism to move in the mucous layer;
2. The previously mentioned urease complex, which may help maintain a neutral micro pH environment in the face of gastric acidity;
3. The VacA protein, which generates vacuoles in eukaryotic epithelial cells;

4. An adhesin binding to the Lewis blood antigen group; and (v) the *cag* pathogenicity island, some of whose encoded proteins help trigger severe inflammatory responses and which, like VacA toxigenicity, is disease associated.

Based on the presence of the *cag* pathogenicity island (PAI), the *H. pylori* isolates are subdivided into two types. Type I strains, containing the *cag* PAI, exhibit increased virulence, since they are predominantly associated with severe gastric disease, whereas type II strains, lacking the *cag* PAI, are more frequently isolated from asymptomatic carriers. It has been demonstrated that some of the proteins encoded by the *cag* PAI trigger severe inflammatory responses in the host. However, the precise function of the gene products of the *cag* PAI and their role in virulence remain to be elucidated. Several other *H. pylori* proteins with known activities, or which are related to similar proteins of known function in other organisms, have been isolated.

While the complete genomic DNA sequence of *H. pylori* 26695 has been reported, many of the proteins inferred from this DNA sequence have no known function, and this DNA sequence clone does not always predict which open reading frames are likely to encode virulence factors or antigens suitable for diagnostic or vaccine studies. A number of studies have begun to address associations of specific *H. pylori* antigens to antibodies in patients with particular gastroduodenal pathologies and of possible autoimmune components to *H. pylori*-associated disease. There is very little information, however, regarding the long-term evolution and clinical implications of these human responses before and after the eradication of *H. pylori* by antibiotic treatment regimens.

Pharmaceutical therapy to treat the *H. pylori* infection involves expensive combinations of various antibiotics, proton pump inhibitors, and bismuth compounds but shows only a limited efficacy (of approximately 80 to 90%) and does not prevent reinfection after successful eradication. In addition, *H. pylori* strains resistant to the most potent antibiotics used in the treatment of *H. pylori* infections, metronidazole and clarithromycin, are emerging rapidly. Considering further that the number of infected people worldwide requiring treatment is far beyond the reach of the antibiotic triple therapy, development of a vaccine seems to make sense for the global control of *H. pylori* infection. It has been shown by various researchers that in animal models of infection protective immunity can be achieved by the coadministration of an appropriate mucosal adjuvant and various *H. pylori* antigens, either separately or in combination, via the orogastric route.

The protective antigens identified include the urease; VacA; CagA, the immunodominant marker protein for the presence of the *cag* PAI; catalase; and HspA and HspB, the *H. pylori* homologs of the heat shock proteins GroES and GroEL. In particular, the *H. pylori* urease gave rise to a high degree of protective immunity in vaccinated animals, and it was reported that 100% protection in *H. pylori*-challenged mice could be achieved by the administration of urease via a live carrier *Salmonella* strain expressing recombinant *H. pylori* subunits A and B. Furthermore, as mentioned previously, it has been demonstrated that therapeutic vaccination with recombinant VacA and CagA eradicates a chronic *H. pylori* infection in mice, demonstrating that the inability of the natural immune response to clear *H. pylori* infection can be overcome. Considering the advantage of an efficacious vaccine, it is important to identify the *H. pylori* proteins that elicit a strong immune response in humans in order to analyze their capability to confer protective immunity. Furthermore, the identification and characterization of immunodominant proteins will contribute to the improvement of serological tests for detecting and monitoring *H. pylori* infections. Another important question is whether there exists a correlation between the presence of antibodies directed against specific *H. pylori* antigens and the particular *H. pylori*-associated gastroduodenal pathology from which a patient is suffering.

In a previous study, proteomics was used to identify common patterns of *H. pylori* antigens that are recognized by sera from patients showing various gastroduodenal pathologies. The proteins from lysed cell pellets of *H. pylori* ATCC 43504 were separated on a series of 2D gels run in parallel with an initial pH gradient of pH 4 to pH 8. The silver-stained gel revealed prominent individual proteins, with several protein "families"-most notably as clusters of bands at approximately 89 (p*I* 6.8), 66, and 58 kDa (p*I* 6.5). The proteins from these 2D gels were transferred to PVDF membranes and incubated with a positive serum pool or a negative serum pool. Western blot data revealed at least 17 spots or groups of spots that were recognized by antibodies in an infected patient serum pool. Transblotted 2D spots from the pH 4 to 8 gel were sequenced by Edman-type amino acid analysis, with the protein within selected spots evaluated further for internal sequence information.

The sequences from these spots were compared with sequences in available databases. Briefly, spots 1 and 2 corresponded to the *H. pylori* urease b subunit and the urease b-associated chaperonin GroEL, respectively. Spot 3 consisted of two proteins: the major species was pyruvate flavidoxin oxidoreductase, and the minor protein species corresponded to the previously described *H. pylori* hypothetical protein 2, or HP0154. Spot 4 corresponded to HP0537, from the *cag* region. Spots 5, 6, and 8 corresponded to flagellin proteins. Spot 7 consisted of two proteins that did not match any previously reported sequences from *H. pylori*. The major component, however, has 90% homology with the *Escherichia coli* TufB protein (possibly HP1205), and the minor component has some sequence homology with various ATPase proton pumps. Spot 9 was homologous to monomine oxidase from various species. Spot 10 corresponded to the neutrophil-activating protein. Spot 11 corresponded to HP1199, a ribosomal protein. Spot 12 had homology with the ClpP protease from various bacteria. The sequencing signals of spots 13 and 14 were too low to be read with confidence. Spots 15 and 16 (major) corresponded to HP0109 (Hsp 70) and HP0589 (ferrodoxin oxidoreductase), respectively. Spots 16 (minor) and 17 corresponded to a protein previously isolated by O'Toole et al.

In the control blot with sera from *H. pylori*-negative persons, only the urease b subunit (spot 1), likely due to cross-reaction with ureases of intestinal bacteria, and the spot 7 proteins showed cross-reactivity. Additional unique proteins were found in a nonequilibrium focusing gel followed by SDS-PAGE, even though fewer proteins, overall, were resolved. Spots 1 through 4 were present in very low quantities; therefore, a clear N-terminal sequence could not be determined with confidence. Spot 5 was the urease b subunit also seen in the pH 4 to 8 2D gels. Likewise, spots 6, 7, and 8 corresponded to urease b-associated chaperonin, flagellin b precursor, and flagellin a protein, respectively, which were also separated on the pH 4 to 8 2D gel. Spot 9 (major) corresponded to HP0027 (isocitrate dehydrogenase), with spot 9 (minor) representing a possible contaminant in the sequencing sample. Spot 10 corresponded to an open reading frame from HP1018, an open reading frame with no known database homologs. Spot 11 corresponded to *H.*

pylori catalase. Spot 12 contained an N-terminal sequence which has been found in several Omp's (Omp 5, 8, 9, 19, and 27). Spot 13 corresponded to HP1350, a putative protease Spot 14 was the previously reported HopC protein, and spot 16 was the urease a subunit. The sequence yields from transblotted spot 15 were low (in the midfemtomole range), suggesting that the protein was blocked. The sequence information derived from spot 15 gave an N-terminal amino acid sequence that did not match any known protein sequences. This suggested that the protein(s) in this spot might be modified at the amino terminus, as sequencing yields were low despite the protein(s) being clearly visible on a silver-stained gel.

These experiments not only illustrate that 2D gel electrophoresis can give a global view of the abundant proteins of *H. pylori*1 but also suggest that the identification of large numbers of proteins and their characterization with defined serum pools raises the possibility of rapid screening for potential vaccine (as well as diagnostic) candidates. Amino-terminal sequencing and/or proteolytic mass spectral mapping on isolated spots allowed for efficient characterization of these potential antigens. Using this approach upfront, downstream peptidomimetic analysis in parallel with libraries of cloned DNA fragments could provide additional information for the construction of specific vaccine clones or diagnostic recombinant "*mosaic*" antigens. This is especially important in the case of pathogens whose genomes have not yet been sequenced. One of the many advantages in using "*proteome*"-type technologies, as here, as opposed to traditional molecular biology (DNA) library approaches, stems from information about likely functionality and utility that comes from initial screening and that is refined as candidate antigens are discovered.

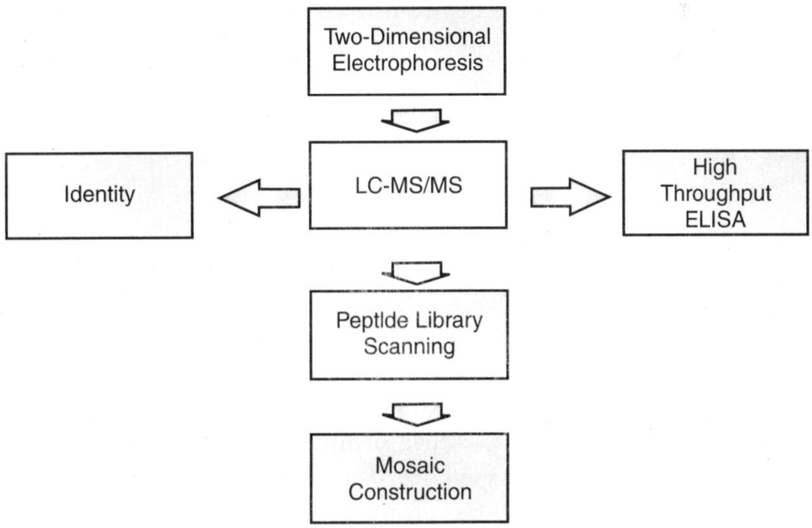

Fig. 12.1. Vaccine development strategy.

MANIPULATIVE ACTIVITY IN *HELICOBACTER PYLORI*

As described, *H. pylori* is a microaerophilic bacterium that exhibits a strict respiratory form of metabolism, which therefore restricts it to an environment of relatively low oxygen tension, such as the gastric mucosa. Microaerophiles are susceptible to redox-active 5-nitroimidazole drugs that are converted to short-lived, DNA-damaging hydroxylamine derivatives by low-potential reductases, including ferrodoxins and flavidoxins, in a series of two-electron transfer reactions. The nitroimidiazole drug metronidazole (MTZ) is a critical component of important combination therapies against *H. pylori*. However, resistance to at least moderate levels of MTZ is very common, ranging up to 90% of isolates in some developing countries, and constitutes a major cause of treatment failures. The development of moderate-level MTZ resistance in *H. pylori* generally depends on mutational inactivation of the *rdx*A (NADPH nitroreductase) gene (HP0954), with higher level resistance resulting from loss of function mutations in additional reductase genes. Loss of metabolic enzymes, even apparently dispensable ones such as these nitroreductases, will often be compensated by changes in other metabolic functions.

Previous physiologic tests suggest that exposure of genetically resistant *H. pylori* strains to sublethal concentrations of MTZ results in changes in levels of some key enzymes, which could enable these strains realize their resistance potential. Proteome technologies may be used to examine the protein profile of a MTZ^R derivative of reference strain 26695 grown in medium with a moderate amount of MTZ. The proteins from lysed cell pellets of *H. pylori* MTZ^R grown in the absence and presence of MTZ were separated on 2D gels run in parallel using an initial pH gradient of pH 4 to pH 9. The Sypro stained gel of the sample grown in the absence of MTZ revealed many prominent individual protein spots. Forty-two individual protein spots were chosen which showed differential regulation between the two gels.

Two-dimensional spots from the pH 4-10 gel were analyzed by *in situ* digestion with Lys-C followed by MALDI-TOF analysis. The identifications of these spots have been described. Of the gel spots identified from the *H. pylori* genomic database, spots 11 and 12, spots 19 and 20, spot 22, and spots 29-42 consistently showed differential expression when *H. pylori* was grown in the presence of metronidazole. Fluorescent intensities of spots 11 and 12 (isocitrate dehydrogenase) appeared to drop by approximately 50% when cells were grown in the presence of metronidazole. Likewise, fluorescence intensities in spots 29-42 appeared to drop by at least twofold when the bacteria was grown in the presence of metronidazole. However, the most obvious difference in expression was observed in spots 19,20, and 22, which appeared to be upregulated when the MTZ-resistant strain of *H. pylori* was grown in the presence of metronidazole. Spots 19 and 20 were identified as alkylhydroperoxide reductase (*ahp*C). Spot 22 corresponded to aconitase B. As to be expected, a considerable number of proteins from various metabolic pathways were repressed by the addition of MTZ to the growth media.

The proteomic approach provided a broader screen, and allowed the detection in particular, of a compensatory increase in expression of a set of proteins encoded by HP 1563, alkyl hydroperoxide reductase (*ahp*C), which has not been reported in *H. pylori.* The question was this: why *ahp*C? Incomplete reduction of molecular oxygen leads to the formation of reactive oxygen species that can damage nearly all subcellular macromolecules in some form or another. Cells are at least partially protected against these reactive oxygen species by superoxide dismutase, which eliminates the O_2^- radical, and catalase, which removes H_2O_2, and peroxidases, which remove alkyl peroxides as well as H_2O_2.

Alkyl hydroperoxidase enzymes reduce hydroperoxide intermediates with electrons donated by NADPH via thioredoxin or other thiol-containing intermediates, and the alkyl hydroperoxidase reductase family seen in *H. pylori* represents a new type of peroxidase that has a conserved cysteine as the primary site of catalysis rather than the selenocysteine of glutathione peroxidase. Antioxidant defense mechanisms are induced as part of a global cellular adaptive response to oxidative stress in many organisms. In previous studies evaluating clinical isolates of isoniazid resistant *Mycobacterium tuberculosis* defective in *katG*, a conserved essential gene, it has been shown that isoniazid-resistant *katG* mutants compensate for the loss of *katG* activity by a second mutation, resulting in hyperexpression of at least one isoform of *ahpC*. *AhpC* in *H. pylori* has been found to upregulated in response to iron deprivation. If this gene is knocked out, the organism will suffer from oxygen toxicity problems.

Two possibilities might explain why there appears to be differential expression of isoforms of *ahpC*: The obvious answer would be phosphorylation; the other, dissociation of an iron–sulfur center similar to *oxyR*. In other words, *ahpC* may be a redox-responsive protein and in that respect may interact with *rdxA*. Regardless, alkyl hydroperoxidase identification could lead to a worthwhile drug target, identified via proteomic analysis.

SURROGATE MARKERS

An area that has been largely unexploited as a potential source for novel antibacterial targets is the complex mechanism surrounding bacterial cell division. This is due largely to the difficulty of devising a high-throughput screen for cell-division inhibitors. Bacterial cell division is dependent on a set of key division proteins that are conserved across a broad range of bacteria and differs significantly from the eukaryotic division apparatus. These proteins were initially identified in *E. coli* as temperature-sensitive mutations that led to filamentous growth (i.e., filamenting temperature sensitive, leading to very long, nonseptate cells). Two of these proteins (e.g., ftsA and ftsZ) are essential to the process of division and have been shown to participate either directly or indirectly in the assembly of a cell-division apparatus. These proteins are localized with other division-associated proteins, at the time of initiation of cell division, to a ringlike structure at the future cell-division site.

The most extensively studied bacterial cell-division protein is the prototubulin gene product ftsZ. This protein, conserved among a very broad range of bacteria, is the earliest protein to assemble at the division site. It has been found to have GTPase activity and has been demonstrated to form GTP-dependent polymers *in vitro*. This *in vitro* assembly mimics the *in vivo* assembly of the ring structure at the bacterial cell-division site. The actin like protein ftsA, also present in all known eubacteria, plays an essential role in septation and cell division. FtsA behaves as a dimer and maintains both ATP binding and ATP hydrolysis activities. The use of proteomics is proposed to determine whether the inhibition of cell division in bacteria leads to cell-cycle-specific changes that can be used to construct an indicator strain to screen for division inhibitors. It is important to note that any changes in the proteome using temperature-sensitive *ftsA* or *ftsZ* mutants must differentiate between changes that are specific for division versus those that are due to general stress responses or the shift to restrictive temperature.

The proteome of a filamenting temperature-sensitive mutant (*ftsA*) of *E. coli* that was unable to divide into daughter cells during growth at 42°C was examined. Comparison of 2D electrophoresis profiles generated from cell lysates of *ftsA* and the wild-type isogenic strain grown at the restricted temperature (42°C) allowed for the identification of protein spots that were specifically up- or down-regulated in the mutant strain. These spots were subjected to in-gel proteolytic digestion followed by mass spectrometry analysis for peptide sequence and identity. From these results, several gene products that were differentially expressed between the two bacterial phenotypes were identified (filamented versus normal morphology). In this experimental setting, *E. coli* strain MC4100 and an isogenic *fts*A temperature-sensitive cell-division strain, constructed via P1 phage transduction, were utilized. The shift to restrictive temperature led to the loss of activity in FtsA resulting in filamentous growth of the cells.

Cell lysates were compared at both permissive (30°C) and restrictive (42°C) temperatures, and several gene products from the MC4100 and *fts*A strains were differentially expressed. In this gel profile, uridine phosphorylase, phosphoglycerate mutase I, and periplasmic ribose binding protein are shown. In this case, it appears that uridine phosphorylase represents a potential candidate for the proposed surrogate marker assay, whereas the other proteins likely do not. Another example is shown a typical surrogate marker response is compared to another cell septation marker, F1 ATP synthase. From this information, Lac reporter fusion vectors can be constructed to test these surrogate marker proteins for utility in a high-throughput cell-based screening assay. From such studies, it may be possible to identify division-specific markers, that could be used in cell-based assays to screen for novel small-molecule inhibitors of the cell-division pathway. Such inhibitors could be developed into potentially novel antibiotics, targeting a pathway specific for and conserved among gram-positive and gram-negative bacterial species.

In a similar manner, known antimicrobial compounds or temperature- sensitive strains in potential targets could be utilized to generate initial proteomic data, thereby establishing guideposts useful for uncovering new structural classes of compounds. This platform could potentially be suitable for evaluation and discovery of other inhibitors of cell processes in both gram-negative and gram-positive bacteria. There have been many published examples of proteomically derived antimicrobial targets for therapeutic intervention such as *Pseudomonas aeruginosa* from chronic lung infections in cystic fibrosis (CF) patients, *Staphlococcus aureus, M. tuberculosis, Francisella tularensis, Haemophilus influenzae, Candida albicans,* and *Salmonella.* It has lately been heartening to see biology and medicine finally starting to take hold in (or, rather, "infect") this new technology. It is important, however, to recognize that as with any technology, there are limitations to proteomic approaches. This point is often lost or obscured by articles touting the "wonders of technology." Therefore, it is critical to note that while proteomics offers a unique approach to the global analysis of cellular proteins, validation of the results can *only* be achieved through studies based on biological phenomena in conjunction with complementary approaches such as molecular biology.

The development of antimicrobial therapies mesh well with the intended goals of modern proteome technologies as there is a plethora of readily available bacterial strains and mutants from various key metabolic pathways to study with relatively little constraint from lack of cost-effective reagents. Although previous biochemical characterization of these mutants has been very challenging and identification of many of their regulatory components and functions remains tricky, many of their genome sequences have been recently deciphered, so the lexicon is in place to enable the proteomic approach. Through proteome technologies, it should be possible in future experimentation to more fully address many of the central issues of bacterial growth, metabolism, and resistance that have gone unanswered. However, let's get real! Proteomics won't do this single-handedly.

13

Lesion in Parkinson's Disease

A fundamental lesion in Parkinson's disease is a marked deficiency in the dopaminergic innervation of the basal ganglia owing to degeneration of neurons in the substantia nigra. Enhancement of dopaminergic transmission restores motor function at least partially. The decrease in dopaminergic activity in the basal ganglia results in a relative excess of cholinergic influence. Therefore, dopaminergic agonists, such as levodopa a precursor of dopamine, and cholinergic (*muscarinic*) antagonists can be combined in the treatment of Parkinson's disease. Parkinson-like syndromes also occur after depletion of central stores by reserpine and after treatment with phenothiazines and other antipsychotic drugs blocking dopamine receptors.

The pathology of Parkinson's disease is typified by the presence of cytoplasmic inclusions (*Lewy bodies*). The formation of these proteinaceous inclusions involves the interaction of several proteins, including a-synuclein, synphilin, parkin and ubiquitin carboxyl-terminal hydrolases. Orr et al. (2002) gave a review on inflammatory aspects of Parkinson's disease and highlighted the cell-to-cell interactions and immune regulations critical for neuronal homeostasis and survival. Parkinson's disease and related synucleinopathies are considered as a new class of nervous system amyloidoses.

PARKINSON-INDUCED COMPOUNDS

Although *in vivo* animal studies have been used to examine the effects of a number of Parkinson-inducing compounds, there is little information on reliable *in vitro* methodologies that can recapitulate the previously observed in *vivo* results. Cardozo (1993) and Smeyne and Smeyne (2002) described details of a method for generating mixed and chimeric neuron/glial cultures of postnatal substantia nigra (SN), independent of other monoaminergic nuclei in the ventral midbrain. Since many toxins do not affect regions of the midbrain except the SN, use of whole ventral midbrain from embryos can dilute any measurement of cell death. By specifically culturing ventrolateral midbrain containing the SN, one can more directly target the effects of dopaminergic toxins. In addition, this method can be used to test potential therapies for amelioration of Parkinson's disease.

Method

Stepwise methodology

1. Postnatal day 2-5 C57Bl/6 mice and/or Swiss-Webster (SWR) mice (Harlan) are placed on ice for 2-3 min to achieve an appropriate plane of anesthesia.
2. Animals are quickly decapitated and ≈ 8-10 brains from P2-P5 C57BL/6 or SWR matings are removed and placed in a freshly prepared dissociation media (= DM containing 12.8 g sodium sulfate; 5.2 g potassium sulfate; 0.036 g calcium chloride· $2H_2O$; 1.18 g magnesium chloride· $6H_2O$; 1.8 g glucose anhydrous; 0.238 g HEPES in 1l deionized water, adjusted to pH 7.4 with NaOH).
3. Brains are placed on their ventral surface and a slice of midbrain rostral to the cerebellum and caudal to the hippocampus is isolated. This removed brain slab is placed flat in DM and either the entire midbrain or the ventrolateral midbrain containing the SN is dissected and minced into small pieces.
4. The minced SN or midbrain is then incubated in papain and DNAase for 30 min at 37°C. A second incubation with fresh papain solution (30 min, 37°C) is followed by three rinses in DM, and one rinse in plating media (PM).
5. Tissue is triturated in 5 ml PM and the cell suspension is added to 2 ml (1 ml of BSA stock and 1 ml ovalbumin stock).
6. The cell suspension is spun 1400 rev/min for 8 min and the pellet is then resuspended in 1.0 ml plating media containing 2% rat serum.
7. Once resuspended, cells are counted using trypan blue (0.4%) to determine cell viability.
8. Cells are adjusted to 1.2×10^6 cells/ml and plated at ≈ 200,000 cells/cm^2 in Lab-Tek 4-well Permanox chamber slides that were previously coated with laminin (200 mg/ml) and poly-D-lysine (200 mg/ml) 1:1 (v:v) and rinsed once with deionized water. Cells are maintained in an incubator at 37°C, 5% CO_2, and fed 2-3 times per week with feeding media complete with 2% rat serum (RS) by exchanging approximately one-fifth of the media (100 μl).

Neuroculture

1. SNpc cells plated on pre-plated SNpc glia are produced using a variation of the previously described methods. SNpc glial feeder layers from P2-5 C57BL/6 or SWR mice are produced using the above described method, but cells are plated at 20,000-50,000 cells/cm^2 and fed with plating media containing 2% rat serum and 8% fetal bovine serum to promote glial proliferation and neuronal death.
2. For astrocyte cultures, this is the final step.
3. To generate mixed neuron/glial cultures, 3-4 weeks after the initial SNpc cells are plated, the glial feeder layers are rinsed once with plating media without serum, and 0.5 ml of plating media with 2% rat serum (RS) are added.
4. SNpc cells from C57BL/6 or SWR mice are isolated and plated at 250,000 cells/well on the previously generated glial feeding layers.

5. Twenty-four hours after plating of neurons onto the glial feeder layers, the cultures are fed with feeding media complete with 2% RS by an exchange of approximately one-fifth of the medium (100 μl) and cytosine β-D-arabinofuranoside (Ara-C, 2 μM) to prevent glial proliferation of the freshly plated cells. Thereafter, Ara-C (10 μM, final concentration) is added at each feeding.

Effect of MPTP

1. 1-Methyl-4-phenyl-1,2,3,6-tetrahydropyridine (MPTP) is added to cultures 7-10 days after the neurons are plated onto glia. The MPTP is prepared by dissolving MPTP in feeding media for a 1-mM stock, then diluting to a 1-μM stock with feeding media complete with 2% rat serum and adding this directly to the cultures. The final concentration of the MPTP in media is 50 nM. Two feedings of MPTP, in 2 days, are necessary to achieve the desired toxic effect.
2. Seven days after MPTP is added, the cultures are rinsed 3 × with TBS, fixed in 4% buffered paraformaldehyde for 10 min, and rinsed 3 × with TBS.

Assessment

1. To determine the number of SN cells, cultures are immunostained for expression of tyrosine hydroxylase. First, endogenous peroxidase activity is quenched by rinsing with 0.3% hydrogen peroxide in methanolin 1 × TBS for 2 × 15 min. Cells are permeabilized with 0.1% Triton X-100, 5% goat serum in TBS for 2 × 15 min. Cultures are covered in 400 μl of a polyclonal antibody directed against tyrosine hydroxylase and incubated in this solution overnight at 4°C. The next day, cultures are rinsed three times with TBS followed by application of secondary antibody (goat α-rabbit) and amplification with Avidin-Biotin. Final visualization of the immuno-positive neurons is made using diaminobenzidine as a chromagen.
2. All TH-positive cells having the previously described characteristics of SNpc neurons (TH-positive cytoplasm surrounding a pale unstained nucleus) from each culture are counted at a magnification of 200 ×. Note: to control for variability between cultures, all MPTP treatments are performed on matched cultures present on one single slide. Once cells are counted in each well, the number of TH-positive cells in the MPTP-treated cultures from each genotype is compared to the adjacent non-treated culture. Since an identical number of cells generated from the same brains was plated on a single slide, these cultures are directly compared to determine the percentage cell loss following MPTP. The percentage cell loss data following MPTP are then pooled from all C57Bl/6, SWR or mixed cultures.

Role of Substantia Nigra

A dopamine-derived neurotoxin, 1(*R*), 2(*N*)-dimethyl-6,7-dihydroxy-1,2,3,4-tetra-hydroisoquinoline [*N*-methyl(*R*)salsolinol] was found to cause parkinsonism in rats and to deplete selectively dopamine neurons in the substantia nigra after infusion in the striatum. This isoquinoline occurs enantio-specifically in the nigra-striatum of human brains. The biosynthesis from dopamine is catalyzed by two enzymes, (*R*)salsolinol synthase and (*R*)salsolinol *N*-methyltransferase. The isoquinoline increases in the cerebrospinal fluid from parkinsonian patients, and the increase is ascribed to high activity of its synthesizing neutral (*R*)salsolinol *N*-methyltransferase, as shown by analyses in lymphocytes. The cell death caused by this neurotoxin in dopaminergic human neuroblastoma SH-SY5Y cells proved to be apoptotic. Apoptosis by this neurotoxin is mediated by intracellular sequential process, loss of mitochondrial membrane potential, activation of caspases and DNA fragmentation.

Neuroprotection to halt progressive cell death of neurons has been proposed as a future therapy for neurodegenerative disorders. In these disorders, such as Parkinson's disease and Alzheimer's disease, apoptosis contributes to neuronal death in most cases. The well-regulated and relatively slow apoptotic process was proposed as a target of neuroprotection.

Apoptosis is induced in neurons by various insults: oxidative stress, metabolic compromise, excitotoxicity, and neurotoxins. Apoptotic signaling is a multi-step pathway induced by opening a mitochondrial mega-channel called *permeability transition* (PT) pore, followed by decline in membrane potential, $\Delta\psi m$, release of apoptosis-inducing factors, activation of caspases and fragmentation of nuclear DNA.

Mitochondrial PT pore is regulated by Bcl-2 protein family, preventively by Bcl-2 and Bcl-xL and in a promotive way by BAX and BAD. Several papers describe the neuroprotective-antiapoptotic action of the anti-Parkinson drug rasagiline.

Maruyama et al. (2004a) described the neuroprotective function of *R*-(-)-1-(benzofuran-2-yl)-2-propylaminopentane, [*R*-(-)-BPAP], against apoptosis induced by *N*-methyl(*R*)salsolinol, an endogenous dopaminergic neurotoxin, in human dopaminergic neuroblastoma SH-SY5Y cells.

Method

SH-SY5Y cells were cultured in Cosmedium-001 tissue culture medium, supplemented by 5% fetal calf serum in an atmosphere of 95% air /5% CO_2.

Assessment

Apoptosis was quantitatively measured by fluorescence-augmented flow cytometry (FACS) with FAC-Scan and CellQuest software. Cells cultured in a 6-well poly-L-lysine-coated culture flask were incubated with or without 1 μM to 1 nM (-)-BPAP analogs at 37°C for 30 min, and then for 24 h with 250 μM *NM*(*R*)Sal in Cosmedium-001 culture medium supplemented with fetal calf serum. The cells were treated with trypsin, gathered, washed with the culture medium and twice with phosphate-buffered saline (PBS). The cells were incubated with 100 nM test drugs solution in an ice-bath for 30 min, washed and suspended in PBS, then subjected to FACS analysis.

Parameters in $\Delta\psi m$

Decline in $\Delta\psi m$ induced by *NM*(*R*)Sal was quantified by measuring the reduction of Rhodamine 123 fluorescence pre-loaded in the cells, as reported by Akao et al. (2002b). To examine the effects of (-)-BPAP analogs, the cells cultured in 6-well poly-L-lysine-coated tissue culture flasks were stained with 5 μM Rhodamine 123 in DMEM for 30 min at 37°C. After being washed twice with PBS, the cells were suspended in DMEM, incubated with 1 μM to 1 nM BPAP derivatives for 30 min, then with 250 μM *NM*(*R*)Sal for 1 h. After being washed and gathered by treatment with trypsin, the cells were suspended in PBS and the fluorescence at 535 nm was measured with excitation at 505 nm in a Shimadzu spectrofluorophotometer, RF-5000.

Calculating bcl-2 mRNA level in the cells treated with BPAP derivatives

SH-SY5Y cells were cultured in the presence of various concentrations (100 nM to 10pM) of (-)-BPAP analogs for 24 h, and mRNA levels of bcl-2 were quantitatively assessed by RT-PCR. The cells were gathered and washed with PBS, and the total RNA was extracted by the phenol/ guanidinium thiocyanate method. cDNA was generated by reverse transcription of 2 μg of the total RNA, and the cDNA fragments were amplified using the PCR primers. The linearity of the relationship of the amount of PCR product to the time of PCR amplification was confirmed under the conditions used in this study. PCR products were analyzed by electrophoresis on 3% agarose gels, and α-actin was used as an internal standard. The amounts of mRNA were quantified using NIH imaging software.

Calculating Bcl-2 levels in the cells treated with (-)-BPAP derivatives

SH-SY5Y cells treated with 1 μM to 1 pM (-)-BPAP analogs for 24 h, and the cells were gathered, washed with PBS and suspended in RIPA buffer (10 mM Tris-HCl buffer, pH 7.5, containing 1% NP-40, 0.1% sodium deoxycholate, 0.1% sodium dodecylsulfate, 150 mM NaCl and 1 mM EDTA 2Na). The lyzed protein (5 μg) was separated by SDS-PAGE using a 10%-20% gradient polyacrylamide gel and electroblotted onto PVDF membranes. After blockage with 5% nonfat milk in PBS containing 0.1% Tween 20, the membrane was incubated overnight at 4°C with anti-human Bcl-2 (100) antibody, or anti-β-actin antibody as control. The membranes were incubated further with alkaline phosphatase-conjugated goat anti-mouse antibody at room temperature. The immunoblots were visualized by use of an enhanced chemiluminescence detection kit, and quantified by computer-assisted image analysis with the NIH imaging software.

Assessment

Experiments were repeated four to eight times, and the results were expressed as the mean and SD. Differences were statistically evaluated by analysis of variance (ANOVA) followed by Sheffe's *F*-test. A p value less than 0.05 was considered to be statistically significant.

Latest methodology

Maruyama et al. (2000) found that dopaminergic neurotoxins, 6,7-dihydroy-1-(3′,4′-dihydroxybenzyl)-isoquinolines, cause different cell death in SH-SY5Y cells: apoptosis was induced by oxidized papaverolines and necrosis by reduced tetrahydropapaverolines. Maruyama et al. (2003) described the anti-apoptotic action of an anti-Alzheimer drug, TV3326 [(N-propargyl)-(3R)-aminoindan-5-yl)-ethyl methyl carbamate], a novel cholinesterase-monoamine oxidase inhibitor.

INVESTIGATING ANTICHOLINERGIC DRUGS

The muscarinic agonists tremorine and oxotremorine induce parkinsonism-like signs such as tremor, ataxia, spasticity, salivation, lacrimation and hypothermia. These signs are antagonized by anticholinergic drugs.

Method

Groups of six to ten male NMRI mice weighing 18-22g are used. They are dosed orally with the test compound or the standard (5 mg/kg benzatropine mesilate) 1 h prior to the administration of 0.5 mg/kg oxotremorine s.c. Rectal temperature is measured before administration of the compound (basal value) and 1, 2 and 3 h after oxotremorine injection. Tremor is scored after oxotremorine dosage in 10-s observation periods every 15 min for 1 h.

Tremor	*Score*
absent	0
slight	1
medium	2
severe	3

Salivation and lacrimation are scored 15 and 30 min after oxotremorine injection.

absent	0
slight	1
medium	2
severe	3

Assessment

Hypothermia

The differences of body temperature after 1, 2 and 3 h versus basal values are summarized for each animal in the control group and the test groups. The average values are compared statistically.

Tremor

The scores for all animals in each group at the three observation periods are summarized. The numbers in the treated groups are expressed as percentage of the number of the control group.

Salivation and lacrimation

The scores for both symptoms for all animals in each group are summarized at the two observation periods. The numbers in the treated groups are expressed as a percentage of the number of the control group.

Assessment

The oxotremorine antagonism has been proven to be a reliable method for testing central anticholinergic activity. The overt isomorphism between the animal model and the symptoms of Parkinson's disease recommend this test for screening of anti-Parkinson drugs. However, the model measures only central anticholinergic activity.

Latest technique

Matthews and Chiou (1979) developed a method for quantifying resting tremors in a rat model of limb dyskinesias. The model involved permanent cannulation of the caudate nucleus for the introduction of carbachol. Tremors were quantified with a small transducer and an electronic data collecting system. The system allows the construction of dose-response curves for tremor inhibition by potential anti-parkinsonism drugs. Johnson et al. (1986) developed a procedure for quantifying whole-body tremors in mice. Displacement of a free floating platform by animal movement created a change in resistance across a strain gauge. Administration of oxotremorine, 2.5 mg/kg, i.p, produced numerous high-frequency, high-intensity peaks within 5 min. Clement and Dyck (1989) constructed and tested a tremor monitor that quantitates soman- and oxotremorine-induced tremors. The device consisted of a force transducer, from which a plastic beaker was suspended containing a mouse. The signal from the force transducer was fed into a tremor monitor and quantitated using the Applecounter from Columbus Instruments.

Coward et al. (1977) recommended *N*-carbamoyl-2-(2,6-dichlorophenyl)acetamidine hydrochloride (LON-954), a tremorigenic agent, as an alternative to oxotremorine for the detection of anti-Parkinson drugs. Rats treated with 3-acetylpyridine show a selective degeneration of neurons in the inferior olive nucleus and the olivo-cerebellar tract with characteristic motor inco-ordination and ataxia. Similar motor dysfunction is seen in patients with olivo-ponto-cerebellar atrophy. To measure the effect of 3-acetylpyridine and the ameliorating effect of TRH agonists in rats the maximal height of vertical jump after stimulation by a foot shock was measured. Stanford and Fowler (1997) used a special technique for measuring forelimb tremor in rats induced by low doses of physostigmine. The rats pressed a force-sensing operandum while a computer measured force output and performed Fourier analyses on resulting force-time waveforms. Several studies indicate that the exposure to the pesticide *rotenone* causes highly selective dopaminergic degeneration and α-synuclein aggregation in rats. *Tremulous jaw movements* in rats induced by tacrine are considered to reflect symptoms in Parkinson's disease. The are suppressed by dopamine agonists.

Symptoms of Parkinson's Disease

MPTP (*N*-methyl-4-phenyl-1,2,3,6-tetrahydropyridine) has been shown to cause symptoms of Parkinson's disease in exposed individuals. The first observation was made in a relatively young, drug-abusing population. This initiated many clinical and experimental studies. When administered to primates this compound causes a partial destruction of basal ganglia and a syndrome that resembles Parkinson's disease.

Method

Burns et al. (1983) treated eight adult rhesus monkeys weighing 5-8 kg over a period of 5-8 days with cumulative intravenous doses of *N*-methyl-4-phenyl- 1,2,3,6- tetrahydro-pyridine (N-MPTP) up to 10-18 mg/kg. These animals showed a parkinsonism-like disorder (akinesia, rigidity, postural tremor, flexed posture, eyelid closure, drooling) which was reversed by the administration of L-dopa. The pathological and biochemical changes produced by N-MPTP are similar to the well established changes in patients with Parkinsonism. The N-MPTP intoxication was applied using marmosets by Nomoto et al. (1985, 1988), Temlett et al. (1989) and by Kebabian et al. (1992) to evaluate potential antiparkinson drugs.

Assessment

The severity of parkinsonian symptoms is rated by trained observers using a scale of 0 (normal) to 17 (maximum severity) that assesses movement (0: normal; 1: reduced; 2: sleepy), checking movements (0: present; 1: reduced; 2: absent), attention and blinking (0: normal; 1: abnormal), posture (0: normal; 1: abnormal trunk; 2: abnormal trunk and tail; 3: abnormal trunk, tail, and limbs; 4: flexed posture), balance and coordination (0: normal; 1: impaired; 2: unstable; 4: falls), reactions (0: normal; 1: reduced; 2: slow; 3: absent) and vocalizations (0: normal; 1: reduced; 2: absent).

Latest technique

Close and Elliott (1991) studied the behavioral effects of anti-Parkinsonian drugs in normal and MPTP-treated marmosets following central microinfusions. Kebabian et al. (1992) tested a selective D_1 receptor agonist with anti-parkinsonian activity in MTPT-treated marmosets. Domino and Sheng (1993) studied the relative potency of some dopamine agonists with varying selectivities for D_1 and D_2 receptors in MPTP-induced hemi-parkinsonian monkeys. Gnanalingham et al. (1995) used MTPT-treated marmosets and found differential effects with D_1 dopamine antagonists as compared with the effects in rats with unilateral 6-OGDA-induced lesions, where 6-OGDA is 6-hydroxydopamine. Doudet et al. (1993) used intravenous administration of ^{15}O-labeled water and 6-(^{18}F)-L-fluorodopa to assess abnormal striatal activity in monkeys after longterm recovery of unilateral lesions of the dopaminergic nigro-striatal system induced by MPTP. Positron emission tomography (PET) data were examined in relation to behavioral and biological parameters, such as cerebral blood flow.

Belluzzi et al. (1994) induced a hemiparkinsonian syndrome in *Macaca nemestrina* monkeys by unilateral infusion of MPTP into the right coronary artery. Raz et al. (2000) recorded hand tremor and simultaneous activity of several neurons in the external and internal segments of the globus pallidus in vervet monkeys before and after treatment with MPTP. Rollema et al. (1989) compared the effects of intracerebrally administered MPP^+ (1-methyl-4-phenylpyridinium) in three species (mouse, rat and monkey) by microdialysis determinations of dopamine and metabolites in the striatum. Asin et al. (1997) tested a selective D_1 receptor agonist in rats previously given unilateral 6-hydroxydopamine injections and in macaques previously given unilateral, intracarotid infusions of MPTP. Lange (1989, 1990) described circling behavior in old rats after unilateral intranigral injection of MPTP. Fuxe et al. (1992) studied the protection against MPTP-induced degeneration of the nigrostriatal dopamine neurons in the black mouse.

There are marked species differences in susceptibility to the neurotoxic effects of MTPT. Grunblatt et al. (2001) described gene expression analysis of mice in the MPTP model of Parkinson's disease using cDNA microarray. Mandel et al. (2003) reviewed the use of cDNA microarrays to assess Parkinson's disease models and the effects of neuroprotective drugs. Hamre et al. (1999) performed a quantitative analysis in seven strains of mice and found that differential strain susceptibility following MPTP administration acts in an autosomal fashion. Sedelis et al. (2001) described behavioral phenotyping of the MPTP mouse model on Parkinson's disease. Muramatu et al. (2003) published an immunocytochemical study on cerebral alterations in an MPTP mouse model of Parkinson's disease.

Clinical Aspect of Reserpine Antagonism

Reserpine induces depletion of central catecholamine stores. The sedative effect can be observed in mice shortly after injection, followed by signs of eyelid ptosis, hypokinesia, rigidity, catatonia, and immobility. These phenomena can be antagonized by dopamine agonists.

Method

Male NMRI mice of either sex weighing 20-25 g are used. They are injected intraperitoneally with 5mg/kg reserpine and tested 24h later. The test compounds are injected 30 min prior to observation. The animals are placed singly onto the floor of a Perspex container (30 × 26 cm, 20 cm high), situated on a Panlab proximity sensor unit. Horizontal movements are recorded for 10 min. Moreover, rearings and grooming episodes are recorded by an experienced observer.

Assessment

Locomotor activity and grooming scores of drug- treated animals are compared with controls treated with reserpine and vehicle only by analysis of variance.

Latest technique

Rats treated with reserpine develop spontaneous orofacial dyskinesia that has features similar to tardive dyskinesia in humans. The incidence of tongue protrusions was recorded to quantify the occurrence of oral dyskinesia.

Investing Nigrostriatal Lesioned Rats

Unilateral lesion of the dopaminergic nigrostriatal pathway in the rat by the neurotoxin 6-hydroxydopamine (6-OHDA) induces hypersensitivity of the postsynaptic dopaminergic receptors in the striatum of the lesioned side. The rats rotate in a direction towards the lesioned side (ipsilateral) when an indirect acting compound such as amphetamine is administered, but to the opposite direction (contralateral) when a directly acting dopamine agonist, e. g., apomorphine, or the dopamine precursor L-dopa is given. Therefore, this test can be used for the study of central dopamine function and the evaluation of dopamine antagonists and agonists, particularly the activity of novel anti-parkinsonian drugs.

Method

Male Wistar rats weighing 200-250 g at the time of surgery are used. The are housed individually in a controlled environment with free access to food and water. The animals are anesthetized with sodium pentobarbital. The head is placed in a stereotaxic device (DKI 900) and positioned according to the atlas of Konig and Klippel (1963). After a sagittal cut is made in the skin of the skull, a 2-mm-wide hole is drilled with an electrical trepan drill. Care is taken not to lesion the meninges. A 30-gauge stainless-steel cannula connected to a Hamilton syringe is aimed at the anterior zona compacta of the substantia nigra (coordinates anterior 4.1, lateral 1.0 and dorso-ventral -2.5 from instrument zero). A total of 8 μg of 6-OHDA in 4γ/l of saline is injected at a rate of 1γ/4 min. After the intracranial injection the wound is closed. The animal is allowed several weeks for recovery and for development of the lesion.

Specially constructed opaque plastic spheres attached to solid-state programming equipment serve as test chambers. The number of full turns, either ipsilateral or contralateral to the lesion, is recorded on an automatic printout counter every 15 min for 1- or 2-h test sessions. To determine the control values for ipsilateral turning, each subject is administered 2.5 mg/kg of d-amphetamine and immediately placed in the circling chamber for 2 h. Control values for contralateral circling are determined by injecting apomorphine at 1 mg/kg s.c. and recording the rat's circling for 1 h.

Test compounds are given i.p or s.c. and the animals placed into the circling chambers. Circling is recorded over a 1-h period.

Assessment

Percent change of drug turns from control turns is recorded. Using various doses ED_{50} values can be calculated.

Latest technique

Etemadzadeh et al. (1989) described a computerized rotometer apparatus for recording circling behavior. The digital pulses derived from the infrared photocell detector induced by the animal rotations were fed directly to a 20-megabyte microcomputer for online recording and were processed further to the Digital Equipment Corporation's VAX computer with the SAS software system for statistical and graphical analysis. Hudson et al. (1993) described a 16-channel automated rotometer system for reliable measurement of turning behavior in 6-hydroxydopamine-lesioned and transplanted rats. The system is preferable to more tedious methods such as videotaping and subsequent manual analysis or various other mechanical systems.

A rotometer differentiating between clockwise and counter-clockwise rotations with computerized evaluation is available from Technical and Scientific Equipment. Carey (1989) tested stimulant drugs as conditioned and unconditioned stimuli in a classical conditioning paradigm using drug-induced rotational behavior in rats with unilateral lesions of dopamine neurons. The production of asymmetry and circling behavior following unilateral, intrastriatal administration of neuroleptics was discussed by Costall et al. (1983). Rotational behavior produced by intranigral injections of bovine and human β-casomorphins in rats was described by Herrera-Marschitz et al. (1989). Perese et al. (1989) created a hemiparkinsonian model in rats in which there is 6-OHDA-induced destruction of the dopaminergic nigrostriatal pathway but sparing of the dopaminergic mesolimbic pathway.

Sauer and Oertel (1994) published a retrograde tracing and immunocytochemical study on the progressive degeneration of nigrostriatal dopamine neurons following intrastriatal terminal lesions with 6-hydroxydopamine in the rat. Garrett and Holtzman (1996) compared the effects of apomorphine, d-amphetamine, cocaine and caffeine on locomotor activity and rotational behavior in rats with unilateral 6-OHDA-induced lesions of the nigrostriatal tract. McElroy and Ward (1995) reported that the high-affinity and selective dopamine D_3 receptor ligand, 7-OH-DPAT, caused 6-OHDA-lesioned rats to rotate in a direction contralateral to the lesioned side similarly to the direct-acting dopamine agonist apomorphine. Haque et al. (1996) directly infused the neurotrophins NT3 and NT4/5 intra-parenchymally in close proximity to transplanted nigral tissue placed in the dopamine-depleted striatum of 6-hydroxydopamine-lesioned rats.

A survey on the unilateral 6-hydroxydopamine lesion model in behavioral brain research was prepared by Schwarting and Huston (1996). DSP-4 [*N*-(2-chloroethyl)-*N*-ethyl-2-bromobenzylamine] is highly toxic to noradrenergic neurons. Srinivasan and Schmidt (2003, 2004) showed that parkinsonian symptoms in 6-hydroxydopamine-induced partial degeneration of substantia nigra in rats are potentiated after depletion of nucleus coeruleus noradrenaline by intraperitoneal treatment with DPS-4. Harro et al. (2003) studied the effect of denervation of the locus coeruleus projections by DSP-4 treatment on [^{3}H]-raclopride binding to dopamine D_2 receptors and D_2 receptor–G protein interaction in rat striatum. Loscher et al. (1996) described a mutant rat strain (*CI*) with abnormal circling behavior reminiscent of 6-hydroxydopamine-lesioned rats. Mele et al. (1997) studied alterations in striatal dopamine overflow by *in vivo* microdialysis during rotational behavior of rats induced by amphetamine, phencyclidine and MK 801. Smith et al. (1996) reported contralateral turning in chronically cannulated rats after stimulation of glutamate receptors by unilateral intrastriatal injections of glutamate receptor agonists, such as kainate or AMPA. Spooren et al. (2000) tested the effects of a prototypical $mGlu_5$ receptor antagonist on rotarod locomotor activity and rotational responses in unilateral 6-OHDA-lesioned rats using an accelerated rotarod and automated rotameter cylinders.

Using the same equipment, Breysse et al. (2002) reported that chronic but not acute treatment with a metabotropic glutamate 5 receptor antagonist reverses the akinetic deficits in rats with 6-hydroxydopamine lesions. Using automated rotameter bowls (TSE Systems), Meissner et al. (2002) found that deep brain stimulation of subthalamic neurons induces contralateral circling in freely moving 6-hydroxydopamine-lesioned rats. With this equipment, Lebsanft et al. (2005) found, that 3,4-methylenedioxymethyl-amphetamine counteracts akinesia enantioselectively in rat rotational behavior. O'Neill et al. (2004) described neurotrophic actions of an AMPA receptor potentiator, which may increase brain-derived neurotrophic factor expression, in rodent models of Parkinson's disease.

Inhibition of sinistrotorsion induced in guinea pigs by injection of physostigmine into the right carotid artery was proposed as a method of screening central anticholinergic activity by De Jonge and Funcke (1962). Behavioral quantification of striatal dopaminergic supersensitivity after bilateral 6-hydroxydopamine lesions in the mouse was reported by Mandel et al. (1992). Worms et al. (1986) and Poncelet et al. (1993) studied turning behavior in mice induced by intrastriatal injection of neuropeptides. Fitzgerald et al. (1992) recommended the chakragati mouse (ckr), a transgenic insertional mutant which displays lateral circling, locomotor hyperactivity and hyper-reactivity, as a model to study aspects of neuropsychiatric disorders associated with dopaminergic abnormalities. Emonds-Alt et al. (1995) injected 0.3 pg senktide into the striatum of gerbils inducing contralateral rotations which were antagonized by intraperitoneal or oral administration of a tachykinin NK_3 receptor antagonist. Vernier and Unna (1963) tested drugs against the tremor induced by stereotactically oriented electric lesions in the region of the subthalamus or the mesencephalic reticular formation in monkeys.

Technique for Elevated Body Swing

Borlongan and Sanberg (1995); Borlongan et al. (1995) proposed the elevated body swing test as a measure of asymmetrical motor behavior of hemi-parkinsonian animals in a drug-free state.

Method

Male, 8-week-old Sprague Dawley rats are anesthetized with sodium pentobarbital (60 mg/kg i.p.) and mounted in a Kopf stereotaxic frame. They are lesioned by injection of 8 μg 6-hydroxydopamine in 4 μl saline containing 0.02% ascorbic acid in the left substantia nigra (AP – 5.0, ML+1.5, DV – 8.0). The solution is injected over a 4-min period and the needle left in place for an additional 5 min before retraction. Seven days after the lesion, behavioral testing is performed.

The animal is allowed to habituate in a Plexiglas box and attain a neutral position having all four paws on ground. The rat is held about 2.5 cm from the base of its tail and elevated 2.5 cm above the surface on which it has been resting. A swing is recorded whenever the animal moves its head out of the vertical axis to either side.

Before attempting another swing, the animal must return to the vertical position for the next swing to be counted. Swings are counted for 60s over four consecutive 15-s segments. The total number of swings made to each side is divided by the overall total number of swings made to both sides to get percentages of left and right swings. The criterion of biased swing is set at 70% or higher. At 30 and 45s, 6-OHDA-lesioned rats exhibit right-biased swings of 70% or higher compared to normal rats.

Assessment

A two-way ANOVA is used to analyze swing behavior data across the 15-s segments.

Studying Skilled Paw in Rats

Montoya et al. (1990, 1991), Abrous et al. (1993), Abrous and Dunnett (1994), Nikkhah et al. (1993), and Barneoud et al. (1996) used the skilled paw reaching test as a model of Parkinson's disease in the rat. The term "staircase test" mentioned in this context has nothing to do with the staircase test described by Thiebot et al. (1973) for evaluating anxiolytic activity in rats. Unilateral injection of 6-OHDA into the medial forebrain bundle results in an impairment of paw reaching on both sides which can be ameliorated by drug treatment or transplantation of a nigral cell suspension. The apparatus has been developed after earlier studies by Whishaw et al. (1986) who investigated the contributions of motor cortex, nigrostriatal dopamine and caudate-putamen to skilled forelimb use in the rat.

Method

Requirements

The apparatus consists of a clear Perspex chamber with a hinged lid. A narrower compartment with a central platform running along its length, creating a trough on either side, is connected to the chamber. The narrowness of the side compartment prevents rats from turning around, so they can use only their left paw for reaching into the left trough and their right paw for reaching into the right trough. A removable double staircase is inserted into the end of the box, sliding into the troughs on either side of the central platform. Each of the steps of the staircase contains a small well, and two 45-mg saccharin-flavored pellets are placed in each well.

Experimental approach

The week before the start of the training period, the rats are deprived of food and their body weight is stabilized at 85% of the weight of non-deprived rats. At the same time, they are gently manipulated and familiarized with the appetitive saccharin-flavored pellets. The animals then begin to learn the paw reaching task. For 4 weeks they are placed in the test boxes once per day for 10-15 min. The number of pellets eaten during the test period indicates the rat's success in grasping and retrieving the pellets; the number of steps from which pellets have been removed provides an index of the attempts to reach the food and how far the rat can reach; the number of missed pellets remaining at the end of the test on the floor of the side compartment indicates a lack of sensorimotor coordination in grasping and retrieving the pellets. In addition to these three parameters, it is noted which forepaw the rat used for the first movement to reach the pellet on each test day. A first choice score of +1 corresponds to the paw contralateral to the lesion, a score of –1, to the paw ipsilateral to the lesion. Because rodents exhibit a "pawedness," it must be noted whether there is a preference for one paw.

Symptoms

The mesotelencephalic system is lesioned by a stereotaxic unilateral injection of 6-OHDA into the medial forebrain bundle under equithesin anesthesia. 6-OHDA is injected in a volume of 1.5 μl and at a concentration of 4 μg/μl of 0.9% saline and 0.01% ascorbic acid twice over 3 min via a 30-gauge stainless steel cannula at the stereotaxic coordinates: L = 1.6 mm, AP = 0 mm, V = - 7.6 mm and L = 1.6 mm, AP = - 1 mm, V = -8 mm. The coordinates AP and L are estimated relative to the bregma, and V is measured from the level of the dura, with the incisor bar set 5 mm above the interaural line. Following each injection, the cannula is left in place for an additional 4 min to allow the diffusion of the neurotoxin away from the injection site. The sham-operated group receives sham lesions by identical injection of ascorbate-saline solution alone.

Clinical approach

The animals are injected i.p. with the test drug or saline 30 min before the unilateral 6-OHDA lesion and 24 h thereafter.

Assessment

Test sessions are performed 4, 5, 7, and 8 weeks after 6-OHDA lesion. The parameters success, attempts and sensorimotor coordination are subjected to a two-way ANOVA with group as the independent measure and weeks as the dependent measure.

Latest technique

Fricker et al. (1996) investigated the effect of unilateral ibotenic acid lesions in the dorsal striatum, placed at anterior, posterior, medial, or lateral loci, in the staircase test of skilled forelimb use. Nakao et al. (1996) studied paw reaching ability in rats with unilateral quinolinic acid lesions of the striatum as an animal model for Huntington's disease. Barneoud et al. (1996) evaluated the neuroprotective effects of riluzole using impaired skilled forelimb use, circling behavior, and altered dopaminergic metabolism of the mesotelencephalic system in unilaterally 6-OHDA-lesioned rats. Fricker et al. (1997) studied the correlation between positron emission tomography, using ligands to the D_1 and D_2 receptors, and reaching behavior in rats with ibotenic acid lesions and embryonic striatal grafts. Grabowski et al. (1993), Marston et al. (1995), and Sharkey et al. (1996) tested drug effects on skilled motor deficits produced by middle cerebral artery occlusion in rats using the paw reaching test. Meyer et al. (1997) described a revolving food pellet test for measuring sensorimotor performance in rats.

Performing Stepping Test in Rats

Schallert et al. (1992), Olsson et al. (1995) and Rosenblad et al. (1997) introduced the stepping test as a clinically relevant unilateral model of Parkinsonian akinesia. The 6-OHDA lesion induced marked and long-lasting impairments in the initiation of stepping movements with the contralateral paw which can be ameliorated by the systemic application of drugs.

Method

6-OHDA lesion surgery

Female Sprague Dawley rats receive two stereotaxic injections of 6-OHDA (3.6 µg/µl in 0.2 µg/ml ascorbate-saline) into the right ascending mesostriatal dopamine pathway using a 10-µl Hamilton syringe at the following coordinates (in mm, with reference to bregma and dura):

1. 2.5 µl at AP –4.4, L 1.2, V 7.8, tooth bar –2.4;
2. 3.0 µl at AP –4.0, L 0.8, V 8.0, tooth bar +3.4 at an injection rate of 1 µl/min.

The cannula is left in place for an additional 5 min before slowly retracted.

Laboratory setup for stepping test

The tests monitoring initiation time, stepping time and step length are performed using a wooden ramp with a length of 1 m connected to the rat's home cage. A smooth-surfaced table is used for measuring adjusted steps. During the first 3 days the rats are handled by the experimenter to familiarize them with the experimenter's grip. During the subsequent 1-2 days the rats are trained to run spontaneously up the ramp to the home cage. The stepping test comprises two parts: first, the time to initiation of a movement of each forelimb, the step length, and the time required for the rat to cover a set distance along the ramp with each forelimb; and second the initiation of adjusting steps by each forelimb when the animal was moved sideways along the bench surface. Each test consists of two tests per day for three consecutive days and the mean of six subtests is calculated.

Experimentation

The rat is held by the experimenter with one hand fixing the hindlimbs and slightly raising the hind part above the surface. The other hand fixes the forelimb not to be monitored. Time is measured until the rat initiates movement with the forelimb not fixed by the experimenter, using 180s as break-off point. Stepping time is measured from initiation of movement until the rat reaches the home cage. Step length is calculated by dividing the length of the ramp by the number of steps required for the rat to run up the ramp. The sequence of testing is right paw testing followed by left paw testing, repeated twice.

Adjusting steps

The rat is held in the same position as described above with one paw touching the table, and is then moved slowly sideways (5s for 0.9m) by the experimenter, first in the forehand and then in the backhand direction. The number of adjusting steps is counted for both paws in the backhand and forehand directions of movement. The sequence of testing is right paw forehand and backhand adjusting stepping, followed by left paw forehand and backhand directions. The test is repeated twice each day.

The paw contralateral to the lesion is passively dragging when the rat is moved in the forehand direction, while the ipsilateral paw performs frequent stepping movements.

Pharmaceutical application

Stepping tests are repeated to determine the baseline weekly after the 6-OHDA lesion. The drug tests are administered for 1 day only. Various drugs can be evaluated in weekly intervals.

Assessment

Results are expressed as means ± SEM. For statistical evaluation, the data are subjected to one-factor analysis of variance (ANOVA) and Fisher post hoc test.

Latest technique

Schallert et al. (2000), Picconi et al. (2003, 2004), and Centonze et al. (2005) described the limb-use asymmetry test to measure motor performance in rats with 6-OHDA-induced lesions. Forelimb use during exploratory activity was analyzed in a transparent cylinder (20 cm diameter and 30 cm height) for 3-10 min depending on the degree of movement maintained during the trial. The number of supporting wall contacts executed independently with the right of the left forelimb was counted. The percentage of wall contacts executed by the impaired forelimb (contralateral to the lesion) was then subtracted from the percentage of contacts with the non-impaired forelimb to obtain a limb-use asymmetry score.

Investing Parkinson's Disease in Transgenic Animal

New insight in the pathobiology of Parkinson's disease has led to the development of transgenic animal models. The most prominent models are related to α-synuclein. The first transgenic mice that express human α-synuclein were generated by Masliah et al. (2000). These mice displayed a progressive accumulation of α-synuclein and ubiquitin-immunoreactive inclusions in neurons of the neocortex, hippocampus, and substantia nigra. These alterations were associated with a loss of dopaminergic terminals in the basal ganglia and with motor impairments. Since then, several studies on transgenic α-synuclein mice have been performed.

Masliah and Hashimoto (2002) discussed the possible role of β-synuclein for development of new treatments for Parkinson's disease using transgenic animal models. Ubiquitin carboxyl-terminal hydrolase L1 (UCH- L1) is a neuron-specific ubiquitin recycling enzyme, which may be implicated in the pathogenesis of Parkinson's disease. Lindsten et al. (2003) developed a model for *in vivo* analysis of the ubiquitin/proteasome system by generating mouse strains transgenic for a green fluorescence protein reporter carrying a constitutively active degradation signal. Loss-of-function mutations in *parkin* are the major cause of early-onset familial Parkinson's disease. To investigate the pathogenic mechanism by which loss of parkin function causes Parkinson's disease, Goldberg et al. (2003) generated a mouse model bearing a germline disruption in parkin. In addition to PARK1 (representing the α-synuclein gene) and PARK2 (representing the UCH-L1 gene), other monogenic forms of Parkinson's disease have been described. The *weaver* mutant mouse has been proposed as another model of Parkinson's disease.

Latest technique

Lo Bianco et al. (2002) studied *a*-synucleinopathy and selective dopaminergic neuron loss in a rat lentiviral-based model of Parkinson's disease. Kirik et al. (2003) described nigrostriatal α-synucleinopathy induced by viral vector-mediated overexpression of human α-synuclein using a primate (adult marmosets) model of Parkinson's disease. Transgenic *Drosophila* models expressing human α-synuclein were described by Feany and Bender (2000), Pendleton et al. (2002), and Scherzer et al. (2003). Brown et al. (2002) published multiplex three-dimensional brain gene expression mapping in a mouse model of Parkinson's disease. Son et al. (2003) described cloning and expression analysis of the Parkinson's disease gene, UCH-L1, and its promoter in zebrafish.

Studying Lesioned Animals

Human embryonic stem cells may potentially serve as a renewable source of cells for transplantation. In Parkinson's disease, embryonic stem-cell-derived dopaminergic neurons may replace the degenerated neurons in the brain. To substantiate this goal, numerous animal experiments were performed. Most authors used rats rendered hemi-Parkinsonian by injection of 6-OHDA into the substantia nigra as recipients. Takagi et al. (2005) reported that dopaminergic neurons generated from cynomolgus monkey embryonic stem cells functioned in the MPTP primate model of Parkinson's disease. However, complications were reported by Blanchet et al. (2003) using a similar approach. In spite of almost 20 years of experimental experience, clinical trials with cell transplantation for Parkinson's disease have had disappointing results.

Recombinant Technology

Recombinant viral vectors derived from adenovirus, adeno-associated virus or lentivirus have been developed into highly effective vehicles for gene transfer to the adult central nervous system. These vectors have been shown to be effective for long-term delivery of glial cell-line derived neurotrophic factor (GDNF) in the nigrostriatal system. GDNF is a strong candidate agent in the neuroprotective treatment of Parkinson's disease. The injection of adeno-associated virus encoding rat glial cell line-derived neurotrophic factor protected nigral neurons in the 6-OHDA-induced degeneration model of Parkinson's disease in rats.

Jollivet et al. (2004) used intra-striatal glial cell line-derived neurotrophic-factor-releasing microspheres. Yasuhara et al. (2005) reported the use of glial cell line-derived neurotrophin-factor-producing cells encapsulated into hollow fibers. Kojima et al. (1997) and Cheng et al. (1998) found that glial cell-line-derived neurotrophic factor protects against MPTP-induced neurotoxicity in mice. Kordower et al. (2000) reported that neurodegeneration was prevented by lentiviral vector delivery of GDNF in primate models (*rhesus monkeys* treated with MPTP) of Parkinson's disease.

14

Nephric Hypertension

Since the classical experiments of Goldblatt et al. (1934) there is clear evidence that the ischemia of the kidneys causes elevation of blood pressure by activation of the renin-angiotensin system. The principle can be used both for acute and chronic hypertension. In rats acute renal hypertension is induced by clamping the left renal artery for 4 h. After reopening of the vessel, accumulated renin is released into circulation. The protease renin catalyzes the first and rate-limiting step in the formation of angiotensin II leading to acute hypertension. The test is used to evaluate antihypertensive activities of drugs.

ACUTE RENAL HYPERTENSION IN RATS

Method

Male Sprague-Dawley rats weighing 300 g are used. The animals are anesthetized by intraperitoneal injection of 100 mg/kg hexobarbital sodium. A PVC-coated Dieffenbach clip is placed onto the left hilum of the kidney and fixed to the back muscles. The renal artery is occluded for 3.5-4 h. 3.5 h following the surgery, the animals are anesthetized by intraperitoneal injection of 30-40 mg/kg pentobarbital sodium. The trachea is cannulated to facilitate spontaneous respiration.

To measure systolic and diastolic blood pressure, the cannula in the carotid artery is connected to a pressure transducer. For administration of the test compound, a jugular vein is cannulated. Following a stable blood pressure state, ganglionic blockade is performed with pentolinium (10 mg/kg i.v.). After obtaining stable reduced blood pressure values, the renal arterial clip is removed. This leads to a rise in blood pressure as a consequence of elevated plasma renin level. Within 15 min a stable hypertension is achieved (control = 100%).

The test substance is then administered by intravenous injection at doses of 10 and 100 µg/kg. Blood pressure is monitored continuously until a renewed increase to the starting level is obtained. Ten-twelve animals are used per compound.

Calculation

Increase in blood pressure after reopening of the renal artery and reduction in blood pressure after administration of the test drug are determined [mm Hg]. Percent inhibition of hypertensive blood pressure values under drug treatment are calculated as compared to pretreatment hypertension values. Duration of the effect is determined [min]. Statistical significance is assessed by the paired *t*-test. A sharp and transient in systemic arterial blood pressure associated with reflex bradycardia can be elicited by injection of 5-hydroxytryptamine, cyanide, nicotine or lobeline into the coronary artery blood stream of dogs. The phenomenon is named the cardiogenic hypertensive chemoreflex and 5-HT proved to be the most powerful agent for its initiation.

ISCHEMIA OF KIDNEY IN MICE

On the basis of the findings of Goldblatt et al. (1934) that ischemia of the kidneys induces hypertension, various modifications of the technique have been described for several animal species. One of the most effective modifications in rats is the so called 1-kidney-1-clip method.

Method

Male Sprague-Dawley rats weighing 200-250 g are anaesthetized with 50 mg/kg i.p. pentobarbital. The fur on the back is shaved and the skin disinfected. In the left lumbar area a flank incision is made parallel to the long axis of the rat. The renal pedicel is exposed with the kidney retracted to the abdomen. The renal artery is dissected clean and a U-shaped silver clip is slipped around it near the aorta. Using a special forceps the size of the clip is adjusted so that the internal gap ranges from 0.25-0.3 8 mm. The right kidney is removed through a flank incision after tying off the renal pedicle. The skin incisions are closed by wound clips. 4-5 weeks after clipping blood pressure is measured and rats with values higher than 150mm Hg selected for the experiments. Blood pressure readings are taken on each of 3 days prior to drug treatment. Drugs are administered orally in volumes of 10 ml/kg. The rats are divided into 4 animals per dose and each animal is used as his own control. Compounds are administered for 3 days and predrug and 2 h postdrug blood pressure readings are taken.

Assessment

Changes in systolic blood pressure are expressed in mm Hg. Activity is determined by comparing treatment blood pressure values with the control blood pressure value (Day 1, predrug blood pressure). Comparisons are made using the paired *t*-test for evaluation of statistical significance. Duan et al. (1996) induced renal hypertension in male Hartley *guinea pigs* by a two-step procedure consisting of ligation of the left caudal renal artery and right nephrectomy. Arterial blood pressure and heart rate were monitored in conscious animals. ACE-inhibitors reduced blood pressure in sham-operated and in renal hypertensive guinea pigs, whereas renin inhibitors were effective only in renal hypertensive animals.

ISCHEMIA OF KIDNEY IN DOGS

Production of hypertension by clamping renal arteries has been first described by Goldblatt et al. (1934) in dogs. Later on, the method has been modified, e. g., as the "wrapping" technique.

Method

Dogs weighing 8-12 kg are anesthetized with i.v. injection of 15 mg/kg thiopental. Anesthesia is maintained with a halothane-oxygen mixture. Under aseptic conditions, a midline abdominal incision is made. One kidney is exposed and wrapped in cellophane and then replaced. The contralateral kidney is exposed. The artery, vein and ureter are ligated and the kidney is removed. The abdomen is closed by sutures and clips. On the day of surgery and for 3 days following, the dogs are given antibiotics. Body temperature is measured twice daily for 4 days following surgery. Six weeks following surgery, blood pressure is measured using a tail-cuff method. For recording, the tail-cuff is attached to a polygraph.

Only animals with a systolic blood pressure higher than 150 mm Hg are considered to be hypertensive and can participate in studies evaluating potential antihypertensive compounds. For the experiment, blood pressure is recorded either by the indirect tail-cuff method or by direct measurement via an implanted arterial cannula. On day 1 readings are made every 2 h, just before, and 2 and 4 h after oral treatment with the potential antihypertensive compound. Drug administration is repeated for 5 days. On days 3 and 5 blood pressure readings are taken before and 2 and 4 h after treatment. At least 3 dogs are used per dose and compound.

Assessment

The starting value is the average of the 2 readings before application of the drug. Each of the following readings is subtracted from this value and recorded as fall of blood pressure at the various recording times. Renal hypertension in rats has been achieved by many modifications of the method such as the technique according to Grollman (1944). The kidney is exposed through a lumbar incision, the renal capsule is removed by gentle traction, and a figure-8 ligature is applied being tight enough to deform the kidney but not tight enough to cut the tissue. Renal hypertension may be induced in the *rat* by encapsulating both kidneys with latex rubber capsules. Moulds are formed from plastic using a rat kidney as a model. The capsules are prepared by dipping the moulds in liquid latex allowing them to dry in the air. Three applications of latex are applied before the capsules are toughened by placing them under warm running tap water. The kidney is exposed by lumbar incision, the renal capsule gently removed and the capsule applied.

BARORECEPTOR IN DOGS

Vasodilator and depressor reflexes, originating in the baroreceptor areas of the carotid sinus and aortic arch, play an important part in the regulation of blood pressure. Stimulation of the afferent buffer fibres exerts an inhibitory influence on the vasomotor center, and their sectioning leads to a persistent rise in blood pressure. In this way, acute neurogenic hypertension can be induced in dogs.

Method

Adult dogs of either sex weighing 10-15 kg are anesthetized using 15 mg/kg sodium thiopental, 200 mg/kg sodium barbital and 60 mg/kg sodium pentobarbital i.v. A femoral vein and artery are cannulated using polyethylene tubing to administer compounds i.v. and record arterial pressure and heart rate, respectively. Left ventricular pressure and dp/dt are recorded via the left common carotid artery (post-deafferentiation) using a Millar microtip pressure transducer. P_{max} is recorded by speeding up the chart paper.

Cardiac output is determined by introducing a Swan-Ganz catheter into the right heart and pulmonary artery via a jugular vein. Five ml of cold 5% dextrose is injected into the right atrium and an Edwards Cardiac output computer is used to calculate the cardiac output from the temperature change in the pulmonary artery. All recordings are made with a polygraph. Both of the carotid arteries are cleared up to the bifurcation of the internal and external carotid arteries. The carotid sinus nerves are isolated, ligated and sectioned and a bilateral vagotomy is performed to produce neurogenic hypertension (mean arterial pressure more than 150 mm Hg). The dog is allowed to equilibrate for approximately 30 min and a bolus of the test compound is administered by intravenous injection. Heart rate, arterial pressure, left ventricular pressure, P_{max} and dp/dt are monitored for 90 min. A minimum of 3 dogs are used for each compound.

Assessment

Changes of the cardiovascular parameters are expressed as percentage of the values before administration of the drug. Neurogenic hypertension through baroreceptor denervation has also been described in rabbits and in rats. The neurogenic hypertension is useful for acute experiments. However, it is less useful for chronic experiments since the elevated blood pressure caused by buffer nerve section is more labile than that caused by renal ischemia.

MINERALOCORTICOID-INDUCED HYPERTENSION IN RATS

Mineralocorticoid-induced hypertension is thought to be due to the sodium retaining properties of the steroid causing increases in plasma and extracellular volume. The hypertensive effect is increased by salt loading and unilateral nephrectomy in rats.

Method

Male Sprague Dawley rats weighing 250-300g are anesthetized with ether. Through a flank incision the left kidney is removed. The rats are injected twice weekly with 20 mg/kg s.c. desoxycorticosteroneacetate in olive oil for 4 weeks. Drinking water is replaced with a 1% NaCl solution. Blood pressure starts to rise after one week and reaches systolic values between 160 and 180 mmHg after 4 weeks. The regimen to induce DOCA-salt hypertension has been modified by many authors. DOCA pellets or implants in silastic devices were used instead of repeated injections. DOCA-salt hypertension can also be achieved without nephrectomy. Using kininogen-deficient Brown Norway Katholiek (BN-Ka) rats, Majima et al. (1991, 1993) showed suppression of rat desoxycorticosterone-salt hypertension by the kallikrein-kinin system. Li et al. (1996) examined small-artery structure on a wire myograph and quantified endothelin-1 messenger RNA by Northern blot analysis in DOCA-salt hypertensive rats after administration of an ACE-inhibitor, a calcium channel antagonist and a nitric oxide synthase inhibitor. Ullian (1997) described the Wistar-Furth rat as a model of mineralocorticoid resistance. These rats developed two-kidney, one-clip hypertension to the same degree as did Wistar rats and reacted to glucocorticoid treatment with a rapid onset of hypertension, but were resistant to the development of DOCA-NaCl hypertension. Studies in DOCA-salt hypertensive *mice* were reported by Gross et al. (1998, 1999), Honeck et al. (2000), Peng et al. (2001).

DIETARY CARBOHYDRATE AND HYPERTENSION IN RATS

Increases in dietary carbohydrate intake can raise blood pressure in experimental animals. The increased intake of either sucrose or glucose was shown to enhance the development of either spontaneous hypertension or salt hypertension in rats. Hwang et al. (1987) first reported that hypertension could be induced in normal rats by feeding a high-fructose diet. Fructose feeding was also found to cause insulin resistance, hyperinsulinemia, and hypertriglyceridemiain normal rats. Dai and McNeill (1995) studied the concentration- and duration-dependence of fructose-induced hypertension in rats.

Method

Groups of 8 male Wistar rats weighing 210-250 g are used. The are housed two per cage on a 12-h light 12-h dark cycle and are allowed free access to standard laboratory diet (Purina rat chow) and drinking fluid. Drinking fluid consists either of tap water or 10%- fructose solution. Body weight, food intake and fluid intake of each rat are measured every week during treatment. Using the tail-cuff method, systolic blood pressure and pulse rate is measured before and every week during treatment. Blood samples are collected before and every second week during treatment for determination of plasma glucose, insulin, and triglycerides.

Assessment

Since maximum effects on the chosen parameters are achieved after 6 weeks, the duration of treatment can be limited to this time. Statistical analysis is performed using a one-way or two-way analysis of variance, followed by the Newman-Keuls test. Reaven et al. (1988, 1989) found an attenuation of fructose-induced hypertension by exercise training and an inhibition by somatostatin treatment. Brands et al. (1991, 1992) found an increase of arterial pressure during chronic hyperinsulinemia in conscious rats. Hall et al. (1995) reported the effects of 6 weeks of a high-fat diet on cardiovascular, renal, and endocrine functions in chronically instrumented conscious dogs. Body weight increased by approximately 16.9kg, whereas MAP, cardiac output, and heart rate increased by 28%, 77%, and 68%, respectively

HEREDITARY HYPERTENSION IN RATS

Inherited hypertension in rats has been described by Smik and Hall 1958; Phelan and Smirk 1960; Laverty and Smirk 1961; Phelan 1968 as *genetically hypertensive* (GH) rats. Okamoto et al. (1963, 1966) reported the development of a strain of spontaneously hypertensive rats from mating one Wistar male rat with spontaneously occurring high blood pressure with a female with slightly elevated blood pressure. By inbreeding over several generations a high incidence of hypertension with blood pressure values of 200mm Hg or more was achieved. These strains were called "Spontaneously hypertensive rats (Akamoto-Aoki)" = SHR or "Wistar-Kyoto rats" = WKY.

Hypertension in these rats is clearly hereditary and genetically determined, thus comparable to primary hypertension in humans. Cardiac hypertrophy and cellular ionic transport abnormalities have been observed. Inbred strains being *salt-hypertension-sensitivey* and *salt-hypertension-resistant* (RD) have been developed by Dahl et al. (1962, 1963), Rapp (1984), Cicila et al. (1993). Inoko et al. (1994) reported the transition from compensatory hypertrophy to dilated, failing left ventricles in Dahl salt-sensitive rats. Two strains of rats with inbred dissimilar sensitivity to DOCA-salt hypertension ("*Sabra strain*") have been separated by Ben-Ishay et al. (1972, 1984).

Another hypertensive strain derived from Wistar rats was produced by brother-sister mating in the group of Bianchi et al. (1974, 1986) at the University of Milan called "*Milan hypertensive strain*" = MHS. These rats show a cell membrane defect resulting in abnormal kidney function. Salvati et al. (1990) studied the diuretic effect of bumetanide in isolated perfused kidneys of Milan hypertensive rats. Furthermore, the "*Lyon*" strains of hypertensive, normotensive and low-blood-pressure rats were developed. These rats show a genetically determined defect in central nervous function.

Spontaneously hypertensive rats which develop failure before 18 months have been selectively bred. Several substrains of spontaneous hypertensive rats were separated by the group of Okamoto et al. (1974) including the stroke-prone strain SHR = SHRSP. These rats have an increased sympathetic tone and show a high incidence of hemorrhagic lesions of the brain with motor disturbances followed by death. A strain of obese spontaneously hypertensive rats has been described by Koletsky (1975), Ernsberger et al. (1993). With new techniques of genetic engineering, *transgenic rats with hypertension* could be created. Increase of blood pressure of spontaneously hypertensive rat is determined by multiple genetic loci. With new technology not only these loci could be defined but also new models in hypertension research and models to detect antihypertensive drugs could be established. Mullins et al. (1990) reported fulminant hypertension in transgenic rats harboring the mouse Re-2 gene.

A rat strain TGR(mREN2)27 as a monogenetic model in hypertension research was described by Peters et al. (1993), Lee et al. (1996), Langheinrich et al. (1996), and Ohta et al. (1996). Bohlender et al. (1996) reconstructed the human renin-angiotensin system in transgenic rats overexpressing the human angiotensin gene TGR(hOGEN) 1623 by chronically injecting human recombinant renin intravenously using Alzet pumps. Zolk et al. (1998) described the effects of quinapril, losartan and hydralazine on cardiac hypertrophy and β-adrenergic neuroeffector mechanisms in transgenic TGR(mREN2)27 rats.

The use of spontaneously hypertensive rats to detect potential antihypertensive compounds is well established. On the basis of available data no preference can be given to a particular strain. The most abundant experience has been gained with the Wistar–Kyoto strain. Transgenic rats with well defined genomes are gaining more importance. Pijl et al. (1994) described streptozotocin-induced diabetes mellitus in spontaneously hypertensive rats as a pathophysiological model for the combined effects of hypertension and diabetes. Rosenthal et al. (1997) used rats of the Cohen-Rosenthal diabetic hypertensive strain to examine the effects of an ACE-inhibitor, an ATII antagonist and a calcium antagonist on systolic pressure and spontaneous blood glucose levels. Holycross et al. (1997) used hypertensive SHHF/ Mcc-facp rats to study plasma renin activity during development of heart failure. Linz et al. (1997) compared the outcome of lifelong treatment with the ACE inhibitor ramipril in young prehypertensive stroke-prone spontaneously hypertensive rats and age-matched normotensive Wistar-Kyoto rats. Lifelong ACE inhibition doubled the lifespan in hypertensive rats matching that of normotensive rats. Studies in *genetically hypertensive mice* were reported by Rosenberg et al. (1985), Hamet et al. (1990) Meneton et al. (2000). Ohkubo et al. (1990) generated transgenic mice with elevated blood pressure by introduction of the rat renin and angiotensinogen genes.

NITRIC OXIDE-INDUCED HYPERTENSION

Tonic basal release of nitric oxide (NO) by vascular endothelial cells controls blood pressure in the basal state. Chronic blockade of NO synthesis in the rat produces systemic hypertension and glomerular damage. This was recommended by Ribeiro et al. (1992) as a model of hypertension. Yang et al. (1996) found an increase of vascular angiotensin II receptor expression after chronic inhibition of NO synthase in spontaneously hypertensive rats. The detrimental sequels of chronic NO synthase inhibition in rats can be inhibited by treatment with ACE inhibitors. Hsieh et al. (2004) reported that NO inhibition by L-NAME accelerates hypertension and induces perivascular inflammation in rats. Hropot et al. (2003) reported that angiotensin II subtype AT_1 receptor blockade prevents hypertension and renal insufficiency induced by chronic NO synthase inhibition in rats.

Method

Male Wistar rats at an age of 7-8 weeks weighing 210 ± 10 g were placed at random in metabolic cages, divided in four to six groups of six to eight rats each. Group 1 (control) had free access to tap water and food. Groups 2-4 were treated with 0.02% L-NAME water solution for 6 weeks in a daily dose of 25 mg/kg. Groups 3 and 4 received the angiotensin receptor antagonists fonsartan (10 mg/kg) or lorsatan (30 mg/kg) for 6 weeks daily per stomach tube. Groups 5 and 6 received fonsartan and lorasatan alone. At the end of the study, 24-h urine samples were collected and retrobulbar blood samples were taken in short inhalation anesthesia. Plasma values of creatine, PRA and electrolytes were determined. For clearance evaluation rats were anesthetized with 50mg/kg thiopentone i.p. In order to determine glomerular filtration rate and renal plasma flow, clearances of inulin and *para*-aminohippurate were performed. After the clearance experiments, the rats were sacrificed and hearts and kidneys removed.

Hearts were perfused in the isolated working heart preparation via the aorta with modified Krebs-Henseleit buffer containing solvents or drugs. After a preischemic period of 20 min, acute regional myocardial ischemia was produced by clamping the left coronary artery close to its origin for 15 min (ischemic period). Thereafter, the clip was removed and changes during reperfusion were monitored (reperfusion period). The following cardio-dynamic and cardio-metabolic parameters were measured: incidence and duration of ventricular fibrillation, left ventricular pressure, contractility (dP/dt max), heart rate and coronary flow; in the coronary effluent lactate dehydrogenase, creatine kinase and lactate; in the myocardial tissue lactate, glycogen, ATP and creatine phosphate.

Left ventricular pressure was measured with a Statham pressure transducer (P 23 DB) which on differentiation yielded LV dP/dt max and heart rate. Coronary flow was determined by electromagnetic flow probes in the aortic cannula. For the determination of lactate release, lactate dehydrogenase (LDH) and creatine kinase (CK) activities in the perfusate samples were taken from the coronary effluent and analyzed spectrophotometrically.

Assessment

Results are presented as arithmetical means ± SEM. A one-way ANOVA was calculated with SYSTAT for Windows followed by multiple pairwise comparisons according to Tukey. Arnal et al. (1993) measured cardiac weight of rats in hypertension induced by NO synthase blockade. Linz et al. (1999) reviewed the interactions between ACE, kinins and NO. Sampaio et al. (2002) reported that hypertension plus diabetes mimics the cardiomyopathy induced by NO inhibition in rats. Rossi et al. (2003) found that chronic inhibition of NO synthase induces hypertension and cardiomyocyte mitochondrial and myocardial remodeling in the absence of hypertrophy.

MONOCROTALINE INDUCED HYPERTENSION

The pyrrolizidine alkaloid monocrotaline, derived from *Crotalaria spectabilis,* is hepatotoxic and pneumotoxic in the rat. A single injection of monocrotaline leads to progressive pulmonary hypertension resulting in right ventricular hypertrophy and cardiac failure. Pathologic changes and hemodynamic changes associated with monocrotaline administration include blebbing of the lung, degeneration and fragmentation of endothelial cells, perivascular edema, extravasation of red blood cells, and muscularization of the pulmonary arteries and arterioles. Rats given monocrotaline develop severe right ventricular hypertrophy often accompanied by ascites and pleural effusions. Amelioration by angiotensin-converting enzyme inhibitors and by penicillamine has been demonstrated.

Method

Treatment of male Sprague Dawley rats weighing 200-225 g with the test drug (angiotensin converting enzyme inhibitor or vehicle) is started one week prior to a single subcutaneous injection of 100 mg/kg monocrotaline up to sacrifice 4, 7, or 14 days later by pentobarbital anesthesia and exsanguination. Heart and lungs are excised from thoracic cavity. After removing atria from the heart, the right ventricle is separated from the left ventricle plus septum which are blotted and weighed separately. Left lung is blotted, weighed, minced and reweighed after drying at room temperature for 14 days. Three pulmonary artery segments, main pulmonary artery, right extrapulmonary artery and an intrapulmonary artery from the from the right lower lobe, are isolated for study of vascular responsiveness.

Cylindrical segments of each vessel are suspended between stainless steel hooks in 10-ml isolated tissue baths containing modified Krebs-Henseleit buffer aerated with 95% O_2/5% CO_2 at 37°C. At the end of each experiment, vessel segments are blotted and weighed and their dimensions measured. Cross-sectional area of each artery is determined from tissue weight and diameter. Arteries are equilibrated for 1 h at 1 g of passive applied load and then are made to contract to KCl (6×10^{-2} M). After washout, the procedure is repeated with applied loads increased by 1 g increments. Responses are normalized to the maximum active force development generated by an artery in each experiment and the data are plotted as a function of applied force.

Changes in isometric force are monitored through force displacement transducers (Grass FT03) and recorded on a polygraph. Responsiveness to contractile and relaxant agonists is assessed in pulmonary arteries from saline- and monocrotaline-treated rats both in verum- and placebo-treated groups. Cumulative concentration-response curves to hypertonic KCl, angiotensin II and norepinephrine are generated sequentially in vessels at resting tone. Arteries are then contracted submaximally with norepinephrine and cumulative concentration-response curves to the vasorelaxants isoproterenol and acetylcholine are determined.

Assessment

Contractions are expressed as active tension development, force generated per cross-sectional area and relaxations are normalized to precontraction tone. Both contractile and relaxation responses are plotted as a function of the negative logarithm of agonist concentration. Differences in mean responses are compared by a *t*-test for grouped data. Molteni et al. (1986) treated rats continuously with monocrotaline in the drinking water at a concentration of 2.4 mg/kg/day for a period of 6 weeks. Test rats received an ACE-inhibitor during this time in the drinking water and controls the vehicle only. At the end of the experiment, hearts and lungs were weighed and examined by light and electron microscopy. Madden et al. (1995) determined L-arginine-related responses to pressure and vasoactive agents in monocrotaline-treated rat pulmonary arteries. Ono et al. (1995) studied the effects of prostaglandin E_1 (PGE_1) on pulmonary hypertension and lung vascular remodelling in the rat monocrotaline model of human pulmonary hypertension.

Yamauchi et al. (1996) studied the effects of an orally active endothelin antagonist on monocrotaline-induced pulmonary hypertension in rats. Gout et al. (1999) evaluated the effects of adreno-medullin in isolated vascular rings from rats treated with monocrotaline (60 mg/kg s.c.) causing pulmonary hypertension and ventricular hypertrophy within 3 to 4 weeks. Kanno et al. (2001) studied the effect of an angiotensin-converting enzyme inhibitor on pulmonary arterial hypertension and endothelial nitric oxide synthase expression in monocrotaline-treated rats. For evaluation of right ventricular hypertrophy as a result of pulmonary arterial hypertension, multislice spin-echo MRI images were acquired at 8-12 time points in a cardiac cycle with respiratory and ECG gating at 2, 3, 4, and 5 weeks after monocrotaline treatment. Kang et al. (2003) reported that a phosphodiesterase-5 inhibitor attenuated monocrotaline-induced pulmonary hypertension in rats.

HYPERDYNAMIC SPLANCHNIC INDUCED HYPERTENSION IN RATS

Portal hypertension is associated with hyperdynamic splanchnic circulation and reduced vascular resistance. Tanoue et al. (1991) developed a method for inducing portal hypertension and esophageal varices in rats - partial ligation of the portal vein after devascularization of the circumference of the left renal vein and complete ligation of the portal vein on the fifth day thereafter. Tsugawa et al. (2000) used this model to study the role of nitric oxide and endothelin-1 in rat portal hypertension.

Method

Male Sprague-Dawley rats were anesthetized with 50 mg/kg Nembutal intraperitoneally. The portal vein was isolated and stenosis created by a single ligature of 3-0 silk placed around the portal vein and a 20-gauge blunt-tipped needle after devascularization of the left renal vein. This devascularization is indispensable in preventing the development of excess collateral vessels, which inhibit the formation of esophageal varices and the portal hypertensive state. The needle was then removed from the ligature. In addition, 3-0 silk was also placed at the area of partial ligation (loose ligation), and both ends were then drawn out through the abdominal wall. Five days after the operation, the ends of the silk that had been placed in the flank were simultaneously pulled to induce complete portal vein ligation. Two weeks later, this portal hypertension model was completed. Portal venous pressure, blood flow volume in the intra-abdominal viscera, plasma NO and plasma endothelin-1 were measured.

Assessment

Results were expressed as mean ± standard deviation. The Student's *t*-test was used to determine significance between portal hypertension rats and sham-operated controls. Portal hypertension by portal vein ligation without devascularization of the left renal vein was used by Lee et al. (1985), Braillon et al. (1986), Oren et al. (1995), Fernandez et al. (1996), Moreno et al. (1996), Connolly et al. (1999), Hilzenrat et al. (1999), Chagneau et al. (2000), Yu et al. (2000), and Sakurabayashi et al. (2002). Dieguez et al. (2002) used a surgical technique based on the development of a triple stenosing ligation to worsen the complications inherent to the prehepatic chronic portal hypertension. Jaffe et al. (1994) and Li et al. (1998) used injection of different sized microspheres into the portal vein of male Wistar rats to induce portal hypertension.

15

Validation of Biomarkers

Clinical medicine is in the midst of a revolution and is continually evolving toward individualized medicine that is based on molecular classification and targeted treatments of diseases. As a result, it has become increasingly desirable to develop new biomarkers with wide clinical applications, including improved resolution of molecular classification of diseases to achieve better precision in diagnosis, monitoring treatment effects or disease progression, and predicting clinical outcomes. Biomarkers are also expected to play an increasingly important role in all phases of drug development and regulatory decision making. Panels of relevant biomarkers may be applicable for predicting the efficacy or safety of a drug and used by regulatory agencies as a basis for approval and market access of drugs.

Recent advances in molecular biotechnology have been one of the major driving forces in transforming the process of biomarker development and its adoption in clinical practice. For example, high-throughput genomic technology, in particular DNA microarray technology, has greatly accelerated the discovery phase of gene expression biomarker development. This technology allows screening of large numbers of genes to identify potential biomarkers whose expression levels are correlated with disease status, clinical outcomes, and treatment regimens. However, before these candidate biomarkers can become useful in clinical settings, they must go through rigorous verification, typically involving both retrospective and prospective clinical studies. These verification studies remain a limiting step in translating biomarker discovery to clinically useful tests in a time- and cost-effective manner. In this chapter, we will focus on the real-time polymerase chain reaction (PCR) technology, in particular TaqMan-based real-time PCR gene expression assays, a simple, robust, and practical tool that has shown great potential in bridging the gaps between biomarker discovery and clinical practice.

REAL-TIME PCR TECHNOLOGY

Quantitative, real-time PCR was first developed in the mid-1990s. This method capitalizes on the fact that there is a quantitative relationship between the amounts of specific transcripts and the amount of PCR products at any given PCR cycle number. For characterizing gene expression, a real-time PCR reaction is typically performed in a two-step reverse transcription-polymerase chain reaction (RT-PCR). In the reverse transcription step, RNA from a biological sample is reverse transcribed into cDNA using a reverse polymerase and random priming, oligo-dT primers or gene-specific primers. The use of gene-specific primers minimizes background priming, whereas the use of random or oligo-dT primers maximize the number of mRNA molecules that can be analyzed from a small sample of RNA. The second step is the polymerase chain reaction. Various methods are available to combine the processes of amplification and detection of the target to permit the monitoring of PCR reactions in real time.

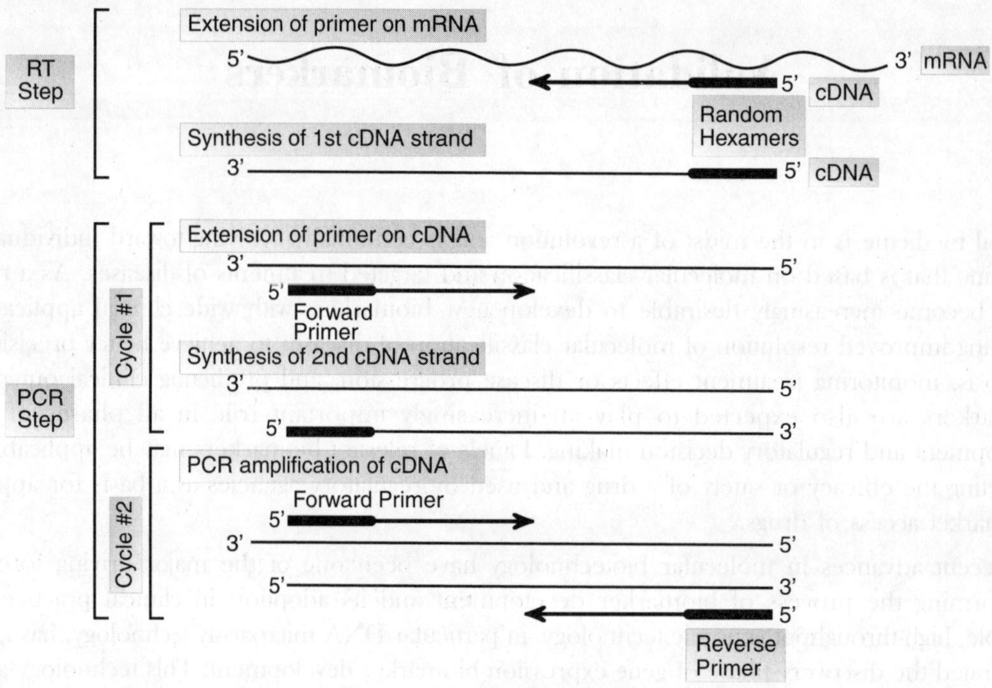

Fig. 15.1. Scheme of two-step real-time PCR reaction.

The simplest method uses fluorescent dyes that bind specifically to double-stranded-DNA, however, the specificity of this method is determined entirely by PCR primers and therefore less optimal. Other methods provide more specificity by using hybridization of fluorescence-labeled probes to the targeted amplicon. These include molecular beacons, hybridization probes, and TaqMan probes that utilize the 5′ nuclease activity of the DNA polymerase to hydrolyze a hybridization probe bound to its target amplicon. We will take the TaqMan-based real-time PCR assay as an example and discuss its chemistry and performance in more detail.

TaqMan Probe

TaqMan probes are single-stranded DNA oligos that contain a reporter dye (6-FAM dye) linked to the 5′ end of the probe and a nonfluorescent quencher (NFQ) at the 3′ end of the probe. When the probe is intact, the proximity of the reporter dye to the quencher results in suppression of the reporter fluorescence, primarily due to Forster-type energy transfer. The nonfluorescent quencher used in TaqMan probes provides greater signal-to-noise ratios and thus increased sensitivity of detection relative to fluorescent dye–quenching adjuncts such as TAMRA. During PCR, the TaqMan probe anneals specifically to a complementary sequence between the forward and reverse primer sites. Only probes that are hybridized to the complementary target are cleaved by the 5′ exonuclease activity of AmpliTaq Gold DNA polymerase. Cleavage separates the reporter dye from the NFQ, resulting in a detectable increase in reporter dye fluorescence at each PCR cycle. This increase in fluorescent signal occurs only if the target sequence is complementary to the probe and is amplified during PCR.

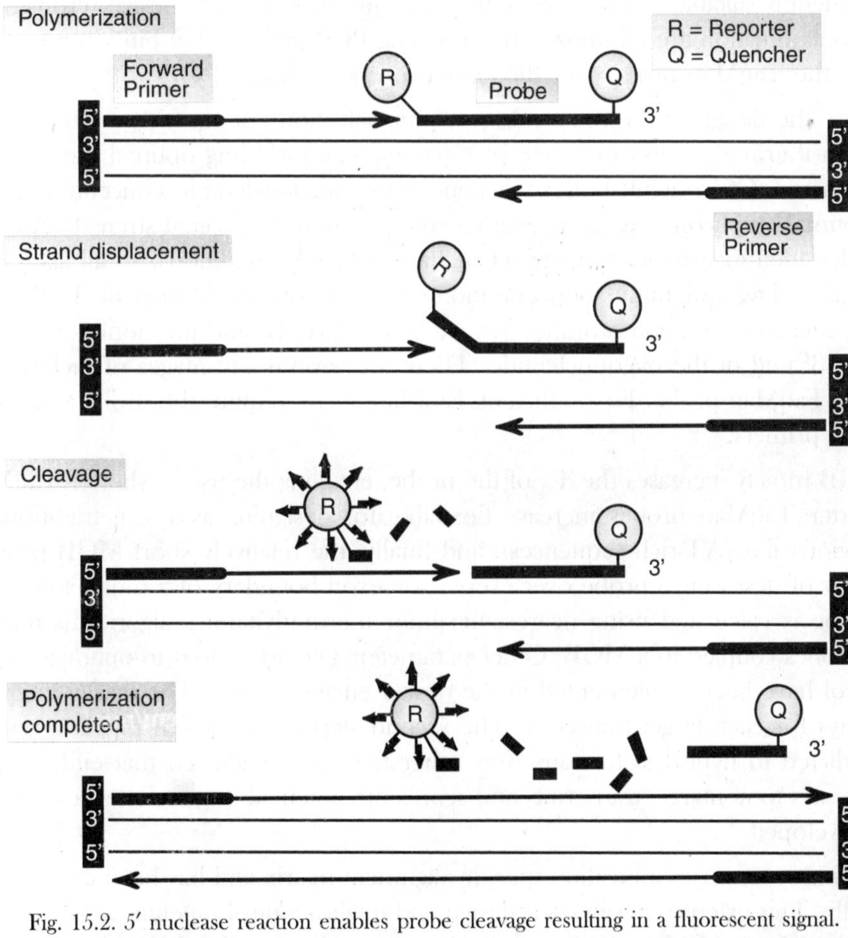

Fig. 15.2. 5′ nuclease reaction enables probe cleavage resulting in a fluorescent signal.

The amount of fluorescence produced from the TaqMan probe is measured at each amplification cycle, providing a real-time measurement of how many amplification targets are produced during each round of PCR. Identification of the PCR cycle when the exponential growth phase is first detectable (C_t) provides highly accurate quantitation of gene expression in the starting samples. The TaqMan real-time PCR assay is sensitive (ability to detect 1 copy in 10 to 100 cells), reproducible (standard deviation <0.25), and has a very large dynamic range (7 to 8 logs).

Specific Assay for Quantification

One of the most critical elements in real-time PCR experimentation is designing highly specific assays for quantifying the transcript of interest. For TaqMan real-time PCR assay, it typically involves design of a single TaqMan probe and two unlabeled oligonucleotide PCR primers. The entire design process is automated using a highly sophisticated assay design pipeline that integrates comprehensive design rules, an extensive array of bioinformatics tools, and assay quality-control strategies. Based on information leveraged from both public and private genome databases, comprehensive sequence analysis is conducted to avoid (mask) SNPs (Single Nucleotide Polymorphisms), repeats, and areas of sequence discrepancy between the public and the private databases to identify suitable locations along mRNA transcripts for assay design. Problems associated with DNA contamination are minimized by designing PCR primers that bind in two distinct exons while placing the TaqMan probe over the exon-exon boundary.

To ensure the design of the most robust assays, selection criteria for primers and probes are based on both thermodynamic and chemistry parameters, including optimal melting temperature (T_m) requirements, GC-content, buffer/salt conditions, oligonucleotide concentrations, secondary structure, optimal amplicon size, primer-dimer minimization, and signal strength. All primers and probes are designed to work in a single set of standard PCR conditions, so all assays can be run with universal cycling and buffer concentrations. In addition, the design of TaqMan probes is improved by incorporating both a minor groove binder (MGB) and the nonfluorescent quencher (NFQ) at the 3′ end of the oligonucleotide. There are several advantages of including the MGB moiety in the TaqMan probe. First, efficient TaqMan assays require the probe to have higher T_m than the PCR primers.

The MGB moiety increases the T_m of the probe, enabling the use of shorter TaqMan probes; second, shorter TaqMan probes increase flexibility for designing assays in traditionally difficult sequence regions (i.e., AT-rich sequences); and finally, the relatively short MGB probes increase the probability of designing a probe over every exon-exon boundary of a multiexon gene. The T_m for each probe is calculated using nearest neighbor thermodynamics algorithms that have been adapted to probes coupled to a MGB. Other parameters that are geared to optimize manufacturing quality control have been implemented in the design engine as well. The design engine generates multiple assays for each target transcript. The second step of the design pipeline removes assays that are predicted to hybridize to transcripts other than the target. To that end, a program that both aligns assays to nontarget transcripts and scores the potential of generating a transcript specific signal was developed.

The scoring system uses a position-specific alignment matrix and has been extensively validated experimentally. This extensively validated design and quality-control pipeline ensures that the assays that are designed are highly specific to their target and also allows the design of assays that discriminate between highly homologous gene families. Currently, there is a comprehensive set of more than 680,000 predesigned TaqMan probe and primer sets available for nearly all exon-exon junctions for transcripts from human, mouse, rat, *Arabidopsis*, *Drosophila*, and *Caenorhabditis elegans.* These predesigned assays are cost-effective and eliminate the labor barriers historically associated with designing and validating individual quantitative RT-PCR assays. The removal of these barriers has dramatically increased the ability of researchers to integrate quantitative real-time PCR technology much earlier and faster in their biomarker screening and validation workflow and greatly reduced the time to results.

Data Processing

Real-time PCR instruments perform data acquisition and analysis by combining amplification, detection, and quantification in an integrated closed-tube system. This avoids problems of contamination and minimizes hands-on time. The entire process can be fully automated, which makes these instruments ideally suited for high-throughput screening applications. For example, the Applied Biosystems 7900HT real-time PCR system can use a robotic plate loading and unloading device to provide a maximum throughput of up to 84 plates at a single time in unattended operation. This instrument uses user-interchangeable thermal cycler blocks that can accommodate 96- and 384-well plates, as well as the new 384- well TaqMan Low Density Array format.

To detect fluorescence signal, samples are illuminated with an argon laser excitation source, and the resulting fluorescent emission is detected by a spectrograph attached to a charge-coupled device (CCD) camera. Emission wavelengths from 500 to 660 nm are monitored allowing the simultaneous detection of multiple fluorophores. Although this system is optimized for TaqMan probe-based and SYBR Green I Dye assay chemistries, it also provides the features and capabilities to accommodate a wide range of other real-time chemistries.

Laboratory Procedures

TaqMan Gene Expression Assays (inventoried: part number 4331182; noninventoried: part number 4351372) consist of a 20X mix of unlabeled PCR primers and TaqMan MGB probe (FAM dye labeled). These assays are designed for the detection and quantitation of specific genetic sequences in RNA samples converted to cDNA. Gene expression quantitation using TaqMan Gene Expression Assays is performed as the second step in a two-step RT-PCR protocol on any ABI PRISM Sequence Detection System Instrument. All TaqMan Gene Expression Assays are optimized to work with either TaqMan Universal PCR Master Mix, No AmpErase UNG (P/N 4324018), TaqMan Universal PCR Master Mix (P/N 4304437), or TaqMan Gene Expression Master Mix Protocol (PN 4371135) and with complementary DNA (cDNA). To prepare the reaction components for a single 20-μL reaction (384- well plate) or a single 50-μL reaction (96-well plate), for singleplex reactions.

How to obtain the PCR reaction mix

1. Prior to use, mix the TaqMan Gene Expression Master Mix thoroughly by swirling the bottle.
2. Calculate the volume of each component needed for all the wells in each assay, based on the number of reactions. Include extra volume to compensate for the volume loss that occurs during pipetting.
3. Cap the tube(s).
4. Vortex the tube(s) briefly to mix the solutions.
5. Centrifuge the tube(s) briefly to spin down the contents and eliminate any air bubbles from the solutions.

How to obtain the PCR reaction plate

1. Transfer the appropriate volume of reaction mixture to each well of a plate:
   (a) MicroAmp Fast Optical 48-Well Reaction Plate: 20 μL per well
   (b) MicroAmp Optical 96-Well Reaction Plate: 50 μL per well
   (c) MicroAmp Fast Optical 96-Well Reaction Plate: 20 μL per well
   (d) MicroAmp Optical 384-Well Reaction Plate: 20 μL per well
2. Cover the plate with MicroAmp Optical Adhesive Film. For MicroAmp 96-Well Reaction Plates, MicroAmp Optical Caps may be used instead. (*Important*: Use a MicroAmp Optical Film Compression Pad with Optical Adhesive Film when using a MicroAmp 96-Well Reaction Plate on the ABI PRISM 7000 Sequence Detection System or the 7900HT Fast Real-Time PCR System.)
3. Centrifuge the plate briefly.

How to obtain the individual tube

1. Set the thermal cycling conditions.
2. In the plate document, select "Standard" mode and enter the correct sample volume (20 μL or 50 μL). (*Important*: Fast thermal cycling conditions are not for use with the TaqMan Gene Expression Master Mix.)

Data analysis the results

Data analysis varies depending on the instrument. Refer to the TaqMan Gene Expression Master Mix Protocol (PN 4371135) and the instrument user guide for information on how to review and adjust threshold values and other details on analyzing your data. Gene expression quantitation using TaqMan Gene Expression Assays should be performed in separate wells (singleplex assay). We recommend that the endogenous control of choice be run in separate wells (singleplex) as this does not require any validation experiments. If performing multiplex experiments, we recommend running multiplex and singleplex assays in parallel to confirm that the C_t values are not affected by multiplex PCR amplification. For additional information regarding relative quantitation of gene expression experiments, refer to the ABI PRISM 7700 Sequence Detection System User Bulletin no. 2 (P/N 4303859).

Storage

Store between -15°C and -20°C; minimize freeze-thaw cycles. The 20X TaqMan Gene Expression Assay may be diluted in TE buffer (Tris-EDTA buffer) (final concentration of TE should be 10 mM Tris-HCl/1 mM EDTA pH 8.0, use RNase-free water).

UTILIZATION

Advances in high-throughput biotechnology, in particular microarray technology, have provided a huge leap forward in the discovery phase of biomarker development. The microarrays' tremendous capability for global gene expression monitoring makes it an ideal starting point for discovering candidate clinical markers and potential therapeutic targets. However, an obvious corollary of increasing the rate at which potential biomarkers are discovered is an increase in the need to validate them in a cost- and time-effective manner.

Biomarker validation generally includes two distinct phases: (i) technical validation to verify candidate biomarkers using an independent technology, and (ii) biological validation to verify the predictive/classification power of the candidate biomarkers in a larger number of samples in both retrospective and prospective clinical studies. The performance capabilities and ease-of-use of real-time PCR chemistries and instrumentation has led to the widespread use of TaqMan Gene Expression Assays as the preferred method for quantifying gene expression as well as for independent validation of candidate biomarkers after microarray analysis.

Validation of Candidate Biomarkers

Although microarrays are an excellent tool for initial target discovery, the reliability of microarray results have been questioned because of the existence of different technologies and nonstandard methods of data analysis and interpretation. To critically examine these issues, large-scale TaqMan Gene Expression Assay-based real-time PCR experimental data sets have been generated and used as the reference to evaluate the performance of various commercial microarray platforms. These studies concluded that microarrays have acceptable sensitivity and accuracy in detecting differential expression, especially for genes with high and medium expression levels and for detecting greater than twofold changes. These studies also characterized some of the limitations of microarrays, in particular the ratio compression phenomena.

Other microarrays limitations revealed by these studies include the significant decrease in overall accuracy of differential expression detection at low expression level and the relatively poor sensitivity in detecting small fold changes (i.e., less than two-fold). Although these limitations have been previously suspected by many, the large-scale TaqMan-based real-time PCR "reference" data set provides a more quantitative view of these limitations for the first time and illustrates why the use of an orthogonal platform such as TaqMan-based real-time PCR to validate data obtained by microarrays is essential. TaqMan real-time PCR assays have been widely applied as an independent validation of candidate biomarkers identified by microarray analysis. For instance, TaqMan Gene Expression Assays have been used to validate potential prognostic biomarkers originally identified with three different microarray platforms. In this study, 20 biopsy samples from early-stage breast carcinomas (T1/T2) were profiled using three different microarray platforms.

Two main clinical-relevant subtypes, luminal A (associated with longer survival) and basal-like (associated with shorter survival), were identified in these carcinomas, substantiating the prognostic value of such expression-defined phenotypes in early breast cancer patients. Of 16,611 common genes among these three platforms, 319 genes were found to be differentially expressed between luminal A and basal-like tumors by all three technologies, which corresponds to an overall consistency of 30% (about 1000 to 1100 genes were identified to be differentially expressed by each platform separately).

A minimal set of 54 genes that best discriminated the two subtypes was identified using the combined data sets generated from the three different array platforms. This set of 54 genes was further validated by TaqMan Gene Expression Assays. Profile correlation analysis showed that the expression profiles across the 20 tumor samples determined by each array platform and by TaqMan assays are highly correlated (median correlation coefficient R > 0.9). The rate of good correlation (Pearson correlation coefficient, R > 0.8) varied from 81% to 88% for the three array platforms. Hierarchical clustering analysis demonstrated not only excellent separation of the luminal A and basal-like subtypes but also that the same tumor samples analyzed by three microarray platforms and TaqMan assays were clustered together by tumor sample rather than by method. Validated by multiple gene expression platforms, this set of 54 predictor genes defines a potential prognostic molecular set for breast cancer.

Validation of the Predictive Value

The second critical aspect of biomarker validation requires validation of the predictive value of the candidate biomarkers from the discovery phase on larger, independent populations of patient samples in a clinical research setting, which usually involves both retrospective study (patients with known outcomes) and prospective study (predict patient outcome). Because clinical studies are often done at multiple sites, a highly robust, easy-to-use, reproducible and standardizable technology is desirable. Difficulties with robustness, reproducibility, and normalization, combined with issues discussed previously, make microarrays a less practical method for this validation phase of biomarker development. TaqMan Gene Expression Assays-based real-time PCR technology is highly reliable, easy to use and interpret, and therefore a much more practical tool for validating biomarkers in clinical research. For example, Lossos et al. demonstrated the use of TaqMan real-time PCR assays for developing gene expression tests that predict survival in diffuse large B-cell lymphoma (DLBCL) patients.

In this retrospective clinical study, gene expression of preselected potential biomarkers was profiled in 66 DLBCL patients with known long-term follow-up histories. A panel of 36 genes was selected as testing targets from several sources of evidence, including reported signal prognostic genes identified for DLBCL and gene lists from three independent microarray studies. Using the TaqMan-based gene expression validation data, the researchers performed a univariate analysis of genes that were correlated with overall survival, either positively or negatively. Six genes were demonstrated to be the strongest predictors: *LMO2, BCL6, FN1, CCND2, CCL3* (also known as *SCYA3*), and *BCL2*. A multivariate model based on the expression levels of the six genes was built and validated by applying it to previously published data from two independent sources. The researchers determined that the six-gene model was highly accurate and independent of the International Prognostic Index (IPI), a well-established predictor of patient outcome. They also further refined the survival predictions of patients within each IPI risk category (low, medium, and high), and by coupling the medium- and high-risk IPI scores with their high-risk expression profile, they were able to identify a group of patients whose survival was especially short (approximately 30% of all patients).

Independently, the genes could not predict how long a patient survived after treatment; however, the six genes taken together were able to predict how long each of the 66 patients survived. This study demonstrated the TaqMan real-time PCR assays to be a more suitable approach for biomarker validation and development in a clinical research setting. To further streamline the work flow of TaqMan real-time PCR and make it a more suitable and robust tool for biomarker development in clinical research setting, TaqMan Low Density Arrays (TLDAs) have been developed to facilitate validation of biomarkers in clinical studies with large cohorts. TLDA is a 384-well micro fluidic card that enables researchers to perform up to 384 simultaneous real-time PCR reactions without the need to use liquid-handling robots or multichannel pipettors. It allows for 1 to 8 samples to be run in parallel against 12 to 384 TaqMan Gene Expression Assay targets that are preloaded into each of the wells on the card. The standardized format allows direct comparison of results between numerous samples from different labs, a critical feature required for clinical testing. The list of genes to be tested by TLDAs is completely customizable making them an ideal solution for biomarker validation in a simple, robust, and standardized format.

Noncoding Regulatory RNAs

MicroRNAs (miRNAs) are an abundant class of endogenous, ~22-nucleotide, noncoding regulatory RNAs that have been shown to have critical functions in a wide variety of biological processes, including development, cell proliferation and death, and oncogenesis. The miRNA profiles have been shown to successfully classify developmental lineage and poorly differentiated tumors, highlighting the potential of miRNA profiling in cancer diagnosis. Furthermore, changes in miRNA gene expression between tumors and their normal counterparts could be exploited, not only for diagnostic and prognostic purposes in the clinic but also to investigate the specific roles of the various miRNAs in tumor development and to identify the genes and pathways that they target. Given that miRNAs regulate gene expression and enter readily into cells, they also have potential as therapeutics to treat cancers and other human diseases. Because of their short size and the sequence similarity between family members, developing sensitive and specific assays for miRNA quantitation is rather challenging. miRNA arrays as well as bead-based approaches have been used for miRNA profiling; however, their detection sensitivity and specificity are typically low.

A novel miRNA quantification method has been developed using stem-loop reverse transcription (RT) followed by TaqMan PCR analysis. Stem-loop RT primers are better than conventional ones in terms of RT efficiency and specificity. These TaqMan miRNA assays are specific for mature miRNAs and discriminate among related miRNAs that differ by as little as one nucleotide. Like standard TaqMan gene expression assays, TaqMan miRNA assays exhibit a dynamic range of seven orders of magnitude. Quantification of five miRNAs in seven mouse tissues showed variation from less than 10 to more than 100,000 copies per cell. This method enables fast, accurate, and sensitive miRNA expression profiling and can identify and monitor potential biomarkers specific to tissues or diseases. We have conducted a study to explore the possibility of using miRNA biomarkers to predict drug potency. A total of 236 TaqMan miRNA assays were used to profile 57 of the NCI-60 cancer cell lines. Using these profiles, linear models were built to predict growth inhibition measurements (GI_{50}) for 119 standard anticancer drugs with which the 57 cancer cell lines were treated. We found that subsets of 7 to 10 miRNAs are sufficient for near optimal prediction (at least 85% accuracy) of drug potency against the NCI-60 cell lines.

Diversified Gene Copies

Gene copy number variation is now recognized as an important type of polymorphism in the human genome. Copy number polymorphisms have been associated with genetic diseases such as cancer, immunologic and neurologic disorders, as well as with variations in drug response. A new application of TaqMan real-time PCR assays has been developed to quantify gene copy number.

Quantitative measurement of gene copy number is achieved by duplexing TaqMan real-time PCR assays, with one FAM dye-based assay designed to detect the genes of interest and one VIC dye-based assay against a reference gene with known two copies per diploid genome (i.e., RNaseP). For example, we have conducted a copy number analysis on five important drug metabolizing enzyme (DME) genes (*CYP2D6, CYP2E1, CYP2A6, GSTM1*, and *GSTT1.*) using TaqMan real-time PCR assays. Survey of 270 DNA samples from the International HapMap Project revealed large variation in gene copy number for these five genes within the HapMap sample panel; furthermore, significant differences in copy number frequency of these genes were found to be associated with different populations. Combined with the TaqMan genotyping assays, the newly developed TaqMan copy number variation assays provide powerful tools to fully characterize genotypes of the genes of interest and their roles in disease or drug response.

16

Heart-Lung Drugs

ISOLATION OF ORGANS

Preparation of Heart-Lung

The isolated heart-lung of the dog was introduced by Knowlton and Starling (1912). Since then, the dog model has been used for many physiological and pharmacological studies. More recently, the rat model has been preferred.

Method

Wistar rats weighing 300-320 g are anesthetized with 50 mg/kg pentobarbitone i.p. Tracheotomy is performed and intermittent positive pressure ventilation is instituted with air. The chest is opened and flooded with ice-cold saline and the heart arrested. Cannulae are inserted into the aorta and the superior (for measurement of central venous pressure) and inferior venae cavae. The heart-lung preparation is perfused with a solution containing rat blood cells from another rat and Krebs-Ringer bicarbonate buffer, with hematocrit and pH of 25% and 7.4, respectively. Concentration of the buffer constituents (mM): NaCl 127, KCl 5.1, $CaCl_2$ 2.2, KH_2PO_4 1.3, $MgSO_4$ 2.6, $NaHCO_3$ 15, glucose 5.5 and heparin. The perfusate pumped from the aorta passes through a pneumatic resistance and is collected in a reservoir maintained at 37°C and then returned to the inferior vena cava. In this model, no other organs except the heart and lung are perfused. Cardiac output is determined by the inflow as long as the heart does not fail. Mean arterial pressure is regulated by the pneumatic resistance. Heart rate is recorded by a bioelectric amplifier and cardiac output is measured with an electromagnetic blood flow meter. Arterial pressure and right atrial pressure are measured with transducers and amplifiers. The heart is perfused initially with cardiac output of 30 ml/min and mean arterial pressure of 80 mm Hg. Test drugs are administered into the perfusate 5 min after start of the experiment.

Assessment

Hemodynamic data within groups are analyzed by two-way analysis of variance (ANOVA) with repeated measures. Recovery time is measured by the Kruskal-Wallis test. The other data are analyzed by one-way ANOVA followed by the Dunnett test for multiple comparisons. Using the Starling heart-lung preparation in dogs, Wollenberger (1947) studied the energy-rich phosphate supply of the failing heart. Shigei and Hashimoto (1960) studied the mechanism of the heart failure induced by pentobarbital, quinine, fluoroacetate and dinitrophenol in dog's heart-lung preparation and effects of sympathomimetic amines and ouabain on it. Imai et al. (1961) used heart-lung preparations of the dog to study the cardiac actions of methoxamine with special reference to its antagonistic action to epinephrine. Capri and Oliverio (1965), Beaconsfield et al. (1974) used the heart-lung preparation of the guinea pig. Robicsek et al. (1985) studied the metabolism and function of an autoperfused heart-lung preparation of the dog.

The *dog heart-lung preparation* was used by Seifen et al. (1988) to study the interaction of a calcium channel agonist with the effects of digoxin, by Somani and Blum (1966) to study blockade of epinephrine- and ouabain-induced cardiac arrhythmias in the dog, by Riveron et al. (1988) to investigate the energy expenditure of an autoperfusing heart-lung preparation, by Namakura et al. (1987) to study the role of pulmonary innervation in an *in situ* lung-perfusion preparation as a new model of neurogenic pulmonary edema, by Hausknecht et al. (1986) to investigate the effects of lung inflation on blood flow during cardiopulmonary resuscitation, by Caffrey et al. (1986) to evaluate the effect of naloxone on myocardial responses to isoproterenol, by Ono et al. (1984) to estimate the cardiodepressant potency of various beta-blocking agents, by Ishikawa et al. (1983) for a graphical analysis of drug effects in the dog heart-lung preparation – with particular reference to the pulmonary circulation and effects of norepinephrine and 5-hydroxytryptamine, by Iizuka (1983) to study the cardiac effects of acetylcholine and its congeners, by Fessler et al. (1988) to investigate the mechanism of reduced LV afterload by systolic and diastolic positive pleural pressure, by Takeda et al. (1973) to study the cardiac actions of oxprenolol. Beaconsfield et al. (1974) used the heart-lung preparation of *guinea pigs* to study the cardiac effect of delta-9-tetrahydrocannabinol. The *rabbit* autoperfusing heart-lung preparation was used by Muskett et al. (1986, 1988). The isolated heart-lung preparation in the *cat* was described by Beaufort et al. (1993). Kontos et al. (1987, 1988) harvested heart-lung blocks from calves.

Work of Langendorff

More than 100 years ago Langendorff (1895) described studies on isolated surviving mammalian hearts using mainly cats as donors. Since then, the method has been improved from the technical site and is nowadays used for studies with guinea-pig, rabbit or rat hearts. In principle, the heart is perfused in retrograde direction from the aorta either at constant pressure or at constant flow with oxygenated saline solutions. Retrograde perfusion closes the aortic valves, just as in the *in situ* heart during diastole. The perfusate is displaced through the coronary arteries flowing off the coronary sinus and the opened right atrium. In this original set-up the ventricles do not fill with perfusate and therefore do not perform pressure-volume work. Parameters usually measured are: contractile force, coronary flow and cardiac rhythm.

Method

Guinea pigs of either sex weighing 300-500 g are sacrificed by stunning. For studies of biochemical parameters in tissue and perfusate, removal of the heart during barbiturate anesthesia and artificial respiration is recommended. The heart is removed as quickly as possible and placed in a dish containing Ringer's solution at 37°C. Associated pericardial and lung tissue are removed. The aorta is located and cut just below the point of its division. A glass or plastic cannula is introduced into the aorta, tied with two threads and perfusion is started with oxygenated Ringer's solution or Krebs-Henseleit buffer. The heart is transferred to a double walled plexiglass perfusion apparatus which is kept at 37°C by the water from a thermostat.

Oxygenated Ringer's solution is perfused at a constant pressure of 40 mm Hg and at a temperature of 37°C from a reservoir. A small steel hook with a string is attached to the apex of the heart. Contractile force is measured isometrically by a force transducer with a preload of 2.5 g and recorded on a polygraph. Coronary flow is measured by a drop counter. Alternatively, flow measurements can be performed using a mechanic-electronic flow meter consisting of a vertical pipe and a magnetic valve. Heart rate is measured through a chronometer coupled to the polygraph. Drugs are injected into the perfusion medium just above the aortic cannula.

Assessment

A reappraisal of the LANGENDORFF heart preparation was given by Broadley (1979) underlining the usefulness to test coronary vasodilating drugs. The value of the LANGENDORFF method can be best assessed by demonstrating a few of its applications in physiology and pharmacology. Direct effects can be measured as well as the antagonism against various physiological and pharmacological agents. A survey on various modifications of the LANGENDORFF-technique and the *isolated working heart preparation* has been given by Ross (1972). Neely et al. (1967) inserted a second cannula into a pulmonary vein or the left atrium. Perfusate from a reservoir flows via this cannula through the mitral valve into the left ventricle. During the systole of the heart, the left ventricle repumps the perfusate through the aorta into the reservoir. The perfusate flowing through the coronary arteries and dripping off from the outside of the heart is collected in a vessel below the heart and recirculated into the reservoir with a roller pump.

Flynn et al. (1978) underlined the difference of this working heart preparation to the original LANGENDORFF method and reported the effects of histamine and noradrenaline on peak left ventricular systolic pressure, contractility, sinus rate, coronary flow, aortic flow, total cardiac output, and external pressure-volume work. Therefore, this method is reported separately. Ishiu et al. (1996) measured simultaneously Ca^{2+}-dependent indo-1 fluorescence and left ventricular pressure on a beat-to-beat basis in Langendorff guinea pig hearts and investigated the changes in Ca^{+2} transient and left ventricular function during positive inotropic stimulation and myocardial ischemia. Hukovic and Muscholl (1962) described the preparation of the isolated *rabbit* heart with intact sympathic nervous supply from the right stellate ganglion. Hendrikx et al. (1994) used the isolated perfused rabbit heart to test the effects of an Na^+/H^+ exchange inhibitor on postischemic function, resynthesis of high-energy phosphate and reduction of Ca^{2+} overload.

Michio et al. (1985) modified the Langendorff method in rabbits to a working heart preparation by cannulating the left atrium. At a pressure of 20 cm H_2O in the left atrium, the heart pumped the solution against a hydrostatic pressure of 100 cm H_2O. Aortic flow, systolic aortic pressure, coronary flow, and heart rate were measured. The influence of an ACE-inhibitor on heart rate, lactate in the coronary effluate and GTP-level in the myocardium after 60-min hypothermic cardiac arrest was studied in working heart preparation of rabbits by Zegner et al. (1996). Gottlieb and Magnus (1904) introduced the so called "*balloon method*".

A small balloon fixed to the tip of a catheter is filled with water and inserted into the left ventricle via one of the pulmonary veins, the left atrium and the mitral valve. The balloon size has to fit the volume of the left ventricle and therefore its size depends on the animal species and body weight. The catheter can be fixed by tying the pulmonary vein stems. Via a three-way valve, the balloon can be extended to a given preload. The beating heart now exerts a rhythmic force to the balloon and thus to the membrane of a pressure transducer. The advantages of this method are that force development and preload can be stated reproducibly in pressure units [mm Hg], left ventricular contraction curves can be used for further calculations, and continuous heart rate recordings can be carried out without any problems when using a rate meter. Sakai et al. (1983) reported a similar method adapted to *mice*. Bardenheuer and Schrader (1983) described a method whereby the balloon is inserted into the left ventricle as described above.

However, isovolumetric pressure in the left ventricle is not measured. Instead, the fluid in the balloon is pumped through the cannula into a closed extra corporal circulation. The fluid is forced into one direction by 2 recoil valves. The balloon is made of silicone material using a Teflon form. The dimensions of the form are derived from casts of the left ventricle of K^+-arrested heart by injection of dental cement. During each heart beat the fluid volume expelled from the balloon corresponding to the stroke volume of the heart, can be recorded by means of a flow meter probe and an integrator connected in series. Preload and afterload can be adjusted independently from each other. The perfusate flow (retrograde into the aorta and through the coronary arteries) is recorded separately. The following parameters were measured in isolated *rat* hearts:

1. LVP (left ventricular pressure) with Statham pressure transducer P 23DB, which on differentiation yielded LV dp/dt_{max} and HR (heart rate). Cardiac output and coronary flow (CF) are determined by electromagnetic flow probes in the outflow system and in the aortic cannula, respectively. Coronary venous pO_2 is measured with a catheter placed in the pulmonary artery by a type E 5046 electrode connected to a PMH 73 pH/blood gas monitor (Radiometer). An epicardial electrocardiogram recording is obtained via two silver electrodes attached to the heart. All parameters are recorded on a Brush 2600 recorder.
2. Myocardial oxygen consumption (MWO_2) [ml/min/g wet weight] is calculated according to the equation:

$$MVO_2 = CF \times (P_a - P_v) \times (c/760) \times 100$$

where CF is the coronary flow [ml/min/g], P_a is the oxygen partial pressure of arterial perfusate (650 mm Hg), P_v is the oxygen partial pressure of the venous effluent perfusate [mm Hg], and c is the 0. 0227 ml O_2/ml perfusate representing the Bunsen solubility coefficient of oxygen dissolved in perfusate at 37°C.

For the determination of lactate dehydrogenase (LDH) and creatine kinase (CK) activities in the perfusate, samples are taken from the coronary effluent. After the experiments, hearts are rapidly frozen in liquid nitrogen and stored at -80°C. Of the left ventricle, 500 mg are taken, put into 5 ml ice-cold $HClO_4$ and disrupted with an Ultra-Turrax. Glycogen is hydrolyzed with amyloglycosidase (pH 4.8) and determined as glucose. Furthermore, ATP and creatine phosphate are measured. Avkiran and Curtis (1991) constructed a dual lumen aortic cannula which permits independent perfusion of left and right coronary beds in isolated rat hearts without necessitating the cannulation of individual arteries. Igic (1996) described a modification of the isolated perfused working rat heart. A special double cannula was designed consisting of an outer cannula that is inserted into the aorta and an inner cannula that is advanced into the left ventricle.

The perfusion fluid flows through the inner cannula into the left ventricle, and is ejected from there into the aorta. If the outer cannula system is closed, the fluid perfuses the coronary vessels and drips off outside the heart. When the outer cannula is open and certain pressure resistance is applied, a fraction of the ejected fluid perfuses coronary vessels and the rest is expelled. Because the inner cannula can be easily retracted into the outer cannula, which is placed in the aorta, the preparation provides an opportunity to use the same heart as a "working" or "non-working" model for investigating functions of the heart. By labeling glucose, lactate, or fatty acids in the perfusate with ^{3}H or ^{14}C, Barr and Lopaschuk (1997) directly measured energy metabolism in the isolated rat heart.

Krzeminski et al. (1991) described a new concept of the isolated heart preparation with on-line computerized data evaluation. Left ventricular pressure was recorded by means of a balloon-catheter, while special suction electrodes obtained the high-amplitude, noise- free electrogram recordings. The coronary effluent partial pressure of oxygen was continuously monitored, which enabled the calculation of myocardial oxygen consumption (MVO_2). The effluent partial pressure of carbon dioxide and pH value were also measured simultaneously. A computerized system of data acquisition, calculation, storage, and end report was described. Doring and Dehnert (1988) described continuous simultaneous *ultrasonic recording* of two cardiac diameters in an isolated perfused guinea-pig heart. For the measurement of the left ventricular transversal diameter the ultrasonic transmitter was positioned at the epicardium at the largest cardiac diameter.

The corresponding ultrasonic receiver was inserted through the right atrium into the right ventricle to approximately the same height as the transmitter. In the right ventricle, which is empty in the isolated perfused LANGENDORFF-heart, it was automatically positioned opposite to the transmitter. Additional transducers were placed both at the heart's base and apex for assessment of the ventricular longitudinal diameter. Several authors used the isolated perfused mouse heart. Bittner et al. (1996) described a work-performing heart preparation for myocardial performance analysis in murine hearts using a modified Langendorff apparatus. Sumaray and Yellon (1998a, b) constructed a specially designed Langendorff apparatus that allows perfusion of the isolated *mouse* heart. These authors reported that ischemic preconditioning reduces infarct size following global ischemia in the murine myocardium.

Brooks and Apstein (1996) measured left ventricular systolic and diastolic pressures in the isovolumically contracting (balloon in the left ventricle) mouse hearts. Sutherland et al. (2003) reviewed characteristics and cautions in the use of the isolated perfused heart of mice. Wang et al. (2001) studied the relationship between ischemic time and ischemia/reperfusion injury in isolated Langendorff-perfused mouse hearts. Tejero-Taldo et al. (2002) reported that α-adrenergic receptor stimulation produces late preconditioning through inducible nitric oxide synthase in mouse heart. Ross et al. (2003) found that the α_{1B}-adrenergic receptor decreases the inotropic response in the mouse Langendorff heart model.

Bratkovsky et al. (2004) measured coronary flow reserve in isolated hearts from mice. Plumier et al. (1995) generated transgenic mice expressing the human heart heat shock protein 70. Upon reperfusion of the hearts after 30 min of ischemia in the Langendorff preparation, transgenic hearts versus non transgenic hearts showed significantly improved recovery of contractile force. Hannan et al. (2000) compared ENOS knockout and wild-type mouse hearts which were perfused in a Langendorff apparatus with Krebs bicarbonate buffer and subjected to 20 min of global normothermic ischemia followed by 30 min of reperfusion. Myocardial function was measured using a ventricular balloon to determine time to onset of contraction, left ventricular developed pressure (LVDP), left ventricular end-diastolic pressure (LVEDP), and heart rate-pressure product (RPP). Sheikh et al. (2001) generated transgenic mice over-expressing fibroblast growth factor (FGF)-2 protein in the heart. An isolated mouse heart model of ischemiareperfusion injury was used to assess the potential of endogenous FGF-2 for cardioprotection.

Clinical Effects

Positive inotropic effects on heart

While negative inotropic substances can be tested in a heart beating with normal force, the evaluation of a positive inotropic compound usually requires that cardiac force is first reduced. Acute experimental heart failure can be induced by an overdose of barbiturates, such as sodium thiopental, or calcium antagonists. This kind of cardiac failure can be reversed by β-sympathomimetic drugs, cardiac glycosides, or increased Ca^{+2} concentration. In this way, the potential β-sympathomimetic activity of a new drug can be measured using isoproterenol as standard. After thiopental-Na treatment, left ventricular pressure (LVP) and dp/dt_{max} decrease considerably, whereas coronary flow is slightly enhanced. β-Sympathomimetic drugs restore LVP and dp/dt_{max} and keep coronary blood flow elevated. Cardiac glycosides increase LVP and dp/dt_{max} and leave coronary flow unchanged.

Negative inotropic effects on heart

The effects of a β-sympathomimetic drug such as isoproterenol at doses of 0.05 to 0.2 μg increasing contractile force as well as heart frequency are registered. After injection of a β-blocker, the effects of isoproterenol are attenuated. The effects of a potential β-blocking agent can be tested comparing the isoproterenol inhibition versus a standard such as propranolol (0.1 mg).

Dilating effect on coronary vessel

The LANGENDORFF heart has been extensively used for assessing the coronary dilating activity of drugs. Rothaul and Broadley (1982) demonstrated the release of coronary vasodilator mediators from guinea pig isolated hearts by a technique employing donor and recipient hearts in series.

Calcium-antagonism

In order to demonstrate the effect of calcium-antagonists, 1 to 5 mg $BaCl_2$ are injected which induce a pronounced spasm of the coronary arteries thereby reducing the coronary flow. Five min later, the test drug is injected. Active compounds have a relaxing effect on coronary arteries indicated by an increase of coronary flow. After this effect has weaned, $BaCl_2$ is injected again and the test drug or a standard drug, e. g. nifedipine, is tested. The increase of coronary flow is expressed as percentage of flow during $BaCl_2$ spasm and compared with the effect of the standard. Using various doses, dose-response curves can be established.

Effect on potassium outflow

Lindner and Hajdu (1968) described a method using the LANGENDORFF heart in which contractile force, coronary flow, and the potassium content in the coronary outflow was determined by flame photometry. Increase in potassium outflow correlates well with the positive inotropic effect.

Hypoxic damage

Lindner and Grotsch (1973) measured the enzymes creatine phosphokinase (CPK), lactate dehydrogenase (LDH), α-hydroxybutyrate dehydrogenase (α-HBDH), and glutamicoxalacetic transaminase (GOT) in the effluent of a guinea pig heart preparation under varying degrees of hypoxia. Potassium content and oxygen tension in the inflowing and outflowing solution were determined. The heart rate, the amplitude of contraction and the rate of coronary vessel perfusion were recorded additionally.

Metabolic observation

Using ^{31}P, studies on metabolism of nucleotides and phosphorylated intermediates of carbohydrates in isolated hearts have been performed.

Influence of compounds on cardiac rhythm

The LANGENDORFF heart preparation is also used to test the influence of compounds on cardiac rhythm. For recording monophasic action potentials, suction electrodes are applied on the heart. Ventricular fibrillation can be induced by simultaneous injection of digitoxin (12.5-25.0 μg) and aconitine (12.5-25.0 μg) into the perfusion fluid. Cardiac glycosides shorten the refractory period, decrease the conduction velocity and increase heterotopic stimulus generation. Aconitine increases markedly heterotopic stimulus generation. Both compounds together induce invariably ventricular fibrillation. Anti-arrhythmic compounds can be tested in this way. Fibrillation is inhibited, at least partially, by 20 μg prenylamine, 10-20 μg quinidine or 20 μg ajmaline. Takeo et al. (1992) described protective effects of anti-arrhythmic agents on oxygen-deficiency-induced contractile dysfunction of isolated perfused hearts.

Hypoxia in isolated rabbit hearts was induced by perfusing the heart for 20 min with Krebs-Henseleit buffer saturated with a gas mixture of 95% N_2 and 5% CO_2 containing 11 mM mannitol. After hypoxic perfusion, the heart was reoxygenated for 45 min with oxygenated buffer containing glucose. Dhein et al. (1989) studied the pathway and time course of the epicardial electrical activation process by means of a computer-assisted epicardial potential mapping, using a matrix of 256 unipolar AgCl electrodes (1 mm spatial and 0.25 ms temporal resolution) in isolated rabbit hearts perfused according to the Langendorff technique. From the activation times of the surrounding electrodes, the direction and velocity of activation for each electrode were calculated, thereby allowing construction of an epicardial vector field. The method was used for the assessment of arrhythmogenic and anti-arrhythmic drug activity.

Ventricular fibrillation

Ventricular fibrillation can be induced in the LANGENDORFF preparation by reducing the glucose content of the perfusion medium to 0.25 g/1000 ml and the KCl content to 0.12 g/1000 ml. After a perfusion period of 20 min, 10 μg epinephrine are injected into the perfusion cannula. Immediately afterwards, the heart is stimulated with a current of 40Hz and 5 mA for 2 min. This procedure is repeated every 10min. Standard conditions are achieved when the fibrillation continues without further electrical stimulation. Hearts treated in this way serve as controls. Other hearts stimulated in the same way are treated with continuous infusion of the test drug or the standard via the perfusion medium. Differences in the incidence of fibrillations are calculated using the χ^2 test.

Modified technique

Balderston et al. (1991) modified the Langendorff technique in rabbit hearts in order to perform electrophysiologic studies. His bundle electrograms were measured with a plunge electrode and allowed atrioventricular nodal physiology to be evaluated directly. Atrial conduction and refractoriness, atrioventricular node conduction and refractoriness, His-Purkinje conduction, and ventricular conduction and refractoriness could be accurately measured. The effects of verapamil and flecainide were described.

Effects of isolated platelets

Lamontagne et al. (1992) isolated platelets from blood of healthy human donors and injected platelets boluses into the perfusion line of the Langendorff preparation of a rabbit heart. In the effluent cyclic GMP was determined as an index for EDRF release.

Ligation of Coronary Artery

In working heart preparations of rats, ischemia can be induced by clamping the left coronary artery close to its origin. After removal of the clip, changes in the reperfusion period can be observed. Prevention of these symptoms can be an indicator of the efficacy of coronary drugs.

Method

The preparation used is a modification of an isolated working heart preparation originally used for guinea pig hearts. Wistar rats of either sex weighing 280-300 g are sacrificed by decapitation. The hearts are removed and dissected free from the epicard and surrounding connective tissue. A cannula is introduced into the aorta from where the coronary vessels are perfused with the non-recirculated perfusion medium according to the Langendorff technique. In the left ventricle a balloon closely fitting the ventricular cavity is placed and connected to an artificial systemic circulation. The fluid in the balloon is pumped through a cannula into the closed extra corporal circulation being forced into one direction by 2 recoil valves. The balloon is made of silicone material using a Teflon form. The dimensions of the form are derived from casts of the left ventricle of K^+-arrested heart by injection of dental cement.

During each heart beat the fluid volume pressed from the balloon, corresponding to the stroke volume of the heart, can be recorded by means of a flow meter probe and an integrator connected in series. Preload and afterload can be adjusted independently from each other. The perfusate flow (retrograde into the aorta and through the coronary arteries) is recorded separately. The following parameters were measured in isolated rat hearts: LVP (left ventricular pressure) with Statham pressure transducer P23 DB, which on differentiation yielded LV dp/dt$_{max}$ and HR (heart rate). Cardiac output and coronary flow (CF) are determined by electromagnetic flow probes in the outflow system and in the aortic cannula, respectively.

Coronary venous pO_2 is measured with a catheter placed in the pulmonary artery by a type E 5046 electrode connected to a PMH 73 pH/blood gas monitor (Radiometer). An epicardial electrocardiogram recording is obtained via two silver electrodes attached to the heart. All parameters are recorded on a Brush 2600 recorder. Myocardial oxygen consumption (MWO_2) [ml/min/g wet weight] is calculated according to the equation:

$$MVO_2 = CF \times (P_a - P_v) \times (c/760) \times 100$$

where CF is the coronary flow [ml/min/g], P_a is the oxygen partial pressure of arterial perfusate (650 mm Hg), P_v is the oxygen partial pressure of the venous effluent perfusate [mm Hg], and c is the 0.0227 ml O_2/ml perfusate representing the Bunsen solubility coefficient of oxygen dissolved in perfusate at 37°C.

Occlusion experiment

For coronary artery occlusion experiment, the isolated working hearts are perfused for a period of 20 min (pre-ischemic period) with modified Krebs-Henseleit buffer at a constant pressure of 65mm Hg. Thereafter, acute myocardial ischemia is produced by clamping the left coronary artery close to is origin for 15 min (ischemic period). The clip is then reopened, and changes during reperfusion are monitored for 30 min (reperfusion period). After coronary artery ligation and reperfusion the hearts develop ventricular fibrillation. From the coronary effluent samples are taken for lactate, lactate dehydrogenase (LDH), and creatine kinase (CK) determinations. After the experiment, glycogen, lactate, ATP, and creatine phosphate in myocardial tissue are measured. The test drugs are given into the perfusion medium either before occlusion or 5 min before reperfusion. For *ex vivo* studies, the rats are treated orally with the test drug 1 h before sacrifice and preparation of the isolated working heart.

Assessment

The incidence and duration of ventricular fibrillation after treatment with coronary drugs is compared with controls. Left ventricular pressure, LV dP/dt max, and coronary flow are reduced after coronary constriction by angiotensin II, whereas enzyme activities in the effluent are increased and the myocardial content of glycogen, ATP and creatine phosphate are decreased. Cardiac protective drugs have the opposite effects. The values of each parameter are statistically compared with controls. Vogel and Lucchesi (1980) described an isolated, blood perfused, *feline* heart preparation for evaluating pharmacological interventions during myocardial ischemia. Ventricular function was measured with a fluid-filled latex balloon within the left ventricle. Vleeming et al. (1989) ligated the left coronary artery in rats after thoracotomy in ether anesthesia. Forty-eight hours after the operation, the hearts were prepared for retrograde constant pressure perfusion, according to the Langendorff technique. Igic (1996) presented a new method for the isolated working rat heart. A special double cannula was designed consisting of an outer cannula that is inserted in the aorta and an inner cannula that is advanced into the left ventricle. The perfusion fluid flows through the inner cannula into the left ventricle, and is ejected from there into the aorta. If the outer cannula system is closed, the fluid perfuses the coronary vessels and drips off outside the heart.

When the outer cannula is open and certain pressure resistance is applied, a fraction of the ejected fluid perfuses the coronary vessels and the rest is expelled. Because the inner cannula can easily be retracted into the outer cannula, which is placed in the aorta, this preparation provides an opportunity to use the same heart as a "working" or "non-working" model for investigating functions of the heart. Pepe and McLennan (1993) described a maintained afterload model of ischemia in erythrocyte-perfused isolated working hearts of rats. Further characterization of the pathophysiological reactions or the isolated working heart was performed by Linz et al. (1999). The external heart power (EHP) [mJ/min/g] was calculated using the formula:

$$EHP_{LV} = \text{pressure} - \text{volume} + \text{acceleration work}$$
$$= [SV(MAP - LAP)]$$
$$+[1/2\, SV \times d \times (SV / \pi r^2 e^2)] HR g^{-1}_{LVwwt}$$

SV indicates stroke volume; *MAP*, mean aortic pressure; *LAP*, mean left arterial pressure; *d*, specific weight perfusate (1.004 g/cm^3); *r*, inner radius of aortic cannula; *e*, ejection time; *HR*, heart rate; LV, left ventricle; LVwwt, left ventricular wet weight.

The function of the left ventricle was altered by changing the aortic pressure (afterload) at constant left atrial filling load (preload). By adjusting the Starling resistance, the aortic outflow could be switched during 1 min from the fixed baseline afterload to a preset higher afterload producing step-wise rises in mean arterial pressure. Lee et al. (1988) studied the effects of acute global ischemia on cytosolic calcium transients in perfused isolated *rabbit* hearts with the fluorescent calcium indicator indo 1. Indo 1-loaded hearts were illuminated at 360 nm, and fluorescence was recorded simultaneously at 400 and 550 nm from the epicardial surface of the left ventricle. The F_{400}/F_{550} ratio was calculated by an analog circuit, which allowed cancellation of optical motion artifact. The resulting calcium transients were registered simultaneously with the ventricular pressure and demonstrated a rapid upstroke and slow decay similar to those recorded in isolated ventricular myocytes. Global ischemia rapidly suppressed contraction, but it produced a concurrent increase in the systolic and diastolic levels of calcium transients, together with an increase in the duration of the peak.

Itter's Experiment

The model of chronic heart failure in spontaneously hypertensive rats described by Itter et al. (2004a) has been used by the same group to study the isolated working heart in rats after chronic infarction.

Method

WKY/NHsd and SHR/NHsd rats at an age of 4 months were randomized into two groups – sham and myocardial infarction (MI). The sham procedure consisted of opening the pericardium and placing a superficial suture in the epicardium of the left ventricle (LV). Chronic heart failure

(CHF) was induced by permanent (8 weeks) occlusion of the left coronary artery 2 mm distal to the origin from the aorta resulting in a large infarction of the free left ventricular wall. Eight weeks after surgery, parameters indicating CHF were measured. Cardiac hypertrophy, function and geometric properties were determined by the "working heart" mode and *in vivo* determinations by MRI and heart weight.

Operative technique

The rats were anesthetized with a mixture of ketamine/xylazine (35/2 mg/kg i.p.). The left ventrolateral thorax was shaved and prepared to create a disinfected surgical access area. When stable anesthesia was achieved the animals were placed on a small animal operation table, intubated and ventilated with room air using a small animal ventilator. The level of anesthesia was deemed as adequate following loss of the pedal withdrawal reflex and absence of the palpebral reflex. Reflexes were evaluated before surgery. The operation took 5 min. The tidal volume was adjusted at 3-5 ml and the ventilation rate was 40 breaths/min. Left thoracotomy was performed via the third intercostal space. The heart was exposed and the pericardium opened. The left main coronary artery was ligated with Perma-Hand silk 4-0 USP near its origin at the aorta (2 mm distal to the edge of the left atrium).

Ligation resulted in infarction of the free left ventricular wall. Ligation was deemed successful when the anterior wall of the left ventricle turned pale. At this point the lungs were hyperinflated by increasing the positive end-expiratory pressure, and the chest was closed. The rats were placed on a heating pad and covered with a layer of unbleached tissue paper. The rats were extubated following return of reflexes. They were continuously monitored until they started moving in their cages. To avoid ventricular arrhythmias, lidocaine (2 mg/kg i.m.) was given before surgery. To prevent acute lung edema, the rats received furosemide (Lasix, 2 mg/kg body weight twice daily for 3 days) via the drinking water. To avoid pain and distress the rats received metamizol treatment (Novalgin, 0.1 mg/kg body weight i.m.) once, directly after the recovery period.

Before killing the animals 8 weeks after MI, non-invasive sequential nuclear magnetic resonance (NMR) measurements of heart geometric properties were done. Thereafter the animals were anesthetized with pentobarbitone (180 mg/kg i.p. Pentobarbital) and subsequently heparinized (Heparin Natrium 500 I.U./100g body weight i.p.). Once stable anesthesia was achieved (stage III 3, reflexes absent), the animals were connected to an artificial respirator via a PE (polyethylene) tube inserted into the trachea and ventilated with room air. A transverse laparotomy and a right anterolateral thoracotomy were performed, and the heart was rapidly removed for the evaluation of its function in the working heart mode. The heart was immersed in physiological buffer chilled to 4°C. The aorta was dissected free and mounted onto a cannula (internal diameter: 1.4 mm) attached to perfusion apparatus.

The hearts were perfused according to the method of Langendorff with an oxygenated (95% O_2/5% CO_2) non-circulating Krebs-Henseleit solution of the following compositions (mM): NaCl, 118; KCl, 4.7; $CaCl_2$, 2.52; $MgSO_4$, 1.64; $NaHCO_3$, 24.88; KH_2PO_4, 1.18; glucose, 5.55; and Na pyruvate, 2.0 at a perfusion pressure of 60 mmHg. Any connective tissue, thymus or lung was carefully removed. A catheter placed into the pulmonary artery drained the coronary effluent perfusate that was collected for the determination of coronary flow and venous pO_2 measurements. The left atrium was cannulated via an incision of the left auricle. All pulmonary veins were ligated close to the surface of the atria. When a tight seal with no leaks had been established and after a 15-min equilibration period, the hearts were switched into the working mode, using a filling pressure (preload) of 12 mmHg in WKY/NHsd and 18 mmHg in SH rats.

The afterload pressure was 60 mmHg in WKY/NHsd and 80 mmHg in SH rats. After validation of the basis parameters the afterload pressure was enhanced in a cumulative manner from an additional 20 mmHg to 140 mmHg. Thereafter the isovolumetric maxima were determined by enhancing the preload pressure in steps of 5 mmHg to 30 mmHg. Flow and pressure signals for computation were obtained from the PLUGSYS-measuring system. Computation of data was performed with a sampling rate of 500 Hz, averaged every 2 s, using the software Aquire Plus V1.21f.

Investigating infarct size

After the evaluation of the external heart work, the total heart weight, and the left and right ventricular weights were determined. The left ventricle was then sectioned transversely into four slices from the apex to the base. Eight pictures were taken of each rat heart, two from each slice. Total infarct size was determined by planimetry of the projected and magnified slices. The areas of infarcted tissue as well as the intact myocardium of each slice were added together and averaged. The infarcted fraction of the left ventricle was calculated from these measurements and expressed as a percentage of the left ventricular mass. The left ventricular perimeter, diameter, infarct scar length, as well as wall thickness and infarct wall thinning were determined as well.

Assessment

The data are given as mean ± SEM. Statistics were performed using the SAS system statistics package with a sequential rejection *t*-test according to Holm (1979).

Bovine Coronary Artery

Eicosanoids can regulate the tonus of coronary arteries. Prostacyclin induces relaxation, whereas thromboxane A_2 causes contraction. Spiral strips from bovine coronary artery can be used for assaying relaxation activity of test compounds.

Method

Freshly slaughtered beef hearts are immersed in cold oxygenated Krebs solution and immediately transported in a thermos flask to the laboratory. The left descending coronary artery and several of its primary branches are cut into spiral strips (about 20 mm long and 2-3 mm wide). The specimens can be stored up to 48 h at 4°C. The artery strips are suspended in a 4 ml organ bath under an initial tension of 2 g and immersed in a Krebs' bicarbonate solution at 37°C being gassed with oxygen containing 5% CO_2 throughout the experiment.

The Krebs solution contains a mixture of antagonists to inhibit any actions from endogenous acetylcholine, 5-hydroxytryptamine, histamine or catecholamines (hyoscine hydrobromide 10^{-7} g/ml, methysergide maleate 2×10^{-7} g/ml, mepyramine maleate 10^{-7} g/ml, propranolol hydrochloride 2×10^{-6} g/ml). The strips are superfused with a solution of the test compounds in concentrations of 0.01, 0.1, 1.0 μg/ml at a rate of 10-20 ml/min with oxygenated Krebs solution containing the mixture of antagonists. Isometric contractions are recorded with Grass force-displacement transducers (type FT 03C) on a Grass polygraph. The strips are superfused with Krebs' solution 3 h prior to the experiment. Standard compounds are 100 ng/ml PGE_2 inducing contraction and 100 ng/ml PGI_2 inducing pronounced relaxation.

Assessment

The relaxation induced by the test compound is expressed as percentage of maximal response to 100 ng/ml PGI_2. Campell and Paul (1993) measured the effects of diltiazem on isometric force generation, $[Ca^{2+}]i$, and energy metabolism in the isolated *porcine* coronary artery. Li et al. (1997) determined the ability of analogues of human α-calcitonin gene-related peptide to relax isolated porcine coronary arteries precontracted with 20 mM KCl.

IN VIVO MYOCARDIAL NECROSIS

Cardiac necrosis can be produced by injection of natural and synthetic sympathomimetics in high doses. Infarct-like myocardial lesions in the rat by isoproterenol have been described by Rona et al. (1959). These lesions can be totally or partially prevented by several drugs such as sympatholytics or calcium-antagonists.

Method

Groups of 10 male Wistar rats weighing 150-200 g are pretreated with the test drug or the standard either s.c. or orally for 1 week. Then, they receive 5.25 and 8.5 mg/kg isoproterenol s.c. on two consecutive days. Symptoms and mortality in each group are recorded and compared with those of rats given isoproterenol alone. Forty-eight hours after the first isoproterenol administration, the rats are sacrificed and autopsied. The hearts are removed and weighed, and frontal sections are embedded for histological examination.

Assessment

Microscopic examination allows the following grading:

Grade 0: no change

Grade 1: focal interstitial response

Grade 2: focal lesions in many sections, consisting of mottled staining and fragmentation of muscle fibres Grade 3: confluent retrogressive lesions with hyaline necrosis and fragmentation of muscle fibres and sequestrating mucoid edema

Grade 4: massive infarct with occasionally acute aneurysm and mural thrombi

For each group the main grade is calculated with the standard deviation to reveal significant differences. The test has been used by many authors for evaluation of coronary active drugs, such as calcium-antagonists and other cardioprotective drugs like nitroglycerin and molsidomine. Yang et al. (1996) reported a protective effect of human adrenomedullin, a C-terminal fragment of adrenomedullin on the myocardial injury produced by subcutaneous injection of isoproterenol into rats.

Myocardial Infarction

Ligation of the left coronary artery in rats as described by Selye (1960) induces an acute reduction in pump function and a dilatation of left ventricular chamber. The method has been used to evaluate beneficial effects of drugs after acute or chronic treatment.

Method

Male Sprague Dawley rats weighing 200-300g are anesthetized with diethylether. The chest is opened by a left thoracotomy, and a thread is inserted near the middle of the lateral margin of the cutaneous wound and carried through a tunnel of the left pectoral muscle around the cranial half of the incision. The heart is gently exteriorized by pressure on the abdomen. A ligature is placed around the left coronary artery, near its origin, and is tightened. Within seconds, the heart is repositioned in the thoracic cavity, and the ends of the musculocutaneous thread are tightened to close the chest wall and enable the animal to breathe spontaneously. The speed of the procedure renders mechanical respiration unnecessary. To evaluate drug effects, the rats are treated 5 min after and 24h after occlusion by subcutaneous injection (standard 5mg/kg propranolol).

Two days after surgery, the rats are anesthetized with 60 mg/kg i.p. pentobarbital and the right carotid artery is cannulated with a polyethylene catheter connected to a pressure transducer. The fluid-filled catheter is then advanced into the left ventricle through the aortic valve for measurement of left ventricular systolic and end-diastolic pressure. After hemodynamic measurements, the heart is arrested by injecting 2 ml of 2.5 M potassium chloride. The chest is opened, and the hearts are isolated and rinsed with 300 mM KCl to maintain a complete diastole. A double-lumen catheter is advanced into the left ventricle through the ascending aorta, the right and left atria are tied off with a ligature, and the right ventricle is opened.

The left ventricular chamber is filled with a cryostatic freeze medium through the smaller of the two catheter lumens and connected to a hydrostatic pressure reservoir maintained at a level corresponding to the end-diastolic pressure measured *in vivo*. The outlet (larger lumen) is then raised to the same level as the inlet to allow fluid in the two lumens to equilibrate. The heart is rapidly frozen with hexane and dry-ice. The hearts are serially cut with a cryostat into 40-µm-thick transverse sections perpendicularly to the longitudinal axis from apex to base. At a fixed distance, eight sections are obtained from each heart and collected on gelatin-coated glass slides. Sections are air-dried and incubated at 25°C for 30 min with 490 µM nitroblue tetrazolium and 50 mM succinic acid in 0.2M phosphate buffer (pH7.6), rinsed in cold distilled water, dehydrated in 95% ethyl alcohol, cleared in xylene, and mounted with a synthetic resin medium. Viable tissue appears dark blue, contrasting with the unstained necrotic tissue.

Assessment

The infarct size can be determined by planimetry and expressed as percentage of left ventricular area, and thickness can be expressed as percentage of noninfarcted ventricular wall thickness. An automatic method for morphometric analysis with image acquisition and computer processing was described by Porzio et al. (1995). Myocardial infarction following coronary artery ligation in Sprague-Dawley rats is a widely used rat model of heart failure. If the left coronary artery is not completely ligated, heart failure may occur as a consequence of chronic myocardial ischemia. Johns and Olson

(1954) described the coronary artery patterns for mouse, rat, hamster and guinea pig. Kaufman et al. (1959), Fishbein et al. (1978, 1980) used various histochemical methods for identification and quantification of border zones during the evolution of myocardial infarction. Sakai et al. (1981) described an experimental model of angina pectoris in the intact anesthetized rat. In anesthetized rats the tip of a special carotid cannula was placed closely to the right and left coronary ostium. Single intra-aortic injections of methacholine or acetylcholine (in the presence of physostigmine) developed a reproducible elevation of the ST segment and the T wave of the electrocardiogram. Coronary drugs were tested to prevent these changes.

Ytrehus et al. (1994) analyzed the effects of anesthesia, perfusate, risk zone, and method of infarct sizing in rat and rabbit heart infarction. Lepran et al. (1981) placed a loose ligature of atraumatic silk around the left anterior descending coronary artery under ether anesthesia in rats. Ten days later, acute myocardial infarction was produced by tightening the ligature. Kouchi et al. (2000) found an increase in $G_{i\alpha}$ protein accompanying progression of post-infarction remodelling in hypertensive cardiomyopathy in rats. G protein α subunits were studied with immunoblotting techniques. The polyclonal antiserum MB1 was raised in rabbits against the carboxyl-terminal decapeptide of retinal transduction (KENLKDCGLF) coupled to keyhole limpet hemocyanine. The MB1 recognized $G_{i\alpha 1}$ and $G_{i\alpha 2}$ but not $G_{0\alpha}$ and $G_{i\alpha 3}$. The membrane fractions were electrophoresed in SDS-polyacrylamide gels and were transferred to nitrocellulose filters.

The filters were incubated with the first antibodies for $G_{i\alpha}$ (MB1) or $G_{s\alpha}$ (RM/1) and then with the second antibody (horseradish peroxidase-conjugated goat anti-rabbit IgG, Amersham). Immunoreactive signals were detected by means of the ECL kit. Liu et al. (1997) found that ligation of the left descending coronary artery in Lewis inbred rats produces an uniformly large infarct with low mortality. The model may be superior to the usual model in Sprague-Dawley rats with a marked variability in infarct size and cardiac dysfunction. Coronary artery ligation induces left ventricular remodeling with cardiomyocyte apoptosis, myocardial fibrosis indicated by morphological studies and by collagen accumulation, which can be prevented by drug treatment. The naturally occurring peptide *N*-acety-seryl-aspartyl-lysyl-proline (Ac-SDKP) is a inhibitor of pluripotent hematopoietic stem cell proliferation and is normally present in human plasma and circulating mononuclear cells. It is cleaved to an inactive form by the NH_2-terminal catalytic domain of ACE. Acute angiotensin converting enzyme inhibition increases the plasma level of *N*-acetyl-seryl-aspartyl-lysyl-proline.

By morphological studies and collagen determinations, Rasoul et al. (2004) found an antifibrotic effect of Ac-SDKP and angiotensin converting enzyme inhibition in hypertension in rats. Similarly, Yang et al. (2004) found that Ac-SDKP reverses inflammation and fibrosis in rats with heart failure after myocardial infarction. Moreover, Azizi et al. (1997) and Le Meur et al. (1998) discussed whether the plasma Ac-SDKP level is a reliable marker of chronic angiotensin converting enzyme inhibition in hypertensive patients. Chen et al. (2004) found inhibition and reversal of myocardial-infarction-induced hypertrophy and heart failure by NHE-1 inhibition. Johns and Olson (1954) described a method of experimental myocardial infarction by coronary occlusion in small animals, such as mouse, hamster, rat and guinea pig. Scholz et al. (1995) described a dose-dependent reduction of myocardial infarct size in rabbits by a selective sodium-hydrogen exchange subtype 1 inhibitor. Gomoll and Lekich (1990) tested the ferret for a myocardial ischemia/salvage model. Varying combinations of duration of left anterior descending coronary occlusion and reperfusion were evaluated.

Ligation of coronary artery

Michael et al. (1995, 1999), and Gould et al. (2001) described the surgical procedure to induce myocardial ischemia in *mice* by ligation of the left anterior descending branch of the left coronary artery. Male C57BL/6 mice 12-16 weeks of age (22.5-30.5 g body weight) were used. Anesthesia was produced by an intraperitoneal injection of pentobarbital sodium (4 mg/ml; 10 μl/g body weight). Mice were placed in a supine position with paws taped to the operating table. With direct visualization of the trachea, an endotracheal tube was inserted and connected to a Harvard rodent volume-cycled ventilator cycling at 100/min with volume sufficient to adequately expand the lungs but not overexpand. The inflow valve was supplied with 100% oxygen.

For studies of the myocardial response to permanent occlusion, ligation of the anterior descending branch of the left coronary artery was achieved by tying an 8-0 silk suture around the artery. The suture was passed under the artery at a position ~1 mm from the tip of the normally positioned left auricle. For studies of the effect of reperfusion after coronary artery occlusion, the ligature was tied at the same location on the coronary artery used for the permanent occlusion. However, to allow subsequent reestablishment of blood flow, occlusion was produced by placing a 1-mm length of polyethylene (PE) tubing (OD = 0.61 mm) on the artery and fixing it in place with the ligature. The artery was then compressed by tightening the ligature, producing myocardial blanching and electrocardiographic (ECG) S-T segment elevation as observed in permanent ligations.

After occlusion for the desired time, blood flow was restored by removing the ligature and PE tubing. The chest wall was then closed by a 6–0 Ticron suture with one layer through the chest wall and muscle and a second layer through the skin and subcutaneous layer. After surgical closing of the chest, the endotracheal tube was removed, warmth was provided by a heat lamp, and 100% oxygen was provided via a nasal cone. The animal was given 0.1 mg/kg butorphanol tartrate as an analgesic, and it became sternally recumbent within 1 h. After surviving the experimental infarct the mice recovered, and this allowed postoperative physiological measurement. Sham-operated mice underwent an identical procedure with placement of the ligature but did not undergo coronary artery occlusion.

Latest technique

Guo et al. (1998) demonstrated the effects of an early and a late phase of ischemic preconditioning in mice. The results demonstrated that, in the mouse, a robust infarct-sparing effect occurred during both the early and the late phases of ischemic preconditioning, although the early phase was more powerful. Guo et al. (2005) found that late preconditioning induced by NO donors, adenosine A_1 receptor agonists, and δ_1-opioid receptor agonists is mediated by inducible NO synthase. Lutgens et al. (1999) reported cardiac structural and functional changes after chronic myocardial infarction in the mouse. Scherrer-Crosbie et al. (1999) described echocardiographic determination of risk area in a murine model of myocardial ischemia. Myocardial contrast echocardiography was performed before and after coronary artery ligation in anesthetized mice by intravenous injection of contrast microbubbles and transthoracic echo imaging. Time-video intensity curves were obtained for the anterior, lateral, and septal myocardial walls. After myocardial ischemia, myocardial contrast echocardiography defects were compared with the area of no perfusion measured by Evans blue staining. Jones and Lefer (2001) described cardioprotective actions of acute HMG-CoA reductase inhibition in the setting of myocardial infarction.

Janssens et al. (2004) reported that cardiomyocyte-specific overexpression of NO synthase 3 (NOS3) improves left ventricular (LV) performance and reduces compensatory hypertrophy after myocardial infarction. The effect of cardiomyocyte-restricted over- expression of one NO synthase isoform, NOS3, on LV remodeling after myocardial infarction in mice was tested. LV structure and function before and after permanent left anterior descending (LAD) coronary artery ligation were compared in transgenic mice with cardiomyocyte-restricted NOS3 overexpression (NOS3-TG) and their wild-type littermates (WT). Before myocardial infarction, systemic hemodynamic measurements, echocardiographic assessment of LV fractional shortening (FS), heart weight, and myocyte width (as assessed histologically) did not differ in NOS3-TG and WT mice. The inotropic response to graded doses of isoproterenol was significantly reduced in NOS3-TG mice. One week after LAD ligation, the infarcted fraction of the LV did not differ in WT and NOS3-TG mice.

Four weeks after myocardial infarction, however, end-systolic LV internal diameter (LVID) was greater, and FS and maximum and minimum rates of LV pressure development were less in WT than in NOS3-TG mice. LV weight/body weight ratio was greater in WT than in NOS3-TG mice. LaPointe et al. (2004) found that inhibition of cyclooxygenase-2 (COX-2) improves cardiac function after myocardial infarction in the mouse. Myocardial infarction was produced by ligation of the LAD coronary artery in mice. Two days later, mice were treated with a selective COX-2 inhibitor, or vehicle in drinking water for 2 weeks. After the treatment period, mice were subjected to two-dimensional M-mode echocardiography to determine cardiac function. Hearts were then analyzed for determination of infarct size, interstitial collagen content, brain natriuretic peptide (BNP) mRNA, myocyte cross-sectional area, and immunohistochemical staining for transforming growth factor (TGF)β and COX-2. Shibuya et al. (2005) reported that *N*-acetyl-seryl-aspartyl-lysine-proline prevents renal insufficiency and matrix expansion in diabetic db/db mice.

Weinberg et al. (2005) found in coronary ligation experiments in mice, that rosuvastatin reduces experimental left ventricular infarct size after ischemiareperfusion injury but not total coronary occlusion. Yang et al. (2005) found that the infarct-sparing effect of A_{2A}-adenosine receptor activation is due primarily to its action on lymphocytes. Chimeric mice were created by bone marrow transplantation from A_{2A}AR-knockout or green fluorescent protein (GFP) donor mice to irradiated congenic C57BL/6 (B6) recipients. In the GFP chimeras, we were unable to detect GFP-producing cells in the vascular endothelium, indicating that bone-marrow-derived cells were not recruited to endothelium at appreciable levels after bone marrow transplantation and/or acute myocardial infarction. Injection of 5 or 10 μg/kg of a potent and selective agonist of A_{2A} adenosine receptor had no effect on hemodynamic parameters but reduced infarct size in B6 mice after 45 min of LAD artery occlusion followed by 24 h of reperfusion. Kanno et al. (2003) found connexin43 to be a determinant of myocardial infarct size following coronary occlusion in mice. Regulation of myocardial connexins during hypertrophic remodeling was reviewed by Teunissen et al. (2004).

Kuhlmann et al. (2006) reported that granulocyte colony stimulating factor (G-CSF), alone or in combination with stem cell factor (SCF), can improve hemodynamic cardiac function after myocardial infarction in mice and reduces inducible arrhythmias in the infarcted heart potentially via increased connexin43 expression and arteriogenesis.

Occlusion of Coronary Artery

The size of infarcts is studied after proximal occlusion of the left anterior descending coronary artery in open-chest dogs. Compounds potentially reducing infarct-size are tested. To delineate the post-mortem area at risk, coronary arteriograms are made after injection of a $BaSO_4$-gelatin mass into the left coronary ostium. The infarct's area is visualized with nitro-blue tetrazolium chloride in myocardial sections.

Method

Dogs of either sex weighing approximately 30 kg are used. The animals are anesthetized by intravenous injection of pentobarbital sodium (bolus of 35 mg/kg followed by continuous infusion of 4 mg/kg/h). The animals are placed in the right lateral position. Respiration is maintained through a tracheal tube using a positive pressure respirator. Arterial blood gases are checked, and the ventilation rate and/or oxygen flow rate are adjusted to achieve physiological blood gas values (P_{O2}: 100-140 mm Hg, P_{CO2}: 32-40 mm Hg, and pH7.47). A peripheral vein (saphenous vein) is cannulated for the administration of test compound. The ECG is recorded continuously from lead II. For recording of peripheral systolic and diastolic blood pressure, the cannula of a femoral vein is connected to a pressure transducer. For determination of left ventricular pressure (LVP), a Millar microtip catheter (PC 350) is inserted via the left carotid artery. Left ventricular enddiastolic pressure (LVEDP) is measured from a high-sensitivity scale. From the pressure curve, dp/dt max is differentiated and heart rate is counted.

Operative technique

The heart is exposed through a left thoracotomy between the fourth and fifth intercostal space, the pericard is opened and the left anterior descending coronary artery (LAD) is exposed. After reaching steady state conditions for the hemodynamic parameters (approx. 45 min), the LAD is ligated just below the first diagonal branch for 360 min. No attempt is made to suppress arrhythmic activity after the ligation. The test substance or the vehicle (controls) is administered by intravenous bolus injection and/or continuous infusion. The schedule of administration may vary. Hemodynamic parameters are registered continuously during the whole experiment. At the end of the experiment, the animals are sacrificed with an overdose of pentobarbital sodium and the heart is dissected.

Coronary arteriograms

Coronary arteriograms are made according to Schaper et al. (1979) to delineate the anatomic post-mortem area at risk. A purse-string suture is placed around the left coronary ostium in the sinus of Valsalva; a cannula is then placed in the ostium and the purse-string suture is tightened. Micronized $BaSO_4$ suspended in 12% gelatin solution (37°C) is injected under increasing pressure (2 min at 100 mm Hg, 2 min at 150 mmHg and 2 min at 200 mm Hg). The heart is placed in crushed ice to gel the injectate. The right ventricle is removed and the left ventricle plus septum is cut into transverse sections (approx. 1 cm thick) from the apex to the level of the occlusion (near the base). From each slice angiograms are made with a X-ray tube at 40 kV to assess the post-mortem area at risk (by defect opacity: reduction of $BaSO_4$-filled vessels in infarct tissue).

Observing infarct tissues

The slices are then incubated in p-nitro-blue tetrazolium solution (0.25 g/L in Sorensen phosphate buffer, pH7.4, containing 100 mM D,L-maleate) in order to visualize the infarct tissue (blue/violet-stained healthy tissue, unstained necrotic tissue). The slices are photographed on color transparency film for the determination of the infarct area. Left ventricle and infarct area, and area at risk are measured by planimetry from projections of all slices with the exclusion of the apex and of the slice containing the ligature.

Assessment

Mortality and the different hemodynamic parameters are determined. Changes of parameters in drug-treated animals are compared to vehicle controls. The different characteristics are evaluated separately. Mean values ± SEM of infarct area and of area at risk are calculated. Statistical analyses consist of regression and correlation analyses and of the Student's *t*-test. Results are considered significant at $p < 0.05$.

Latest technique

Nachlas and Shnitka (1963) described the macroscopic identification of early myocardial infarcts by alterations in dehydrogenase activity in dogs by staining the cardiac tissue with Nitro-BT [2,2′-di-p-nitrophenyl-5,5′ diphenyl-3,3′-(3,3′-dimethoxy-4,4′-biphenylene) ditetrazolium chloride] yielding a dark blue formazan in viable muscle but not in necrotic muscle fibers. Chiariello et al. (1976) compared the effects of nitroprusside and nitroglycerin on ischemic injury during acute myocardial infarction in dogs. Black et al. (1995) studied the cardioprotective effects of heparin or *N*-acetylheparin in an *in vivo* dog model of myocardial and ischemic reperfusion injury. The hearts were removed after 90 min of coronary occlusion and a 6 h-reperfusion period. Area at risk was determined by the absence of Evans blue dye after perfusion of the aorta in a retrograde fashion and infarct zone by the absence of formazan pigment within the area at risk after perfusion of the circumflex coronary artery with triphenyltetrazolium chloride.

Reimer et al. (1985) tested the effect of drugs to protect ischemic myocardium in unconscious and conscious dogs. In the conscious model, dogs of either sex weighing 10-25 kg were anesthetized with thiamyal sodium (30-40 mg/kg i.v.) and underwent thoracotomy through the 4th intercostal space. Heparin-filled polyvinyl chloride catheters were positioned in the aortic root, the left atrium via the left atrial appendage, and a systemic vein. A mechanical adjustable snare type occluder was placed around the proximal left circumflex coronary artery above or below the first marginal branch, so that temporary occlusion resulted in cyanosis of at least 75% of the inferior wall. The catheters and snare were either exteriorized or positioned in a subcutaneous pocket at the back of the neck. Penicillin, 1000,000 units, and streptomycin, 1.0 g, were given i.m. for the first 4 postoperative days, and at least 7 days were allowed for recovery from surgery.

Dogs were fasted overnight prior to the study. After exteriorization and flushing of the catheters, 30-40 min were allowed for the animals to adjust to laboratory conditions. Morphine sulfate, 0.25

mg/kg, i.m., was given 30 min before occlusion, and an additional 0.25 mg/kg, i.v., was given 20 min later. Heart rate and aortic and left atrial pressures were monitored continuously. Permanent coronary occlusion was produced by a sudden one-stage tightening of the snare occluder. Drugs were administered by continuous i.v. infusion over 6 h.

Hemodynamic measurements were taken 5 min before occlusion and 10, 25, 105, 180, and 360 min after occlusion. Raberger et al. (1986) described a model of *transient myocardial dysfunction in conscious dogs*. Mongrel dogs, trained to run on a treadmill, were chronically instrumented with a miniature pressure transducer in the left ventricle and a hydraulic occluder placed around the circumflex branch of the left coronary artery. Two pairs of piecoelectrical crystals for sonomicrometry were implanted subendocardially to measure regional myocardial functions.

Comparable episodes of regional dysfunction of the left coronary artery area during treadmill runs were found after partial left coronary artery stenosis induced by external filling of the occluder. Hartman and Warltier (1990) described a model of *multivessel coronary artery disease* using conscious, chronically instrumented dogs. A hydraulic occluder was implanted around the left anterior descending coronary artery (LAD) and an Ameroid constrictor around the left circumflex coronary artery (LCCA). Pairs of piecoelectric crystals were implanted within the subendocardium of the LAD and LCCA perfusion territories to measure regional contractile function. A catheter was placed in the left atrial appendage for injection of radioactive microspheres to measure regional myocardial perfusion. Bolus injections of adenosine were administered daily via the left atrium to evaluate LAD and LCCA coronary reserve.

After stenosis by the Ameroid constrictor, radioactive microspheres were administered to compare regional perfusion within normal myocardium to flow in myocardium supplied by the occluded or stenotic coronary arteries. Holmborn et al. (1993) compared triphenyltetrazolium chloride staining versus detection of fibronectin in experimental myocardial infarction in pigs. Klein et al. (1995) used intact pigs and found myocardial protection by Na^+/H^+ exchange inhibition in ischemic reperfused hearts. Klein et al. (1997) measured the time delay of cell death by Na^+/H^+ exchange inhibition in regionally ischemic, reperfused porcine hearts. Garcia-Dorado et al. (1997) determined the effect of Na^+/H^+ exchange blockade in ischemic rigor contracture and reperfusion-induced hypercontracture in *pigs* submitted to 55 min of coronary occlusion and 5 h reperfusion.

Myocardial segment length analysis with ultrasonic microcristals was used to detect ischemic rigor (reduction in passive segment length change) and hypercontracture (reduction in end-diastolic length). Symons et al. (1998) tested the attenuation of regional dysfunction in response to 25 cycles of ischemia (2 min) and reperfusion (8 min) of the left circumflex coronary artery in *conscious swine* after administration of a Na^+/H^+ exchange inhibitor. .

The animals were instrumented to measure arterial blood pressure, regional myocardial blood flow (colored microspheres), systolic wall thickening in the normally perfused left anterior descending and left circumflex coronary artery regions (sonomicrometry), left circumflex coronary artery blood flow velocity (Doppler) and reversibility to occlude the left circumflex coronary artery (hydraulic occluder). Etoh et al. (2001) studied myocardial and interstitial matrix metalloproteinase activity after acute myocardial infarction in pigs.

Hemodynamic Measurements

Severe left ventricular failure is induced by repeated injections of 50 μm plastic microspheres into the left main coronary artery of anesthetized dogs. Hemodynamic measurements are performed under theses conditions testing drugs which potentially improve cardiac performance. The test can be used to evaluate the influence of drugs on myocardial performance during acute ischemic left ventricular failure in dogs.

Method

Dogs of either sex weighing approximately 30 kg are anesthetized by an intravenous bolus injection of 35-40 mg/kg pentobarbital sodium continued by an infusion of 4 mg/kg/h. The animals are placed in the right lateral position. Respiration is maintained through a tracheal tube using a positive pressure respirator and controlled by measuring end-expiratory CO_2 concentration as well as blood gases. Two peripheral veins are cannulated for the administration of narcotic (brachial vein) and test compounds (saphenous vein). The ECG is recorded continuously in lead II (Einthoven).

Experimental approach

For recording of peripheral systolic and diastolic blood pressure, the cannula of the right femoral vein is connected to a pressure transducer. For determination of *left ventricular pressure* (LVP), a Millar microtip catheter is inserted via the left carotid artery. Left ventricular enddiastolic pressure (LVEDP) is measured on a high-sensitivity scale. From the pressure curve, dp/dt max is differentiated and heart rate (HR) is counted. To measure right ventricular pressure, a Millar microtip catheter is inserted via the right femoral vein. Systolic, diastolic and mean *pulmonary artery pressure* (PAP), mean pulmonary capillary pressure, and cardiac output are measured by a thermodilution technique using a Cardiac Index Computer and a balloon-tip triple lumen catheter with the thermistor positioned in the pulmonary artery via the jugular vein.

The heart is exposed through a left thoracotomy between the fourth and fifth intercostal space, the pericard is opened and the left circumflex coronary artery (LCX) is exposed. To measure coronary blood flow, an electromagnetic flow probe (Hellige Recomed) is placed on the proximal part of the LCX. Polystyrol microspheres with a diameter of 52.5 ± 2.24 μm are diluted with dextran 70, 60 mg/ml and saline at a concentration of 1 mg microspheres/ml (1 mg = approx. 12,000 beads). For administration of microspheres, an angiogram catheter is inserted into the left ostium via the left femoral artery.

Termination of embolization

The microspheres are injected through the angiogram catheter into the left ostium initially as 10 ml and later as 5 ml boluses about 5 min apart. The microsphere injections produce stepwise elevations of LVEDP. Embolization is terminated when LVEDP has increased to 16-18 mm Hg and/or PAPm has increased to 20 mm Hg and/or heart rate has reached 200 beats/min. The embolization is completed in about 70 min and by injection of an average dose of 3-5 mg/kg microspheres. Hemodynamic variables are allowed to stabilize after coronary embolization for at least 30 min.

Stepwise treatment

The test substance or the vehicle (controls) is then administered by intravenous bolus injection or continuous infusion, or by intraduodenal application. Recordings are obtained:

1. Before embolization
2. After embolization
3. Before administration of test compound
4. 5, 30, 45, 60, 90, 120 and, eventually, 150 and 180 min following administration of test drug.

At the end of the experiment, the animal is sacrificed by an overdose of pentobarbital sodium.

Assessment

Besides the different directly measured hemodynamic parameters, the following data are calculated according to the respective formula:

stroke volume [ml/s],

$$SV = \frac{\text{cardiac output}}{\text{heart rate}}$$

tension index [mmHgs],

$$IT = \frac{BPs \times \text{heart rate}}{1000}$$

Coronary vascular resistance [mm Hgmin/ml],

$$CVR = \frac{BPm \times RAPm}{CBF}$$

total peripheral resistance [dyns/cm^5],

$$TPR = \frac{BPm \times RAPm}{\text{cardiac output}} \times 79.9$$

Pulmonary artery resistance [dyn s/cm^5],

$$PAR = \frac{PAPm - PCPm}{\text{cardiac output}} \times 79.9$$

right ventricle work [kgm/min],

$$RVW = (PAPm - RAPm) \times \text{cardiac output} \times 0.0136$$

left ventricle work [kgm/min],

$$LVW = (BPm - LVEDP) \times \text{cardiac output} \times 0.0136$$

left ventricular myocardial oxygen consumption [ml O_2/min/100 g],

$$MVO_2 = K_1\,(BPs \times HR) + K_2 \frac{(0.8BPs + 0.2BPd) \times HR \times SV}{BW} + 1.43$$

$K_1 = 4.08 \times 10^{-4}$

$K_2 = 3.25 \times 10^{-4}$

BPs = systolic blood pressure [mm Hg]

BPd = diastolic blood pressure [mm Hg]

BPm = mean blood pressure [mm Hg]

CBF = coronary blood flow in left circumflex coronary artery [ml/min]

RAPm = mean right atrial pressure [mm Hg]

PAPm = mean blood pressure A. pulmonalis [mm Hg]

PCPm = mean pulmonary capillary pressure

HR = heart rate [beats/min]

SV = stroke volume [ml]

BW = body weight [kg]

Changes of parameters in drug-treated animals are compared to vehicle controls; statistical significance of the differences is calculated with the Student's *t*-test. Mean embolization times, doses of microspheres and number of microsphere applications are evaluated.

Latest technique

Gorodetskaya et al. (1990) described a simple method to produce acute heart failure by coronary vessel embolization with microspheres in rats.

Prolonged Ischemia

Damage to the mammalian heart produced by prolonged ischemia and reperfusion can be reduced by "preconditioning" the myocardium via a brief cycle of ischemia and reperfusion prior to the protracted ischemic event. Ischemic preconditioning has been shown to decrease infarct size and increase recovery of post-ischemic ventricular function, and to reduce leakage of cellular marker proteins indicative for cardiac myocyte death.

In addition, preliminary preconditioning also attenuates cardiac arrhythmia associated with subsequent occlusion and reperfusion. The mechanistic basis of this phenomenon is under discussion. Adenosine receptor involvement in myocardial protection after ischemic preconditioning in rabbits has been shown by Baxter et al. (1994).

Adenosine (A_1 receptor) antagonists have been demonstrated to block the protection produced by preconditioning, and short term administration of adenosine was shown to simulate the protective effects of ischemic preconditioning. These observations together suggest that adenosine is generated by the short preconditioning ischemia. Other recent pharmacological studies indicate the involvement of the ATP-sensitive potassium channel.

Recent investigations indicate that an increase of NO production after ACE inhibitors may be a part of the protective mechanism. Moreover, the involvement of prostanoids and bradykinin in the preconditioning process has been discussed. Gho et al. (1994) found a limitation of myocardial infarct size in the rat by transient renal ischemia, supporting the hypothesis that the mechanism leading to cardiac protection by ischemic preconditioning may not only reside in the heart itself.

Method

New Zealand rabbits of either sex weighing 2.5-3.5 kg are initially anesthetized with an intramuscular injection of ketamine (50 mg/ml)/xylazine (10 mg/ml) solution at a dose of 0.6 ml per kg body weight. A tracheotomy is performed to facilitate artificial respiration. The left external jugular vein is cannulated to permit a constant infusion (0.15-0.25 ml/min) of xylazine (2 mg/ml in heparinized saline) to assist in maintaining anesthesia and fluid volume. Anesthesia is also maintained by i.m. injections (0.4-0.6 ml) of ketamine (80 mg/ml) and xylazine (5 mg/ml) solution. After the xylazine infusion is started, animals are respired with room air at a tidal volume of 10 ml/kg and a frequency of 30 inflations per min. Thereafter, ventilation is adjusted or inspiratory room air is supplemented (5% CO_2/95% O_2) to maintain arterial blood chemistry within the following ranges: pH 7.35-7.45, P_{CO2} 25-45 mm Hg, P_{O2} 90-135 mm Hg.

The right femoral artery and vein are isolated and catheterized for measurement of arterial pressure and administration of drugs, respectively. A thoracotomy is performed in the fourth intercostal space, and the lungs are retracted to expose the heart. The pericardium is cut to expose the left ventricle, and a solid-state pressure transducer catheter is inserted through an apical incision and secured to enable measurement of pulsatile left ventricular pressure. The maximal rate of increase in left ventricular pressure (LV dP/dt max) is determined by electronic differentiation of the left ventricular pressure wave form. A segment of 4-0 prolene suture is looped loosely around a marginal branch of the left main coronary artery to facilitate coronary occlusion during the experiment. Needle electrodes are inserted subcutaneously in a lead II configuration to enable recording of an ECG in order to determine heart rate and help confirm the occurrence of ischemia (ST segment elevation) and reperfusion of the myocardium distal to the coronary occlusion.

Continuous recording of pulsatile pressure, ECG, heart rate, and LVdP/dt are simultaneously displayed on a polygraph and digitized in real time by a personal computer. Hemodynamic data are condensed for summary and later statistical analysis. Ischemic preconditioning is induced by tightening the prolene loop around the coronary artery for 5 min and then loosening to reperfuse the affected myocardium for 10 min prior to a subsequent 30 min occlusion. After surgical preparation, and prior to 30 min of occlusion, rabbits are randomly selected to receive ischemic preconditioning, no preconditioning, or ischemic preconditioning plus treatment with test drugs. After 30 min of occlusion, the ligature is released and followed by 120 min of reperfusion. Occlusion is verified by epicardial cyanosis distal to the suture, which is usually accompanied by alterations in hemodynamics and ECG. Reperfusion is validated by return of original color. Systemic hemodynamics are summarized for each experimental period.

The experiment is terminated after 120 min of reperfusion, and the heart is excised for determinations of infarct size and area at risk. Immediately before the animal is sacrificed, the marginal branch of the left coronary artery is reoccluded and India ink is rapidly injected by syringe with a 18-g needle into the left ventricular chamber to demarcate blackened normal myocardium from unstained area at risk. After the rabbits are sacrificed, the heart is removed and sectioned in a breadloaf fashion from apex to base perpendicular to the long axis. The right ventricle is removed from each slice leaving only the left ventricle and septum. After each slice is weighed, the portions are washed and incubated in a phosphate buffered saline solution of triphenyl tetrazolium chloride

(1 g/ml, Sigma) for 10-15 min. Salvaged myocardium in the area at risk stains brick red, whereas infarcted tissue remains unaltered in color. Slices are then placed between sheets of Plexiglas and the areas (normal, risk, infarct) of each slice are traced on a sheet of clear acetate. Traces are then digitized and analyzed using computerized planimetry to compare the relative composition of each slice with respect to normal tissue, area at risk, and infarcted myocardium.

Planimetry is performed with a computerized analysis system, e. g., Quantimet 570C image analysis system. Surface areas of normal tissue, area at risk, and infarcted myocardium on both sides of each slide are averaged for the individual slide. The contribution of each slide to total infarcted and area at risk (%) and area at risk as a percentage of total left ventricular mass for the entire left ventricle is prorated by the weight of each slice. By adding the adjusted contributions from each slice to infarcted tissue, area at risk, and left ventricular mass, a three-dimensional mathematical representation of total myocardial infarct size and risk zone can be calculated for each rabbit, and a mean tabulated for each treatment group for statistical comparison.

Assessment

All data are presented as mean ± SD. Systemic hemodynamic data are analyzed by ANOVA using Statistica/W software. Means are considered significantly different at $p < 0.05$.

Latest technique

Li et al. (1990) found in dog experiments that preconditioning with one brief ischemic interval is as effective as preconditioning with multiple ischemic periods. In contrast, Vegh et al. (1990) found in other dog experiments that two brief preconditioning periods of coronary occlusion, with an adequate period of reperfusion between, reduce the severity of arrhythmias. Yang et al. (1996) found a second window of protection after ischemic preconditioning in conscious rabbits which minimizes both infarction and arrhythmias. Late preconditioning against myocardial stunning in conscious pigs together with an increase of heat stress protein (HSP) 70 was described Sun et al. (1995). Szilvassy et al. (1994) described the anti-ischemic effect induced by ventricular overdrive pacing as a conscious rabbit model of preconditioning. Rabbits were equipped with right ventricular electrode catheters for pacing and intracavital recording and polyethylene cannulae in the left ventricle and right carotid artery to measure intraventricular pressure and blood pressure. One week after surgery in conscious animals, ventricular overdrive pacing at 500 beats/min over 2, 5, or 10 min resulted in an intracavital S-T segment elevation, shortening of ventricular effective refractory period, decrease in maximum rate of pressure development and blood pressure, and increase in left ventricular end-diastolic pressure proportional to the duration of stimulus. A 5-min preconditioning ventricular overdrive pacing applied 5 or 30 min before a 10-min ventricular overdrive pacing markedly attenuated ischemic changes, whereas a 2-min ventricular overdrive pacing had no effect. The ventricular overdrive pacing induced preconditioning effect was lost in atherosclerotic rabbits, however, delayed cardiac protection could be induced in these animals.

Magnetic Resonance Imaging Studies

Magnetic resonance imaging (MRI) is the preferred technique for the visualization of lesions in the brain and spinal cord of patients with MS. It visualizes the resonance signals of tissue protons when they are placed in a time-varying strong magnetic field. The most frequently used parameters measured in MS are the spin-lattice relaxation time (T_1) and the spin-spin relaxation time (T_2). MRI is routinely used as a tomographic imaging technique, where anatomical pictures are created of 1-mm-thick tissue sections.

The contrast differences between brain structures in most MRI techniques are determined by the different densities and diffusion of protons, as well as differences in relaxation times. T_2 images are sensitive to water and, because all pathological alterations in MS brains are associated with altered distribution of tissue water (edema), this technique is highly useful for visualization of the spatial distribution of lesions. Contrast in T_1 images is determined mainly by different lattice densities.

Dense structures, such as compact white matter, have low T_1 values, whereas relatively loose structures, such as grey matter or lesions, have higher T_1 values. To distinguish inflammatory active from inactive lesions, the paramagnetic dye gadolinium-DTPA is intravenously injected (0.1-0.3 mmol kg^{-1}) and, in areas of increased blood-brain barrier permeability, leaks into the brain parenchyma, causing local enhancement of the T_1-weighted signal intensity. A third important MRI technique in MS is magnetization transfer ratio (MTR) imaging. The MTR of a given tissue is defined as the ratio of free protons versus protons bound to tissue macromolecules. MRI has emerged as a highly accurate and quantitative tool for the evaluation of cardiac function. Al-Shafei et al. (2000a, 2002b) performed MRI analysis of cardiac cycle events in diabetic rats and tested the effect of angiotensin-converting enzyme inhibition.

Method

Diabetes was induced in Wistar rats at an age of 7, 10, and 13 weeks. The rats were anesthetized using 1%-2% halothane in oxygen, their blood glucose levels were checked. They were then given a single intraperitoneal injection of streptozotocin 65 mg/kg body weight. The control rats received sham injections of the citrated buffer when they were 7 weeks old. One diabetic group was treated with 2 g/l captopril in the drinking water For *MR imaging*, rats were anesthetized using 1%-2% halothane in oxygen, weighed and their systolic blood pressures measured non-invasively using a rat tail blood pressure monitor both before and after imaging sessions to confirm physiological stability.

Electrocardiographic (ECG) monitoring used shielded subcutaneous electrodes and a Tektronix 2225 oscilloscope. The cine imaging protocols were performed with the anesthetized animal placed in a specially designed home-built half-sine-spaced birdcage radiofrequency (RF) probe unit contained within a cylindrical plastic holder fitted within a gradient set of internal diameter 11 cm. The RF probe unit was made up of a half-sine spaced birdcage RF probe of internal diameter of 4.5 cm with open ends, an RF shield consisting of a cylinder of copper gauze surrounding and sliding over the birdcage, a tuning capacitor and a coaxial cable to carry the RF.

The assembly included ECG leads, attachment plugs for the ECG leads and a unit to anchor anesthetic delivery tubes near the nose of the animal. All experiments used a 2T Oxford Instruments (UK) superconducting magnet with a horizontal internal bore of 31 cm. A gated cine protocol synchronized line acquisition to set times following alternate electro-cardiographic R waves. This acquisition was then repeated at the same slice position at 12 equally incremented times through the cardiac cycle. This sequence in turn was repeated for each of the 128 lines to generate each 128 × 128 image, which itself was acquired twice for signal averaging. The preceding procedure was in turn repeated 12 times to obtain signal-averaged images for every one of the 12 contiguous transverse slices examined. Each imaging session therefore required (128 × 12 × 2 × 2) times the cardiac cycle duration. The effective repeat time (TR) was approximately 13 ms. The short echo time (TE) of 4.3 ms reduced motion artifacts and ensured good contrast between blood and myocardium.

Assessment

The image data were transferred from the MRI console using in-house hardware and software to remote UNIX workstations for quantitative analysis using in- house software based on CaMReS libraries. Itter et al. (2004) used non-invasive MRI techniques in a model of chronic heart failure in spontaneously hypertensive rats. Bryant et al. (1998) and Franco et al. (1999) described MRI and invasive evaluation of development of heart failure in transgenic *mice* with myocardial expression of tumor necrosis factor-α.

Wiesmann et al. (2002) reported analysis of right ventricular function in healthy mice and a murine model of heart failure by*in vivo* MRI. Kraitchman et al. (2003) described quantitative ischemia detection during cardiac magnetic resonance stress testing by use of fast harmonic phase MTI (FastHARP) in dogs. Reddy et al. (2004) discussed the feasibility of a porcine model of healed myocardial infarction by integration of cardiac MRI with three-dimensional electroanatomic mapping to guide left ventricular catheter manipulation. Pelzer et al. (2005) reported that the estrogen receptor-α agonist 16α-LE2 inhibits cardiac hypertrophy and improves hemodynamic function in estrogen-deficient spontaneously hypertensive rats. Improved left ventricular function upon 16α-LE2 treatment was also observed in cardiac MRI studies.

Cardiac Allograft Rejection

Acute *cardiac allograft rejection* continues to be the cause of graft loss and contributes to the morbidity and mortality after cardiac transplantation. Endomyocardial biopsy is used routinely for cardiac transplant rejection surveillance. A sensitive and non-invasive method for detecting rejection is desirable. Kanno et al. (2001) developed a rat model of heterotopic heart and lung transplantation for MRI experiments. Allotransplantations were performed with syngeneic transplantations serving as controls. MR images were obtained with a gradient echo method.

Method

All rats used in the experiments were male, 2-3 months of age, and weighed 220-250 g each. Animals were housed individually and provided with food and water ad libitum. Inbred Brown Norway (BN; $RT1^n$) and DA ($RT1^a$) rats were obtained from Harlan Sprague Dawley.

Heart and lung transplantation

Under anesthesia with injection of 35 mg/kg body weight of sodium pentobarbital IP, 500 U/ kg body weight of heparin was injected. In the syngeneic group, an en bloc donor heart and lung were taken from a BN rat and transplanted to another BN rat. In the allogeneic group, a graft from a DA rat was transplanted to a BN rat. This group was divided into two groups: one group was treated with 3 mg/kg per day ciclosporin (CsA), and the other group was not given CsA. Graft survival was monitored every day by palpating contraction of the transplanted heart. Operative procedures have been described by Kanno et al. (2000). In brief, after the chest wall of the donor rat was opened, the left lung was ligated and excised.

The azygos vein with the left superior and right superior venae cavae was ligated and divided. The descending thoracic aorta was transected, and 10 ml of cold University of Wisconsin solution was infused into the inferior vena cava until the fluid draining from the aorta was clear, followed by ligation and division of the inferior vena cava. The ascending aorta was dissected and transected at the portion between the left common carotid artery and the left subclavian artery, followed by ligation and division of the right brachiocephalic artery and the left common carotid artery.

After removal of the heart and lung from the donor, the right lung was washed 3 times through the bronchus with UW solution containing penicillin G. The grafts were then placed into cold UW solution for ≈ 5 min until transplantation. Next, the left inguinal portion of the recipient rat was opened and dissected to make enough space for the transplanted organs. The left lower part of the abdominal wall was opened in a transverse fashion from the left femoral vessels to the midline.

The abdominal organs were retracted to the right, and both the aorta and the inferior vena cava just beyond the bifurcation were dissected. The vessels were clamped, and an appropriate opening of the aorta was made to receive the aorta of the graft in an end-to-side fashion. Rhythmic heartbeats commenced spontaneously as the heart and the lung regained circulation after removal of the clamp. After hemostasis of the surgical field, the abdominal wall was sutured, with care taken not to kink or obstruct the aorta of the graft.

MRI experiments

MRI measurements were carried out on a 4.7-T/40-cm Bruker AVANCE DRX MR instrument equipped with 15-cm, 10-gauss/cm shielded gradients. *In vivo* MR images of transplanted heart-lung were obtained over a period of 24 h after infusion of dextran-coated ultrasmall superparamagnetic iron oxide (USPIO) particles. The imaging sequence consisted of a gradient echo sequence, triggered to ECG and ventilator (60 strokes/min, 10 ml/kg), with TR/TE 500/10 ms, flip angle equal to Ernst angle, slice thickness 1 mm, field of view 6.0 cm, data matrix size 256 × 130 (zero-filled to 256 × 256), and scan time 5 min. ECG leads were placed on both of the hind limbs of the rat with the transplant to pick up the heartbeat from the transplanted heart more effectively.

The change of MRI signal intensity was measured in whole ventricular wall in each transplanted heart. The MR signal intensity of the heart was normalized to that of the leg muscle, because USPIO particles are not readily taken up by muscular tissue, according to Gellissen et al. (1999) Dextran-coated USPIO particles were synthesized according to the method of Palmacci and Josephson (1993) with slight modifications The MR relaxivities R_1 (spin-lattice relaxation rate constant, $1/T_2$, per mol of Fe in USPIO) and R_2 (spin-spin relaxation rate constant, $1/T_2$, per mol of Fe in USPIO) measured at 4.7 T were 3.8×10^4 and 9.1×10^4 (mol/l)/s, respectively. For *in vivo* studies, dextran-coated USPIO particles were dialyzed against PBS solution and diluted to a concentration of 18 μmol Fe/ml, and 0.8 ml of the suspension (i. e., ≈ 3 mg Fe/kg body weight) was injected intravenously for each study.

At 6 days after transplantation, dextran-coated US-PIO particles were injected intravenously as mentioned above, and the animals were subjected to MRI. Then 24 h later, these animals were again placed inside the magnet and scanned. The regions of interest were defined manually with Bruker software. MR signal intensity in the entire ventricular wall in the plane was measured. After injection of USPIO particles at postoperative day (POD) 6, animals with allotransplants were given CsA for 4 (POD 7 to 10) or 7 (POD 7 to 13) days and reinjected with USPIO particles on POD 14.

Clinical studies

After an MR experiment was completed, the transplanted hearts were extirpated, fixed in 3.7% formaldehyde, and embedded in paraffin for 5-μm sections. Hematoxylin-eosin staining and Perl's Prussian blue staining were performed in the Transplantation Pathology Laboratory of the University of Pittsburgh Medical Center. Histological analysis for pathological grading of heart rejection, which is based on the criteria established by the International Society for Heart and Lung Transplantation, was also performed by this laboratory in a blinded manner. Monoclonal anti-rat macrophage antibody was used as a primary antibody for macrophages. Immunohistochemistry was carried out with the ABC staining system according to the manufacturer's protocol.

Assessment

The results are presented as mean ± SD. The results were analyzed by ANOVA with StatView software (SAS Institute). A value of $P < 0.05$ was considered to be statistically significant.

EX VIVO TECHNIQUE FOR PLASTIC CASTS

Prolonged administration of coronary drugs has been shown to increase the number and size of interarterial collaterals of dogs and pigs after coronary occlusion. An increased rate of development of collateral arteries was observed after physical exercise in dogs, as well as after chronic administration of coronary dilating drugs. An even more effective stimulus for collateral development is an acute or gradual occlusion of one or several major coronary branches. Filling the arterial coronary bed with a plastic provides the possibility to make the collaterals visible and to quantify them.

Method

Dogs weighing 10-15 kg are anesthetized with pentobarbital sodium 30 mg/kg i.v. They are respirated artificially and the thorax is opened. After opening of the pericard, Ameroid cuffs are placed around major coronary branches. Gradual swelling of the plastic material occludes the lumen within 3-4 weeks. The dogs are treated daily with the test drug or placebo. After 1 week recovery period they are submitted to exercise on a treadmill ergometer. After 6 weeks treatment, the animals are sacrificed, the heart removed and the coronary bed flushed with saline. The liquid plastic Araldite is used to fill the whole coronary tree from the bulbus aortae. The aortic valves are glued together in order to prevent filling of the left ventricle. Red colored Araldite is used to fill the arterial tree. The venous part of the coronary vasculature can be filled with blue colored Araldite from the venous sinus. The uniformity of the filling pressure, the filling time, and the viscosity of the material are important. Polymerization is complete after several hours. Then, the tissue is digested with 35% potassium hydroxide. The method gives stable preparations which can be preserved for a long time.

Assessment

Plastic casts from drug treated animals are compared with casts from dogs submitted to the same procedure without drug treatment. The procedure allows impressive demonstration of the formation of arterial collaterals. The results of post mortem Araldit impletion agree with the functional results of experimental coronary occlusion. Boor and Reynolds (1977) described a simple planimetric method for determination of left ventricular mass and necrotic myocardial mass in postmortem hearts.

17

Pharmaceutical Applications

As the pace of pharmaceutical drug discovery quickens and greater numbers of preclinical candidates are identified using combinatorial and other high throughput methods, the demand on safety assessment assays increases. As most *in vitro* toxicology assays are, at best, medium throughput, it is readily apparent that rapid *in silico* assessment protocols must be developed and validated for their use in the early discovery phase. No strangers to the increased demand for accurate safety assessments of candidate compounds and the additional constraints imposed by limited resources, regulatory agencies have long been at the forefront of utilizing and championing computational methods.

As regulatory databases of safety information are populated and legacy data incorporated, methods to utilize this data to extract meaningful information must be developed and validated. As this is not intended to be an exhaustive review of all *in silico* tools for toxicology assessment, the reader is referred to a number of recent articles which do an outstanding job of summarizing the algorithms, benefits and shortcomings of many of the commercial packages available.

CANDIDATE COMPOUNDS

During the development process, there are many competing factors which influence the success of a candidate compound. Physical properties such as lipophilicity, solubility and permeability are closely monitored during the development process and *in vitro* and *in silico* models are continuously developed and refined on a per chemotype basis. Empirical metrics such as Lipinski's rule of five are routinely used in profiling the drug likeness of candidate compounds. As we strive to increase the success rates of compounds both in the pre-clinical and post-NCE, we must be increasingly stringent on the criteria used to determine if a compound will be successful. Safety assessment is one of the most inflexible of these criteria.

Treating safety liabilities as simply another property, in this case one to be minimized, has been a paradigm shift in the pharmaceutical industry over the last decade. What remains, though, is that differing stages of the development process have dissimilar needs for such a design algorithm.

Ideally, one would have an assessment protocol in place which would correctly partition all of the harmful from the innocuous compounds in both the training and test/validation sets (TS and VS, respectively). Furthermore, the most usable model is one which can successfully extrapolate beyond either the training or validation sets when encountering novel compounds in a *prediction set* (PS). This type of model exists only in an idealized context, as neither the data nor our understanding of the biology is usually of sufficient detail to allow us to construct models of such quality. As such, we are often left with three classes of models, those with high overall predicitivity (concordance), high specificity (ability to partition the inactive subset) or high sensitivity (ability to partition the active subset).

Most automated algorithms strive to achieve the first goal, usually achieving a high TS concordance, most easily achieved by skewing the model to identify the majority population in the TS. Subjective algorithms, usually involving human intervention and biasing have been used to achieve high sensitivity or specificity, almost always at the expense of concordance. A simple question arises: why would one want to bias a model toward high sensitivity or specificity? Three issues determine the form of the biasing: (a) the predictive ability of the confirmatory assay; (b) the stage in the discovery pipeline of the assessment being performed; and, (c) the cost of each assay performed compared to the cost of a missed assessment. If the second tier assay has a high specificity, then a simple approach to achieve high overall process concordance is to have a first-tier, *in silico* method which has high sensitivity (the compounds assessed in the confirmatory assay are those in which the *in vitro* assay has the highest predictive ability). As models become more widely used by bench scientists throughout a research organization, the needs of the various research entities using these models are an influencing factor in how to appropriately bias the models.

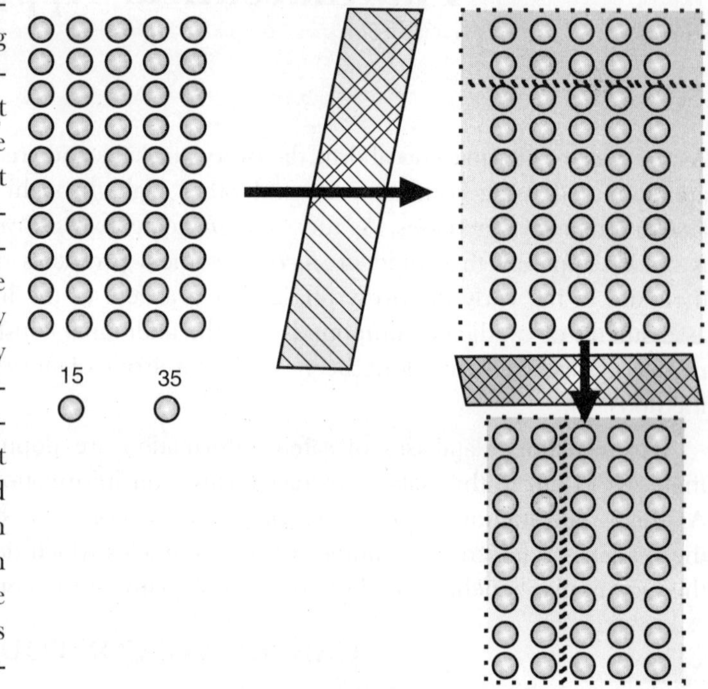

Fig. 17.1. Illustration of a tiered analysis system geared toward correctly partitioning the toxic from the non-toxic in which the first phase has a high sensitivity (93%) and the second phase has high specificity (85%). The overall concordance of this approach is 86%, with an overall sensitivity of 80% and specificity of 89%.

Scientists involved in the final stages of safety assessment (e.g. before an NCE filing) or formulation tend to be more cautious with models that have a minimal rate of false negative assessments. Scientists involved in scaffold development or lead identification, would typically desire a model with as few false positive assessments as possible, so as not to limit the exploration of chemical space. An additional motivation for biasing a model would be if the confirmatory assay were prohibitively expensive or were of low throughput, leading to the need for a model with few false positives. As an example we will consider carcinogenicity modeling. The assay approved by the Food and Drug Administration (FDA) to evaluate the *in vivo* carcinogenicity of a compound is the two year rodent assay with cohorts of male and female rats.

In order to build a model which is not simply an exercise in statistics, a cost-benefit analysis could be conducted as follows: if the cost (either in real dollars or in opportunity cost due to halting a compound's development) of excluding a compound from an assay is higher than the cost of losing a compound in a later stage of development due to carcinogenicity, the model should have high true negative rate. It would be this pool of compounds which would be carried forward and taken into the *in vitro/vivo* assays at a later stage of development.

Traditional Quantitative Models

Our ability to model a liability is not merely dependant on the ability of the modeler or the selection of the optimal algorithm, set of descriptors or subset of data for a given endpoint. More so than with traditional quantitative structure activity models, models in the realm of predictive toxicology must be acutely dependant on endpoints for which they are developed (e.g. models for teratogenicity, in which the phenotype characterized as teratogenic is due to an *unknown* number of biological mechanisms, building a predictive model becomes extremely difficult). Even within an endpoint with a limited number of mechanisms, local models (e.g. the mutagenicity of thiophenes) can often be combined to give a higher overall success rate and interpretability than a single, global model to explain an entire endpoint (e.g. mutagenicity). Another factor which determines the tractability of a problem is the level of experimental detail available on the biological entities being studied.

Modeling of phase I drug metabolism has been greatly impacted by the high resolution x- ray crystal structure of the human cytochrome P450 2C9 and 3A4 enzymes, allowing researchers to carry out high level *ab initio* quantum mechanical calculations in order to predict possible sites of oxidative metabolism upon docking ligands into the active site of the enzyme. As our understanding of this system increases, our ability to predict the inhibition of the enzyme by pharmaceuticals will also increase. As such, the endpoints which have been successfully modeled are varied; the greatest number of, and most mature, models exist for mutagenicity, rodent carcinogenicity (both genotoxic and non-genotoxic) and occupational toxicities (eye, skin and respiratory irritation/sensitization and lachrymation). As the quantity of high quality, controlled data increases, the predictive ability of the models developed from the data and the insights provided by those models will also increase.

Specificity and Sensitivity

There are currently two general approaches used to develop models which assess a compound's potential toxicological liabilities, *quantitative structure-activity relationship* (QSAR) derived models and *expert system* (ES) derived knowledge bases. Hybrid approaches which employ QSAR methods to determine the components rules of ES databases have also been developed. Each of these methods has had varying degrees of success in modeling toxicological endpoints. Local models of activity are often best classified by QSAR derived models (e.g. mutagenicity of anilines). Global models of a toxicological endpoint are often best represented by an expert system (e.g. mutagenicity of the world drug index) due to the general nature of the rules (e.g. aromatic compounds with good leaving groups often form DNA adducts). Both of these approaches may also be combined so that several class-specific models create a larger, global model. QSARs develop an abstract representation of a series of compounds and seek to extract the features (descriptors) which best correlate with an observed activity. Developing a virtual depiction of a molecule is accomplished by the calculation of molecular indices (e.g. counting the number of rotatable bonds, halogens or heavy atoms) and properties (e.g. LUMO energy, polarizability or logP). These *descriptors* are then correlated with the experimental data using multivariate statistical methods (e.g. multiple linear regression, principle component analysis or Shannon entropy methods) to keep only those descriptors with strong resolving power.

QSAR models can be used to build structural models of theoretically ideal ligands (such as Comparative Molecular Field Analysis and pharmacophore models) given the current data. In addition, this class of model can give a quantitative estimate as to the likelihood of a novel compound having the activity of interest. It is readily apparent that QSARs benefit most from two things: (a) a large pool of high quality data; and, (b) a single endpoint assay with few mechanisms of action. As the number of molecules modeled increases, the resolving power of the selected descriptors increases. Likewise, if the assay readout to be modeled is a result of multiple mechanisms, the likelihood of finding simple descriptors with a correlation to activity is low, as the signal from any single mechanism is low.

Conversely, QSAR models perform extremely well when assays generate a large volume of data through a single mechanism of action, as most high and medium throughput assays are based on single mechanisms. Expert systems are built upon a body of knowledge obtained from a specialists and/or literature sources. This information, which can be as varied as "rule of thumb" relationships to experimentally determined relationships (e.g. kinasesubstrate pairs) or structure activity data from compounds assayed using the same protocol, is collated, evaluated and then entered into a rules-based system which describes these cause-effect relationships (i.e. if substructure is aniline and metabolic activation present then genotoxic else nongenotoxic).

Perhaps the best known such relationships in predictive toxicology are the Ashby-Tenant structural alerts for carcinogenicity which have been the basis of numerous expert system databases to identify unwanted activity in compounds. Because each of the rules of an expert system form a component of a decision tree, the collection of rules can be easily combined into modeling global endpoints. Expert systems alone do not provide percent likelihood estimates of an outcome occurring; they merely indicate that it is a possibility (i.e. binary relationship). In order to quantitatively predict an event, predicate logic methods must be applied, as in the DEREK system. The methods described above, QSARs and ESs, can be combined to give greater predictive ability or can each be utilized in the development of the other.

A QSAR model can be developed using fragment-based descriptors (e.g. partitioning a molecule at its rotatable bonds) and then these fragments scored for predictive ability. This database of statistically significant fragments can then be used as the expert system. Such methods have been widely utilized by regulatory agencies and are routinely part of the safety assessment process. One such method, the MultiCASE algorithm, will be used extensively in this work to develop novel fragments for the variety of datasets discussed within. The goal of this paper is not to introduce a novel modeling methodology for toxicology or to present yet another model for a given set of data, but to highlight what can be done with available technologies in order to boost the productivity and predictive accuracy of currently available *independent vendor software* (IVS) methods.

To accomplish this, we will directly compare the performance of these differing methods across a diverse set of chemotypes using a controlled dataset consisting of results from the SOS Chromotest, a gentoxicity screening assay for mutagenicity. Compound libraries for secondary and aromatic amines, thiophenes and polycyclic aromatic compounds have been assayed at Bristol-Myers Squibb. QSAR models for many of these data sets have been developed and published previously and compared to DEREK, Topkat and MultiCASE. We have used the MultiCASE algorithm to extract structural fragments with statistical resolving power and will discuss how this new knowledge could be reincorporated into the overall toxicology assessment paradigm to enhance its overall predictive ability.

PROCEDURE

Descriptive Relationships

The data employed in this publication have been presented elsewhere and we will give only a description of the methods used to choose compounds and calculate descriptive relationships. All libraries were selected, purchased and assayed at Bristol-Myers Squibb. The compounds comprising these datasets of aromatic amines, thiophenes and polycyclic aromatic compounds were selected based upon their availability from Sigma-Aldrich, in pure form, and their drug-likeness, as profiled using an expanded Lipinski filter.

Of the multitude of compounds which could conceivably qualify for this study, the most chemically diverse subset was selected. Chemical diversity was based upon clustering of the Kier-Hall electrotopological indices, molecular weight, and various atom and ring counts. Compound selection was then further enhanced by the selection of specific groups to develop local structure-activity relationships (i.e. *para* versus *ortho* substitution about a ring).

Bacterial Reverse Mutation Assay

The SOS Chromotest was used in this paper as an alternative method to the widely accepted bacterial-reverse-mutation assay, also known as the Ames test. The SOS test has been used extensively with many different chemical classes. A review of published genotoxicity data between 1982 and 1992 indicated that, for the 1776 compounds evaluated, the SOS Chromotest had 90% concordance with the Ames mutagenicity test. The assay is a simple and rapid test for mutagenicity that requires only a few milligrams of compound. These benefits along with the high Ames test concordance, make the SOS Chromotest a good tool for the purposes in this paper. The SOS Chromotest assay is a colorimetric assay that measures induction of a lacZ reporter gene in response to DNA damage.

The SOS pathway plays a leading role in *E. coli* response to damage of nuclear material. SOS induction is used as an early monitor for DNA damage because this pathway is sensitive to a broad spectrum of genotoxic substances. E.coli were modified with a *lacZ* reporter gene fused to an SOS gene, *sfiA,* with the endogenous *lac* sequence deleted, so all β-galactosidase activity is dependent upon *sfiA* induction. The strain was made more sensitive to genotoxic substances by increasing cell envelope permeability (*rfa* mutation) and by eliminating the excision repair pathway (*uvrA* mutation).

In addition, constitutive expression of alkaline phosphatase, an SOS independent gene, was included as a control for cytotoxicity. In response to DNA damage the SOS repair genes are induced resulting in production of ,-galactosidase, the gene product of lacZ. The assay readout is the fold increases in gene induction as determined by measuring p-galactosidase activity using o-nitrophenyl-d-galactopyranoside (ONGP).

Bioinformatic Approaches

ADAPT

ADAPT is a neural network based QSAR modeling environment. Briefly, relationships are derived using a probabilistic neural network which evolves the optimal descriptor set to partition active from inactive compounds (i.e. mutagens from non-mutagens). The ADAPT software package is a product of the Jurs research group and can be compiled to run on many platforms. Model construction using ADAPT follows several distinct phases: structure geometry optimization and modeling, descriptor generation, training and prediction set formation, objective feature selection, model building and model validation. The exact procedure has been published in the works describing the specific models and will not be discussed here. The ADAPT system has been used to model a variety of toxicologically relevant endpoints.

Toxicological endpoint

DEREK is an expert system created specifically for toxicological endpoints and assessments. Originally developed and implemented by Schering Agrochemicals, DEREK is currently available on two platforms from two sources. The IRIX instance of DEREK is available from the Lhasa Group at Harvard University. The PC version is available from Lhasa, Ltd. UK and with an expanded rules base from the IRIX version. As such, it was used for all work presented here (version 5.0 and 6.0). PC DEREK (referred to hereafter as DEREK) covers a broad range of endpoints with its greatest strengths in carcinogenicity, mutagenicity (both genotoxic and non-genotoxic), skin irritation and sensitization. The rules set in DEREK is researched, generated and validated through a collaborative effort involving researchers from academia, industry and the staff of Lhasa. These SARs represented by these rules are encoded as substructures (with the ability to handle exceptions) which are subsequently located in a novel compound. Novel SARs can be incorporated by the user via an ISIS Draw input method.

Modified version

"BMS DEREK" is a heavily modified version of DEREK 5.0 for PC. The source code was ported to UNIX with the DEREK engine processing the compounds. BMS DEREK differs from DEREK 5.0 in that it contains thirteen new mutagenicity rules along with six carcinogenicity substructural alerts. The recategorized rules were found to give the DEREK higher overall sensitivity with a minimal impact on concordance as evaluated on a BMS validation set.

Other expert system

At its operational core, MultiCASE is a QSAR-derived expert system. The range of activity can be rescaled either automatically by the application or by the user. The descriptors generated are connected structural fragments which are from two to ten non-hydrogen atoms in length (referred to as *biophores*). Once a library of all possible biophores has been constructed, various statistical tests are applied to determine the resolving power of given biophores to correlate to an activity. These biophores are then assigned the average activity of all of the molecules in which they are contained (the CASE algorithm). This procedure is then recursively applied to each biophore to find additional biophores which serve to modify activities (activating and deactivating groups) with their effect being quantified in a manner similar to the above (the "Multi" in MultiCASE). MultiCASE then assesses a compound by giving it a score, calculated as the sum of all biophores present, with additional weight being applied for bioavailability.

Biostatistical Application

The overall agreement between a predcitve method and the experimental data is given by the concordance:

$$\text{Conc} = \frac{\text{PA} + \text{NA}}{\text{PA} + \text{PD} + \text{ND} + \text{NA} + \text{PI} + \text{NI}}$$

The ability of a predictive method to correctly assess the positive compounds is given by the sensitivity of a method:

$$\text{Sens} = \frac{\text{PA}}{\text{PA} + \text{ND} + \text{PI}}$$

The ability of a predictive method to correctly assess the negative compounds is given by the specificity:

$$\text{Spec} = \frac{\text{NA}}{\text{PD} + \text{NA} + \text{NI}}$$

The probability of an *in silico* method to correctly assess a compound by random chance is the frequency of success given the known distribution of positive and negative compounds in the data set. It is defined as:

$$\text{Rand} = \frac{(\text{PA} + \text{ND} + \text{PI})^2 + (\text{NA} + \text{PD} + \text{NI})^2}{(\text{PA} + \text{ND} + \text{PI} + \text{NA} + \text{PD} + \text{NI})^2}$$

Another method used to evaluate categorical classification models is the κ statistic. In the notation used here, it is defined as:

$$\kappa = \frac{P_o - P_e}{(1 - P_e)}$$

with P_o equivalent to the concordance as defined above:

$$P_o = \frac{\text{PA} + \text{NA}}{\text{PA} + \text{ND} + \text{PI} + \text{NA} + \text{PD} + \text{NI}}$$

and P_e as a random weighted by the methods success sensitivity and specificity:

$$P_e = \frac{PA * (PA + ND + PI) + NA * (NA + PD + NI)}{(PA + ND + PI + NA + PD + NI)^2}$$

κ can best be thought of as a concordance adjusted for the random success rate of a model. Models with values for κ above 0.5 are considered to have overall predictive ability when classifying compounds into bins.

Multiple Models

Results from ADAPT models which have been published prior to this work are reported here. Briefly, the compounds were taken into ADAPT after their structures were optimized using CORINA and MOPAC. The data were then partitioned into multiple TS and VS using a bagging algorithm. Multiple models were thus generated and the overall consensus model is the "*majority rules*" vote of the individual models. This way all compounds were predicted at least once. The DEREK expert system evaluates compounds using substructure-based searches. Standard and BMS modified DEREK were both used to predict the mutagenic activity of all compounds without any specific models being built using this data (the entire data set was treated as the PS).

In order to make an equivalent comparison, both the MultiCASE A2I *Salmonella* mutagenicity module and a model trained in an equivalent fashion to the ADAPT models, were used. Briefly, the data were partitioned *exactly* as described in the original ADAPT based models (e.g. identical compound distributions were utilized for the TS and VS). The activities were then scaled automatically in MultiCASE so that the activity range would be between zero (inactive) and 99 (most active) with the active-inactive cutoff at twenty-nine units (corresponding to the experimental cut-off in activity of 1.5 fold activity in the SOS Chromotest assay). Studies varying the cutoff value by ± ten percent showed a minimal impact on the sensitivity and concordance of the resulting models. The models thus generated were then allowed to each vote for the membership of a compound in a given class ("mutagenic" or non-mutagenic").

INDIVIDUAL MODELS

In all cases, the ADAPT QSAR consensus model (a majority rules voting of several individual models) had the best concordance/κ value, ranging from 95%/0.88 (thiophenes) to 72%/.45 (aromatic and secondary amines). The MultiCASE *Salmonella* mutagenicity module, the A2I database, performed better than random for only the thiophene data set, with a concordance of 65%, but had a low κ value (0.39). In the remaining two data sets, it had the worst overall performance, with a concordance of 24% in the polycyclic aromatic compounds and a concordance of 32% in the aromatic and secondary amines. In two of the three data sets, BMS DEREK had the highest sensitivity, ranging from 100% (thiophenes and aromatic and secondary amines) to 90% (secondary and aromatic amines), but always had a low κ value.

The polycyclic aromatic compounds were not classified well by the expert systems approach, chiefly due to the lack of rules for heterocyclic compounds. The high sensitivity reported in the BMS DEREK models is offset by their having amongst the lowest specificity, ranging from 0% (thiophenes) to 35% (polycyclic aromatic compounds). Analysis of the specific performance of the expert system based approaches has been reported elsewhere and will only be touched upon here.

Stable Epoxide

The mutagenic potential of a thiophene is linked to formation of a stable epoxide. If the epoxide is stabilized at C2, then the likelihood of mutagenicity increases. Likewise, in systems which stabilize the formation of the *S*-oxide, nucleophilic attack at C2 is further stabilized leading to an increased possibility of reactivity with protein and/or DNA. The thiophene data set was the smallest of all the data sets, and was thus the only model for which a consensus model was not built. As is evident from the table, there was very little representation in the PS of any of the biophores derived. Furthermore, while the overall positive accuracy of these biophores is quite high (69%), the biophore corresponding to the primary alcohol was derived from a small sample (N = 2) of marginal activity (29 CU, the cutoff between active and inactivity).

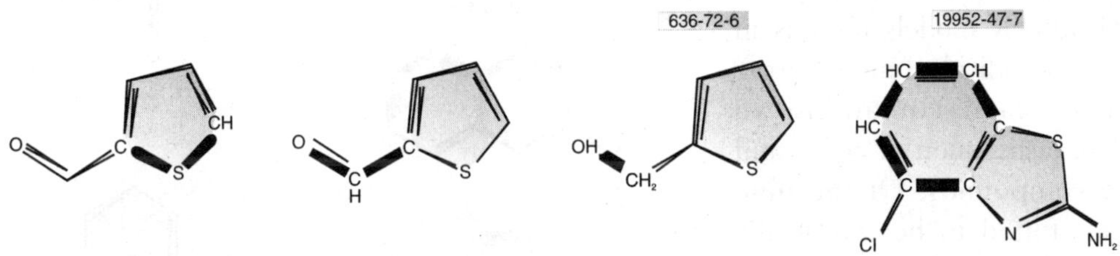

Fig. 17.2. Statistically significant biophores derived from the thiophene dataset highlighted within compounds from the TS in which they are contained.

While the MultiCASE algorithm found this a significant biophore, the author would not utilize this in any further development of structural rules for thiophene (or any other) genotoxicity. Removal of this biophore from the model increases the positive accuracy to 91.7%. These four biophores do successfully select the genotoxic compounds from the PS as is reflected in the very high sensitivity (83%), but do not do an acceptable job of partitioning out the non-genotoxic compounds (36% specificity). This leads to an overall concordance of 50% and a κ of 0.33, worse than random.

Polycyclic Compounds

The polycyclic aromatic compounds used in this study were representative of many different classes of polycyclic compounds. Using the bagging approach employed previously in analyzing this data set, three model triplets were generated, each containing an equivalent distribution of active and inactive compounds. Of the nine significant fragments, four were found in more than one model and all were based upon at least three molecules in the TS (with a median of 4.7 TS molecules/fragment). The positive accuracy of the MultiCASE derived biophores ranged from a high of 80% to a low of 0%. Unlike the thiophene model above, this consensus model had high specificity (87%), but low sensitivity (28%), failing to identify 33 of the 46 genotoxic compounds in the data set, giving an overall concordance of 77%. Interestingly, most of the biophores identified by this approach contain a nitrogen atom, either as part of the heterocycle or as a primary amine. Amines can undergo P450 (1A2) activation to form the reactive nitrenium species to yield the ultimate N-hydroxylated form.

Polycyclic aromatic amines generally exhibit greater mutagenic potency than other aromatic amines provided that the polycyclic system does not exceed six or more rings in size. The extended aromatic system offers a greater degree of stabilization to the nitrenium ion and may allow for a more efficient activation due to the planarity of the structure. The aromatic and secondary amines are considered in the following section at length.

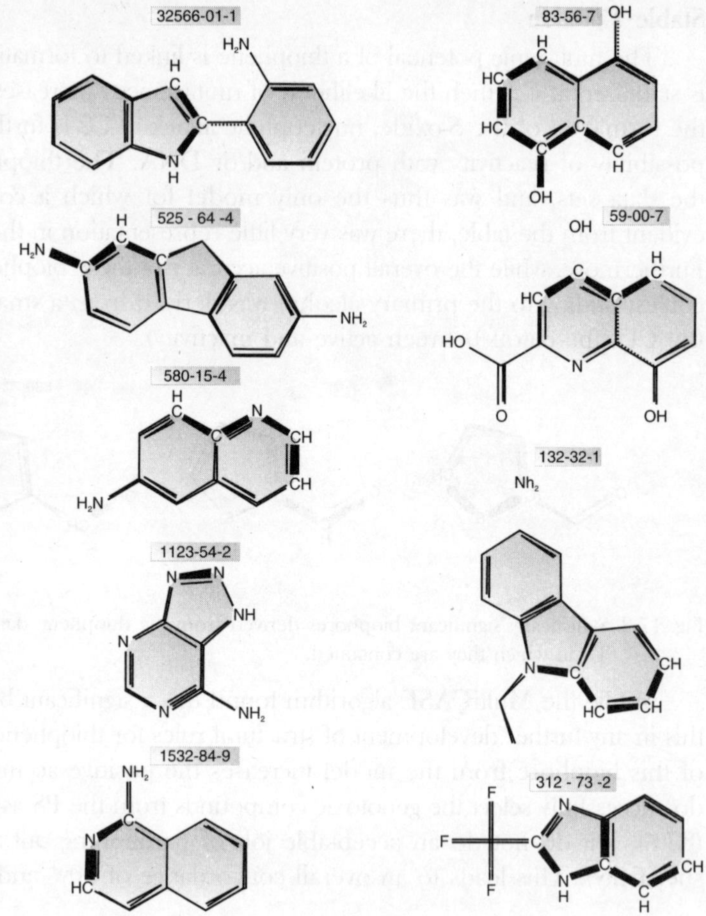

Fig. 17.3. Statistically significant biophores derived from the polycyclic aromatic compounds data set highlighted within compounds from the TS in which they are contained.

Generation of Biophores

The aromatic and secondary amines were chosen to be fairly diverse. Using the data subsetting approach employed to develop the ADAPT QSAR models for this in data set, nine individual models were generated, each containing an equivalent distribution of active and inactive compounds. Of the nine fragments found to be statistically significant, six were derived by more than one model and all were based upon, on average 9.5 TS molecules/fragment, with a minimum of 2.8 TS molecules/fragment. The positive accuracy of these biophores ranged from a high of 80% to 0%.

The MultiCASE consensus model had the second highest specificity (72%) and concordance (69%) of all of the models reported for this data set. The low positive accuracy of the fragments is reflected in the overall sensitivity of the consensus model (58%). The biophores generated by this fragment based approach are the only ones from three models which show a correlation between overall positive accuracy and activity of the underlying molecules upon which the fragments are based. The Pearson correlation coefficient is 0.79, indicating a statistically significant correlation between these two quantities. In the polycyclic aromatic compounds the overall positive accuracy and activity correlation coefficient is negative and in thiophenes, there is a correlation coefficient of 0.45, indicating an uncorrelated system. The correlation is a product of several features of this data set.

The overall population of this set is larger than the other datasets (334 compounds; polycyclic aromatic compounds: 277 compounds; thiophenes: 140 compounds), with a correspondingly larger number of active compounds. This leads to several benefits in model creation. First, as the pool of TS molecules is larger, the number of two to ten atom fragments generated and associated modulating fragments is larger, allowing for a more robust fragment selection process. The mean activity of the significant fragments generated is approximately ten percent higher than that of the other data sets, with more fragments having activities in the moderate range (forty to sixty CASE units).

CONCLUDING REMARKS

The evaluation of independent vendor solutions for prediction of the genotoxic potential of drug-like compounds is presented here. The availability of genotoxicity data from a single lab on several series of congeneric compounds which had been previously assessed by DEREK, ADAPT and MultiCASE made these attractive starting points. We utilized the fragment based approach as implemented in the MultiCASE algorithm to develop a library of genotoxic substructures to analyze these three congeneric series of compounds: thiophenes, polycyclic aromatic compounds and a set of secondary and aromatic amines. The substructural fragments generated for these data sets were found to vary widely in their predictive ability. For compounds with PS coverage greater than five compounds, the positive accuracy ranged from a high of 80% to a low of 0%). Except for fragments generated from the secondary and aromatic amines, the predictive ability was not correlated with either representation in the PS, the number of training set compounds giving rise to a specific fragment, or the mean activity of the fragment.

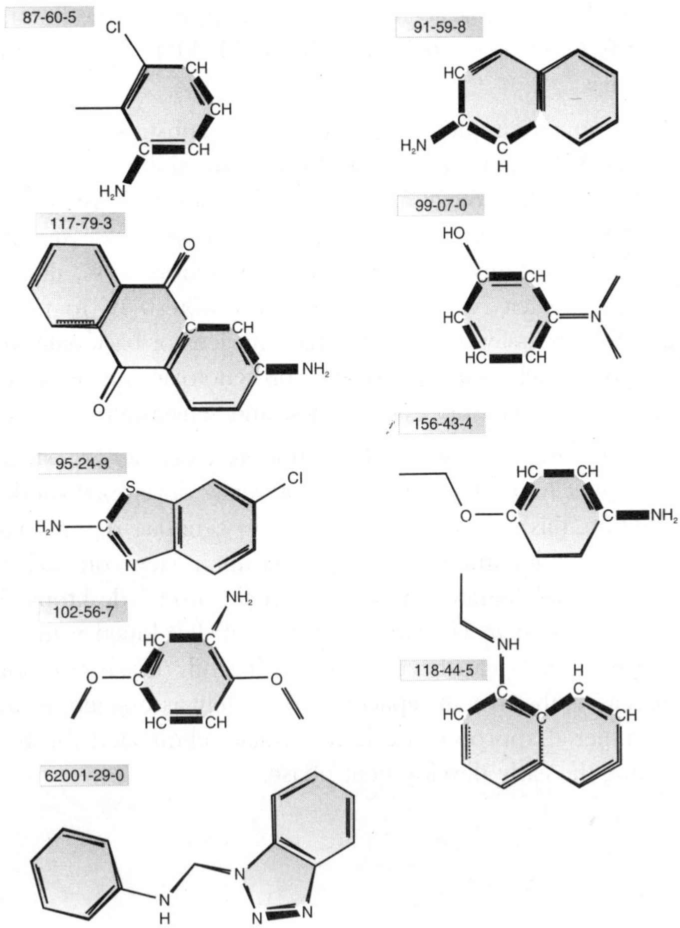

Fig. 17.4. Statistically significant biophores derived from the secondary and aromatic amines dataset highlighted within compounds from the TS in which they are contained.

In the one case with correlation, the number of training set compounds per fragment was found to have significant correlation with activity (correlation coefficient of 0.79). Developing a QSAR using solely substructural fragments was never as predictive as adding additional QSAR factors, as is evidenced by the difference in predictive ability between ADAPT and MultiCASE as ADAPT uses not only path-based substructural fragments but a variety of additional descriptors. The authors would like to provide a cautionary note to researchers utilizing the power of ensemble based modeling methods (such as many of the ADAPT models presented here, machine-learning algorithms, neural networks and the Random Forest algorithm) in that it is very easy to over-train a model, making it no more than a mapping of the training set. There are over sixty descriptors for the set of 334 compounds in the ADAPT ensemble models for the Secondary and Aromatic Amines.

A much simpler model, with nine substructural-fragments as descriptors, developed using MultiCASE has nearly identical predictive ability. The system described here was implemented in an internal toxicology assessment web portal at Bristol-Myers Squibb. This system allows any authorized user on a discovery team or in safety assessment to analyze both synthesized and virtual compounds at any phase in the development process, allowing them to prioritize testing in a second tier genotoxicity assay (the SOS Chromotest) or to provide supporting information for other genotoxicity assays (i.e. *in-vitro* micronucleus or bacterial-reverse-mutation assay). Other toxicological endpoints such as hepatotoxicity, non-genotoxic and genotoxic carcinogenicity, environmental toxicity (respiratory, skin and eye irritation and sensitization and lachrymation) can also be added.

The ability to query IVS models (such as the standard MultiCASE carcinogenicity and mutagenicity models) as well as internally developed models has been added to the web tool. In addition, this system allows the user to generate reports based on the assessment and provides all supporting documentation, including minireviews on each of the mutagenic endpoints with model statistics and literature references at the users' desktops. While no toxicology model will be as accurate as an *in vitro* model system, if developed with a clear goal in mind *and* used within its scope, models can be used to intelligently triage compounds from the testing queue (thereby increasing the overall capacity of the safety assessment assays). At Bristol-Myers Squibb, using this multi-tiered approach, we have virtually eliminated the mutagenic liability as a cause of attrition during the early development phase.

18

Toxicology Biomarkers

Toxicology biomarkers are structural or functional measurements that correlate with an adverse change in the physiology of an organism. A familiar example is elevated serological alanine aminotransferase (ALT) activity as a biomarker of hepatic injury. Toxicology biomarkers such as ALT help us select compounds with desirable margins of safety in the *lead optimization* (LO) phase of drug development, and allow us to identify target-organ toxicities that should be monitored in patients during clinical drug development. Unfortunately, useful clinical and pre-clinical biomarkers are not available for many compound-induced pathologies. Promising new molecular techniques that can address this gap by finding candidate biomarkers include genomics, proteomics, and metabonomics.

Successful application of these technologies to develop toxicology biomarkers could increase the quality and number of clinical candidates selected during LO, and improve our ability to monitor drug safety in patients. In this chapter we present a case history using toxicogenomics in which DMBT1 was developed as a transcriptional biomarker of bile duct hyperplasia induced via oval cell proliferation, a toxicologic pathology for which biomarkers are needed.

COMMON TOXICOLOGIC PATHOLOGY

Bile duct hyperplasia (BDH) is a relatively common toxicologic pathology that occurs in both rodent and non-rodent species. Two distinct processes can lead to chemical-mediated BDH: cholestasis and oval cell proliferation. The pathogenesis involving cholestasis is exemplified by α-naphthylisothiocyanate (ANIT). A single oral dose of ANIT leads to a rapid decrease in biliary function within 24 hr, measurable as decreased biliary taurocholate transport and decreased elimination of erythritol from systemic circulation, a compound normally excreted via the bile. Biliary epithelial cell necrosis ensues ~48 hr after exposure to ANIT, leading to sloughing of epithelial cell debris into bile ducts, causing complete obstruction of many ducts.

A compensatory proliferative response 4-7 days after treatment with ANIT increases the number of bile ducts and biliary epithelial cells per bile duct, resulting in BDH. BDH pathogenesis involving oval cell proliferation is less clearly defined, in large part because of the complex biology of oval cells. Oval cells are thought to be progenitor cells that can differentiate into both parenchymal (i.e., hepatocytes) and non-parenhymal (i.e., biliary epithelial) hepatic epithelial cell types. Under normal conditions in the mammalian liver oval cells occur at low abundance, but under pathological conditions they can become a prominent feature, particularly following severe liver injury.

Oval cells are recognizable by the ovoid shape of their nuclei, a large nuclear:cytoplasm ratio, and basophilia using H&E stain. Their proliferation can be induced with compounds such as acetylaminofluorene, in transgenic mice that overexpress SV40 large T antigen or ErbB-2, and by genetic disruption of genes such as Tg737. In each of these models, oval cells proliferate around the portal triad and expand into the periportal region. Under severe conditions basophilic oval cells delineate periportal regions. Concurrent with oval cell proliferation in these models is the proliferation of bile ducts, and proliferation of biliary epithelial cells per bile duct. Progression of these lesions to cholangiocarcinoma can occur in animals that live sufficiently long.

BILE DUCT HYPERPLASIA (BDH)

BDH can progress to cholangiocarcinoma with low margin of safety, a scenario that can lead to costly, late stage compound terminations and increased risk to patient safety. On the other hand, progression of BDH to neoplasia is not a certainty; at least one blockbuster medication is taken chronically by patients to reduce cholesterol, in spite of the fact that this compound caused BDH at 3 months or less in both rats and dogs. Interpretation of the significance of BDH therefore requires consideration of a number of variables, including duration of exposure, margin of safety, intended patient population, and availability of biomarkers to monitor patient safety. Many compounds cause BDH after just a few days of once daily dosing. In such instances it is reasonable to expect that with greater duration of exposure BDH will progress in severity and become observable at lower doses, leading to decreased margin of safety in GLP studies that support clinical trials.

Because progression of BDH with extended treatment contributes to compound attrition during drug development, methods to identify compounds that do not cause BDH, or with improved margin of safety, are desirable. Ideally this would be an *in vitro* screen that is predictive of the *in vivo* response. Considerable effort has been made to develop *in vitro* culture techniques for biliary epithelial cells, including their use in screening assays. This work has improved our understanding of oval cells and other hepatic progenitor cell populations, but an *in vitro* assay to screen compounds for their ability to cause BDH has yet to be reported. Several factors are likely operative in limiting the success of *in vitro* BDH assays, including failure of progenitor cells to proliferate and differentiate in culture, and inability of primary cells to metabolize xenobiotics.

In addition, since BDH occurs in some instances due to signals emanating from physical obstruction of the bile duct or bile cannaliculi, culture models may not accurately replicate some pathophysiological events that are important in the etiology of BDH. The desire to identify compounds that do not cause BDH, and the lack of adequate models to screen compounds *in vitro*, requires testing additional compounds *in vivo*. This strategy can be successful when the incidence of BDH in the structural series is reasonably low. If, however, the incidence of BDH in the structural series is high, or the lesion does not appear until prolonged exposure, the need to screen multiple compounds *in vivo* can result in significant delays or even program termination. Thus, drug developers need an alternative strategy, such as a biomarker that predicts outcome of long-term exposures. Such biomarkers would be useful not only for pre-clinical toxicology testing, but also to support clinical drug development. Currently there is a substantial difference in the availability of *in vivo* biomarkers of BDH for the two mechanisms that lead to compound-induced BDH.

Changes in liver function associated with the cholestasis mechanism can be monitored pre-clinically and clinically using serological biomarkers such as total bilirubin concentration, alkaline phosphatase activity, and γ-glutamyl-transpeptidase activity. Consequently, development of compounds that cause BDH through a cholestatic mechanism is supported by well-validated serological biomarkers. In contrast, BDH induced via oval cell proliferation is a toxicological endpoint for which drug developers do not have biomarkers to support clinical or pre-clinical drug development. Instead, we are limited to histology of biopsy tissue to monitor this pathogenesis. As a result, development of compounds that cause BDH is more challenging for molecules that cause oval cell proliferation than for compounds that cause cholestasis. Biomarkers to monitor or predict oval cell proliferation would therefore be of value to drug developers, both to screen compounds during LO and to minimize risk to patients during clinical development.

RECOGNITION OF BIOMARKER

Our objective was to identify a serum-based protein biomarker of oval cell mediated BDH that could be incorporated into a clinical pathology panel. Our execution of this strategy began with gene expression analysis using microarrays to identify transcripts that were induced by compounds that cause BDH via a mechanism involving oval cell proliferation. Bioinformatic analyses were used to identify differentially expressed transcripts that encode a secretable protein. We then evaluated the specificity and sensitivity of these biomarkers by measuring their transcript and/or protein concentrations in livers of animals treated with compounds that cause BDH or other hepatotoxicities. A number of technical challenges were anticipated. Principal among these was distinguishing gene expression changes occurring in biliary epithelial and oval cells compared with those in other cell types in the liver.

Biliary epithelial cells constitute <1% of liver volume in mammals, therefore we expected >95% of the RNA isolated from rat liver to be unrelated to the phenomenon of interest. Furthermore, many toxicologic pathologies that alter hepatic gene expression may be coincidental with compound-induced BDH, such as cytotoxicity, hepatocyte regeneration, inflammation, apoptosis, and altered metabolic enzymes. We therefore needed a chemical tool that would maximize BDH and minimize unrelated toxicological phenomena, combined with a strategy to increase the proportion of transcripts isolated from proliferating bile ducts and oval cells. Methapyrilene treatment of rats was selected as a model for BDH because it caused a moderate increase in BDH without causing extensive hepatocellular injury or inflammation that would complicate microarray analysis.

Chronic treatment of F344 rats with methapyrilene produced marked proliferation of oval cells, severe BDH, and progression to cholangiocarcinoma. Our goal was to identify early markers of this process. Rats were administered methapyrilene hydrochloride (MP) at 200 mg/kg at 0 and 48 hr and sacrificed at 96 hr. Histological examination showed oval cell proliferation and BDH, with some evidence of hepatocellular apoptosis, necrosis, and inflammation. In order to minimize the contribution of hepatocytes to the gene expression profile, a collagenase perfusion technique was developed to remove the hepatocyte population. The technique is similar to that used to prepare primary cultures of hepatocytes, but instead of harvesting the hepatocytes, we retained the biliary tree for microarray analysis.

In this way we enriched for biliary epithelial cells by eliminating ~80% of the wet weight of the liver due to loss of hepatocytes. Importantly, the bile ducts isolated using the collagenase perfusion were largely devoid of hepatocytes and had normal histology. RNA samples were prepared from the biliary trees of vehicle-treated and MP- treated rats and processed for analysis using Affymetrix U34A microarrays. The transcript levels of hundreds of genes were altered by treatment with MP. Since our goal was to identify a secreted biomarker, the transcript pool was further analyzed to identify the fraction of transcripts that encoded proteins containing motifs typical of secreted proteins, and for which there was evidence for preferential expression in epithelial cells. One such transcript is called ebnerin. Ebnerin is expressed in secretory duct epithelial cells of the von Ebner gland (VEG) of the rat tongue. It is a large protein (170-210 kDa) that contains repeats of two protein domains, SRCR and CUB, with high homology to secreted proteins and trans-membrane receptors, respectively.

The VEG secretes many proteins and other biological molecules into the lumenal surface along the apical region of taste buds. Of interest to us was the observation that ebnerin was detectable in secretions from VEG, human tear fluid, and in the medium of cell cultures. In earlier studies, ebnerin transcripts were not detected in rat tissues other than tongue, including liver, lung, kidney, colon, testes, brain, and spleen. However, it is common that transcripts and proteins that are first identified under one experimental condition are later discovered to have additional structures and functions in other tissues and other biological processes. We therefore considered it a possibility that ebnerin could be made by, and secreted from, oval and/or biliary epithelial cells, and as such might be detectable in serum and/or GI contents of animals treated with compounds that cause BDH through oval cell proliferation.

Biological Development

Examination of the primary structure of ebnerin transcripts in GenBank and the published literature helped us deduce that a group of transcripts encoded by the same gene had been studied in different species under different names, including ebnerin, gp340, hensin, CRP-ductin, and DMBT1. The transcripts varied significantly in size from 4.5 to 8 kb as a function of species, tissue, and stage of biological development. Splice variants lacking exons at the 5' end of the transcript were identified as a distinguishing feature of the ebnerin transcript. QPCR using TaqMan can be used to distinguish subtle variations in nucleic acid sequence. We therefore designed two sets of TaqMan primers and probes to quantify two different regions of the transcript: one for an exon toward the 5' end of the transcript and one for an exon toward the 3' end.

Of these two exons, only the 3' exon is associated with ebnerin transcripts, whereas both exons are associated with DMBT1 transcripts. Similar levels of expression were observed for both

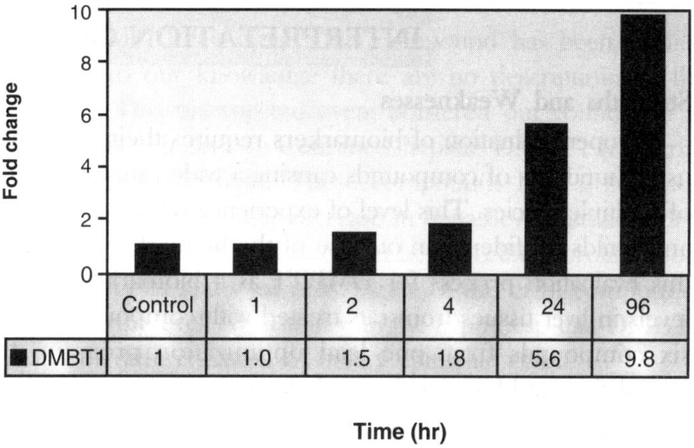

Fig. 18.1. Quantitation of DMBT1 transcript in livers of rats treated with methapyrilene hydrochloride at 200 mg/kg at 0 and 48 jr amd sacrificed at 96 hr.

exons, demonstrating that the livers of both vehicle and MP-treated rats primarily expressed a transcript that is more properly named deleted in malignant brain tumor 1 (DMBT1), than ebnerin. DMBT1 transcript induction measured using TaqMan (~10 fold) was significantly greater than that detected using microarrays. This points to tradeoffs related to the use of microarrays. The single hybridization procedure that permits detection of thousands of transcripts requires that we accept decreased accuracy and narrower linear dynamic range for individual transcripts. Complementary techniques, such as TaqMan, are therefore used in parallel with microarrays to improve qualitative and quantitative assessments of individual transcripts. Investigation of the kinetics of induction of DMBT1 in livers of rats treated with MP revealed a 2-fold increase 4 hr following treatment, greater than 5-fold after 24 hr, and approximately 10-fold at 96 hr following a second administration. DMBT1 transcript induction was therefore delayed relative to the kinetics of compound exposure. This differential pattern of drug exposure and response suggests that the cause of induction is not a direct effect of methapyrilene; rather, it results from a cascade of events initiated by the compound through an as yet unknown mechanism.

Confirmation of Induction

Further confirmation of the induction of DMBT1 by MP was demonstrated by immunoblotting for DMBT1 protein. An immunodetectable band at the appropriate molecular weight was observed in rat liver from MP-treated animals but not from vehicle-treated animals. The band was faint, consistent with the estimate that biliary epithelial cells constitute just 0.14% of the total volume of the liver. The apparent molecular weight of DMBT1 in our experiments (164 kDa) was similar to that reported previously for the expressed cDNA for ebnerin, but was smaller than what was detected in VEG of rat tongue.

Splice variants and differences in extent of protein glycosylation are both likely to contribute to the apparent difference in molecular weight of DMBT1 in these different tissues. DMBT1 protein has been localized by other investigators and ourselves to biliary epithelia and oval cells, as well as epithelial cells in other tissues. The proteins encoded by transcripts of this gene appear to play a significant role in terminal differentiation of epithelial cells of the prostate and vas deferens and of kidney epithelial cells cultured *in vitro*, including polarity reversal of the intercalated cells taken from rabbit kidney. It is likely that DMBT1 plays a role in oval cell differentiation into biliary epithelial cells, and may involve protein-protein interactions with extracellular matrix proteins such as galectin-3.

and by driving the development of *in vitro* screens. A key challenge in reaching these objectives is identifying panels of predictive toxicology biomarkers. In this chapter we demonstrated the use of toxicogenomics to identify a transcriptional biomarker of oval cell-mediated bile duct hyperplasia. Once identified, the specificity and sensitivity of the lead candidate biomarker, DMBT1, was evaluated in livers of rats treated with more than 30 different compounds comprising hepatotoxicities that ranged from BDH and hepatocyte proliferation, to phospholipidosis, hepatocellular vacuolation, apoptosis, and inflammation.

Since its identification, DMBT1 has been used successfully as an early indicator of oval-cell mediated BDH in preclinical toxicology studies *in vivo,* and as a molecular endpoint to screen compounds for their ability to induce BDH *in vitro.* These successes are evidence that biomarkers such as DMBT1, combined with additional toxicology biomarkers being discovered using the strategies described in this chapter, as well as other methods, are tools that will increase the efficiency of the drug discovery process, through improved methods of drug candidate identification and development.

Index